International Academic Conference "The State, Trends and Problems of Development of the Oil and Gas Potential of Western Siberia" (IAC OIL AND GAS 2017)

AA006794

IOP Conference Series: Earth and Environmental Science
Volume 181

Tyumen, Russia
28 – 29 September 2017

ISBN: 978-1-5108-6939-4
ISSN: 1755-1307

Printed from e-media with permission by:

Curran Associates, Inc.
57 Morehouse Lane
Red Hook, NY 12571

Some format issues inherent in the e-media version may also appear in this print version.

Copyright© (2017) by the Institute of Physics
All rights reserved. The material featured in this book is subject to
IOP copyright protection, unless otherwise indicated.

Printed by Curran Associates, Inc. (2018)

For permission requests, please contact the Institute of Physics
at the address below.

Institute of Physics
Dirac House, Temple Back
Bristol BS1 6BE UK

Phone: 44 1 17 929 7481
Fax: 44 1 17 920 0979

techtracking@iop.org

Additional copies of this publication are available from:

Curran Associates, Inc.
57 Morehouse Lane
Red Hook, NY 12571 USA
Phone: 845-758-0400
Fax: 845-758-2633
Email: curran@proceedings.com
Web: www.proceedings.com

Table of contents

Volume 181

International Academic Conference "The state, trends and problems of development of the oil and gas potential of Western Siberia"

28–29 September 2017, Tyumen, Russia

Accepted papers received: 30 July 2018
Published online: 16 August 2018

Preface

Papers

IOP Conf. Series: Earth and Environmental Science **181** (2018) 011001 doi:10.1088/1755-1315/181/1/011001

PREFACE

International Academic Conference "The state, trends and problems of development of the oil and gas potential of Western Siberia" (IAC OIL AND GAS 2017) is an annual conference which offers a place for participants to present and to discuss their innovative recent and ongoing research and theoretical work and/or their applications.

The first "IAC OIL AND GAS" was held in Tyumen, Russia in 2017. The second conference (IAC OIL and GAS) will be held in late 2018. The second conference (IAC OIL and GAS) will be held in late 2018. The organizer will be the company "NeftePAK" or Tyumen industrial University. The conference provides useful and convenient platform for researchers and engineers to present their latest research results and exchange ideas in a wide range of advanced technologies in the oil and gas industry. The conference participants were from leading oil and gas organizations and institutions.

64 reports were submitted. After a thorough review, 42 articles were accepted. Finally, 36 papers are collected in this IAC OIL AND GAS 2017 conference proceeding. This process covers a wide range of issues in the field of oil and gas Geology, drilling of oil and gas wells, geological 3D modeling of hydrocarbon deposits, as well as hydrogeology.

Content from this work may be used under the terms of the Creative Commons Attribution 3.0 licence. Any further distribution of this work must maintain attribution to the author(s) and the title of the work, journal citation and DOI.
Published under licence by IOP Publishing Ltd

Peer review statement

All papers published in this volume of *IOP Conference Series: Earth and Environmental Science* have been peer reviewed through processes administered by the proceedings Editors. Reviews were conducted by expert referees to the professional and scientific standards expected of a proceedings journal published by IOP Publishing.

Content from this work may be used under the terms of the Creative Commons Attribution 3.0 licence. Any further distribution of this work must maintain attribution to the author(s) and the title of the work, journal citation and DOI.
Published under licence by IOP Publishing Ltd

IOP Conf. Series: Earth and Environmental Science **181** (2018) 012001 doi:10.1088/1755-1315/181/1/012001

Structure of the hydrogeological field of the Krasnoleninsky arch

R N Abdrashitova and Yu I Salnikova

Industrial University of Tyumen, 38 Volodarskogo st., Tyumen, 625000, Russia

E-mail: ritte@list.ru

Abstract: The study is devoted to the problem of formation of the hydrogeological field of the deep oil and gas bearing horizons of the Krasnoleninsky Arch. Analysis of seismic data shows the presence of a large concentration of discontinuous violations, which can be traced from the basement and have a different penetration depth in the sedimentary cover. It is shown that significant fluctuations in the parameters of the composition of groundwater and hydrostatic pressure at the depth of the Lower-Middle Jurassic hydrogeological complex are associated with the receipt of deep high-temperature fluids from discontinuous disturbances that are formed under the influence of neotectonic processes. The pressure deficit can reach 9 MPa, an average of 1.6 MPa, and fluctuations in mineralization values range from 3 g/l to 12 g/l. Data have been obtained indicating the dependence of the occurrence of hydrocarbonate-sodium type waters with low mineralization from the position of faults (this type of water is found in samples taken from wells located at a distance of no more than 3 km from the recorded rupture violation). Chlorcalcium type, according to V.A. Sulin, with mineralization normal for these depths is determined in samples taken also from the Lower-Middle Jurassic complex, but from wells located at a distance of more than 2.5-3 km from the revealed disruption. Evidence of the deep fluids' effect on the complex rocks are provided (the presence of such authigenic minerals as pyrite, siderite, dickite, carbonates, and kaolinite in rocks of the complex).

1. Introduction

The relevant task of modern oil and gas hydrogeology is the development of a methodology for studying the hydrogeology of the deep oil and gas bearing horizons of the West Siberian megabasin, the development of more effective practical recommendations for carrying out work on maintaining reservoir pressure (reducing the amount of calcite deposits, etc.) and calculating underground water reserves. Since almost all aspects of groundwater formation at the depths of the Mesozoic basin of Western Siberia are directly related to the development of the doctrine of naftidogenesis.

The study of the mechanisms of groundwater movement at the depths of oil and gas bearing horizons is one of the urgent problems of modern oil and gas hydrogeology. V.M. Matusevich, A.A. Kartsev, S.B. Vagin, V.I. Dyunin and many others considered these issues for the conditions of Western Siberia. The most important aspects of oil and gas hydrogeology abroad are studied by Toth J. [1, 2] who considered water-pressure systems and their role in the formation of hydrocarbon reserves, the mechanisms of groundwater movement and hydrocarbon accumulation in the case of Alberta in Canada. The complexity of the problem lies in the fact that we can judge the patterns of motion based on indirect information only, by combining data on lithology, petrography, tectonics, geohydrotemperature conditions and hydrogeology. For each field, such an analysis should be carried out individually.

Content from this work may be used under the terms of the Creative Commons Attribution 3.0 licence. Any further distribution of this work must maintain attribution to the author(s) and the title of the work, journal citation and DOI.
Published under licence by IOP Publishing Ltd

The basic hydrogeological data obtained during the work on the calculation of groundwater reserves for the purpose of maintaining reservoir pressure or works for the utilization of industrial effluents involves determining the macro- and micro-component composition of water samples, measuring the reservoir pressure and temperatures. There are often situations when the data in wells located close enough to each other (even within 5-10 km) is several times different. For example, within the Lower-Middle Jurassic hydrogeological complex in the Talinskoye oilfield, samples were obtained with a salinity of 3 g/l and 12 g/l under the same paleohydrogeological conditions for the formation of the complex. Also, reservoir pressures within a single hydrogeological complex in one exploration area can vary significantly (from over hydrostatic to lower hydrostatic). This kind of data that are "knocked out" of the general picture because of the absence of complex analysis are most often rejected, sometimes despite their frequent occurrence. On the example of the results described below, we wanted to show that this is not always correct.

First of all, it is necessary to understand that studies of hydrogeological conditions at the depths of oil and gas bearing horizons should be carried out taking into account the fact that the formation of deep fluids is often accompanied by phase transitions in the fluid-rock system, which leads to a significant change in the filtration-capacitive properties of fluid-bearing rocks, change in chemical and gas composition of fluids, etc. [3]

2. Research

On the example of the Krasnoleninsky arch, we carried out a comprehensive analysis of information on hydrogeological conditions (hydrogeochemistry, geohydrology, hydrogeodynamics) and geological structure, structure of the basement, the presence of faults, autogenic processing of rocks.

The main provisions, based on which the analysis was performed, are as follows:

1) The Krasnoleninsky arch refers to an elution lithostatic water-pressure system according to the classification of water-pressure systems by A.A. Kartsev, S.B. Vagin [4] with the additions of V.M. Matusevich and O.V. Bakuev [5]. Elision is the process of squeezing buried waters from clay rocks to sandy as the sedimentary basin is filled and the geostatic load increases - is considered one of the main processes that determined the structure of the hydrogeological field of the research area. According to this theory, the elision waters create superhydrostatic pressures and contribute to the formation of the hydrocarbonate-sodium type of waters by V.A. Sulin;
2) Three hydrogeological basins are distinguished in the section: Cenozoic, Mesozoic and Paleozoic and, respectively, seven hydrogeological complexes: Oligocene-Quaternary, Turonian-Oligocene, Aptian-Albian-Cenomanian, Neocomian, Upper Jurassic, Lower-Middle Jurassic, Triassic-Palaeozoic (Matusevich, 1998). A feature of the geological structure of the research area is the completely clay composition of the Neocomian complex with a capacity of more than 750 m, which led to the attribution of the Krasnoleninsky arch area to an elision water-pressure system. The tracking in the study area of three megacyclites should also be noted: the Triassic-Aptian, Aptian-Oligocene and Oligocene-Quaternary;
3) The object of research is the Lower-Middle Jurassic hydrogeological complex within the fields of the Krasnoleninsky Arch, composed of an uneven alternation of sandstones, siltstones with interlayers of argillite-like clays;
4) Analysis of seismic data shows the presence of a large concentration of faults that start from the foundation and have a different penetration depth into the sedimentary cover.

3. Results and discussion

As it was mentioned above, the research area refers to the Krasnoleninsky arch. In particular, the lower-middle Jurassic hydrogeological complex of the Talinskoye oil field was considered in the most detail. The area is characterized by the development of a large number of disruptive violations. In case they had no effect on the hydrogeochemical, hydrogeodynamic and hydrogeothermal conditions, proceeding from the paleohydrogeological conditions in the Lower-Middle Jurassic complex, the chloralkalcium type of waters would be widely distributed, according to V.A. Sulin, with mineralization of 15-20 g/l and higher. The hydrogeodynamic field would be homogeneous and would be characterized by the development of normal and superhydrostatic pressures.

But the picture of the hydrogeological field is quite different. The hydrocarbonate-sodium type of water (according to V.A. Sulin) is widespread, the mineralization varies from 2 g/l to 16 g/l. Ion-salt composition - hydrocarbonate-chloride sodium.

The hydrodynamic field is extremely inhomogeneous, the areas of the pre-hydrostatic pressures alternate with the areas of superhydrostatic ones. The pressure deficit can reach 9 MPa, with an average of 1.6 MPa. Areas with superhydrostatic pressures (SHSP) are fixed locally. For the Talinskoye oil field (as for other Krasnoleninsky arch fields), the development of such conditions when vertical fluid migration prevails over the lateral one, since the pressure drops in the section exceed those on the laterals by 1-2 MPa, is the most possible [6].

We believe that it is possible to explain such contrasting conditions from the position of the geodynamic concept in modern hydrogeology [7], according to which, underground waters form, move, and also transform their composition under the conditions of functioning of geofluidal systems (GFS). V.M. Matusevich and his co-authors interpret the concept of GFS as a complex block structure, where the elements are structural-lithological blocks and their boundary zones [7]. In our opinion, at the present stage of the development of the West Siberian geosyneclise, these zones, which are represented by discontinuous disturbances, places of increased or decreased permeability of rocks (barriers), determine the directions of movement of the underground fluids at the depths of the aquifers of the Jurassic and older ages of the Krasnoleninsky arch. Similar conditions are described by P.W. Huntoon [8] for the Wyoming Artesian Basin.

Fig. 1 shows the seismic picture of the lower part of the Mesozoic hydrogeological basin [9] and indicates the hydrogeochemical characteristics of water samples taken from appropriate exploratory wells. Hydrocarbonate-sodium type of waters with low salinity is recorded near the disruption of disturbance (no more than 3 km). Chlorcalcium type, according to V.A. Sulin, with mineralization normal for these depths is determined in samples taken also from the Lower-Middle Jurassic complex, but from wells located at a distance of more than 2.5-3 km from the revealed disruption. The obtained data are summarized in Table 1.

The foundation of the sedimentary cover in this region is complicated by numerous intrusive bodies of the Paleozoic age, represented by gabbro, gabbro-diorites, serpentinites. These bodies are located at the intersection of large faults, which are the boundaries of geoblocks. In these zones, increase in basement temperatures from 110 ^{0}C to 145 ^{0}C is fixed. Disruptive disturbances are a kind of "roots", which can also be traced in the form of boundary zones between structural-lithological blocks (Figure 1). We believe that the reduced mineralization and the hydrocarbonate-sodium type of water are the result of the penetration of deep high-temperature fluids through these zones, which diluted stratal sedimentary waters. In the study area, there are no brines, so, most likely, deep fluids have a low salinity.

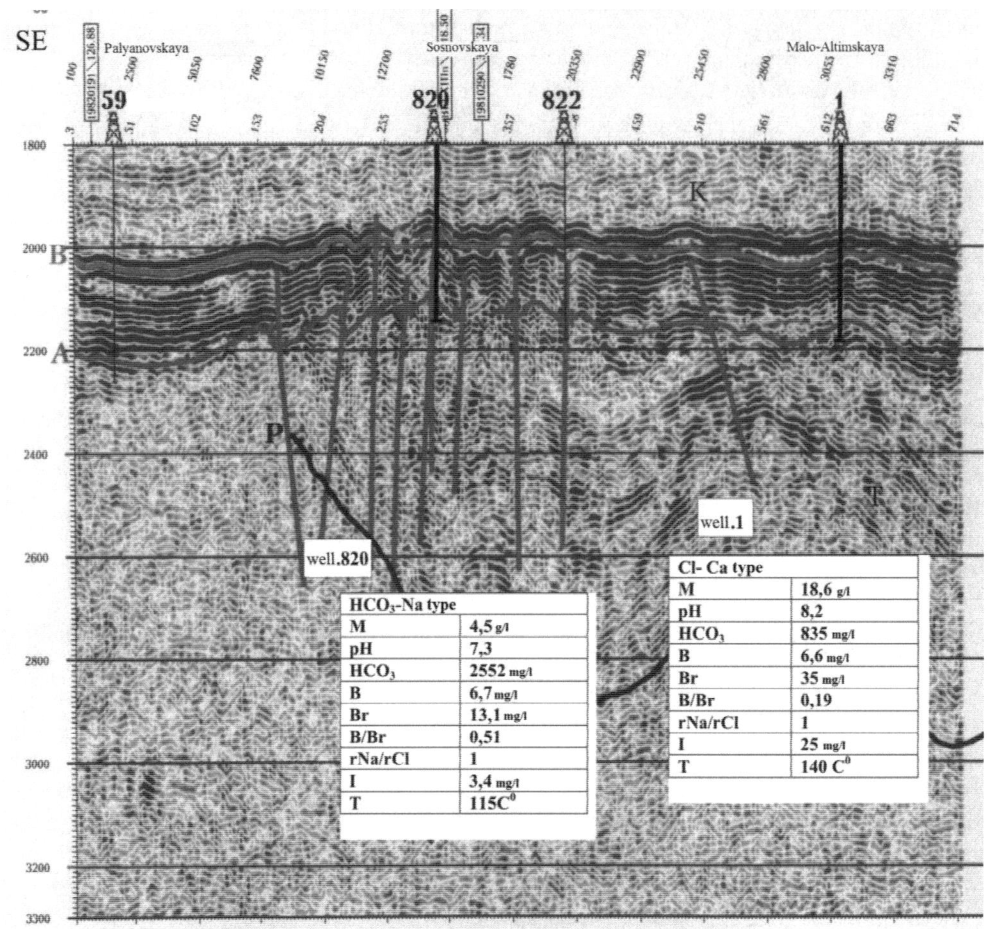

Figure 1. Seismic profile of the lower part of the Mesozoic hydrogeological basin.

This is confirmed by several facts below. For example, the coefficient B/Br presented in Table 1 (a value of up to 2.2, an average of 1.5), which indicates the genesis of deep waters. Values of B/Br - coefficients approaching 1, or exceeding these values, should indicate a sharp increase in the temperature of the feed solution, i.e. on the arrival of deep fluids [10]. A detailed explanation of this phenomenon is given in the articles of V.A. Vsevolozhsky and T.A. Kireeva.

Table 1. Generalized hydrogeological characteristics.

Index	Fracture zone	No faults and fracture
Reservoir pressure, MPa	22 - 27	25 - 27
Water type according to V.A. Sulin	Hydrocarbonate-sodium	Hydrocarbonate-sodium, sometimes chloralcium
Mineralization, g/l	3 - 15, 9	5 - 18
basement roof temperature, ^{0}C	100-145	93 - 140
B/Br	0.34 – 2.2	0.19 – 0.63
SiO_2, mg/l	4 - 28	7 - 99
I, mg/l	1.7 – 6.1	2.6 – 15.2
rNa/rCl	0.87 – 2.46	0.96 – 1.96
Formula of ionic-salt	(Na+K)95 Ca3 Mg2	(Na+K)96 Ca3 Mg1

composition	Cl74 $HCO_3$26	Cl90 $HCO_3$10
Limits of changes in the ionic composition	(Na+K)(92-97) Ca(1-6) Mg(1-3) Cl(70-75) HCO_3(25-30)	(Na+K)(96-98) Ca(2-4) Mg(0-1) Cl(88-93) HCO_3 (7-12)

Numerous studies of products of hydrothermal activity have also been carried out in the area of work [11], such authigenic minerals as pyrite, siderite, dickite, carbonates, kaolinite, etc. were found. Zubkov M.Yu. [6], on the basis of reconstructions of the mechanism of formation of secondary porosity and collectors in the Jurassic sediments of the Talinsky field of the Krasnoleninsky arch, showed a sequence of hydrothermal mineral formation. The author provides numerous evidences of hydrothermal processing of rocks due to the introduction of deep fluids, conditionally calling them hydrothermal He notes that the separation of various types of hydrothermal minerals occurs repeatedly, which is determined by the thermodynamic conditions and composition of the hydrothermal fluid. The same fluids can be a "force", the power of which can lead to hydraulic fracturing of rocks, and thus the formation of vertical migration channels (block boundaries in the GFS) of groundwater, which facilitate their mixing and transformation of the composition.

The contrast of the hydro-geodynamic field, the alternation of piezominimums and piezomaximums within the Krasnolenysky arch in the Lower-Middle Jurassic complex, can in many ways be related to the manifestation of two oppositely directed flows of pore fluids [6]. Using the example of the basal productive deposits of the Talinskaya area, it has been shown that high-temperature, weakly mineralized hydrothermal fluids that have acidic composition intensively dissolve unstable minerals, forming at their expense a secondary capacity, which must necessarily lead to an improvement in the reservoir properties of the rocks, which they enter.

4. Conclusion

An analysis of the hydrogeological conditions of the lower part of the Krasnoleninsky section shows that at the late stages of the development of the West Siberian megabasin, the main reason for the transformation of the hydrogeochemical and hydrogeodynamic fields is the neotectonic processes. Their action leads to the frequent occurrence of the hydrocarbonate-sodium type of waters by V.A. Sulin with low mineralization, occurrence of areas with lower- and superhydrostatic reservoir pressures. In the area of research, the products of hydrothermal activity of deep fluids are widespread, often changing the reservoir properties of rocks, both by forming channels of vertical migration, and vice versa - hydrodynamic screens due to secondary mineral formation.

References

[1] Toth J 1987 Petroleum hydrogeology – a new basic in exploration *World Oil* **205** pp 48–50
[2] Toth J 1987 Petroleum hydrogeology – a potential application of groundwater science *J. of the Geol. Society of India* **29**(1) pp 172–9
[3] Dyunin V I 2000 *Hydrodynamics of Deep Horizons of Oil-and-Gas Bearing Basins* (Moscow: Nauchnyi mir)
[4] Kartsev A A, Vagin S B and Matusevich V M 1986 *Hydrogeology of Oil-and-Gas Bearing Basins* (Moscow: Nedra)
[5] Matusevich V M and Bakuev O V 1986 Geodynamics of water-pressure systems of the West Siberian oil-and-gas bearing basin *Soviet Geology* 2 pp 117-22
[6] Zubkov M Yu 2017 Tectonohydrotermal processes in Jurassic sediments of Western Siberia *Geology of Oil and Gas* 1 pp 64-82
[7] Matusevich V M, Rylkov A V and Ushatinskiy I N 2005 *Geofluidal Systems and Problems of Oil and Gas Content of the West Siberian Megabasin* (Tyumen: TSOGU)
[8] Huntoon P W 1986 Fault severed aquifers along the perimeters of Wyoming artesian basin *Ground Water* **23** pp 176–81
[9] Akinshin A V and Kornev V A 2013 Increase in the determination reliability of permeability

coefficient of complexly built reservoirs on the example of the Vikulovian deposits of the Krasnoleninsky arch *Natural and Technical Science* 5 (67) pp 159-61
[10] Vsevolozhsky V A and Kireeva T A 2009 To the problem of formation of inversions of hydrogeochemical zonality *Bulletin of Moscow University* 5 pp 19-25
[11] Garipov O M and Lukin A E 1995 Features of regional and superimposed catagenesis of the basal sediments of the Krasnoleninsky arch *The Main Directions of Research in the Oil Industry* (Tyumen) pp 54-7

IOP Conf. Series: Earth and Environmental Science **181** (2018) 012002 doi:10.1088/1755-1315/181/1/012002

Drilling automation based on rigs equipped with the top drive system

A M Andriyanov[1], P A Andriyanov[1] and M Bur[2]

[1] Industrial University of Tyumen, Tyumen
[2] LLC "Bentek", Tyumen

E-mail: andriyanov@list.ru

Abstract: The article overviews the issues of increasing labour productivity due to the introduction of drilling automation systems into the technological process.

1. Introduction

At present, billions of dollars are being invested in the oil and gas industry, and the intellectual forces of many countries are working on the improvement of technology [1]. In this connection, the problem arises of increasing the efficiency of capital investments. One of the ways to improve efficiency is automation of technological processes.

The least studied area of the oil and gas industry is drilling automation. Drilling automation helps to increase labour productivity by improving technological processes, optimizing the rate of penetration. In addition, drilling automation should improve the efficiency of environmental protection, labour protection and compliance with safety regulations. Formalization and algorithmization of the drilling process can be considered as a way to consolidate best practices and accumulated knowledge and, subsequently, as a basis for unification and standardization.

When choosing a drilling drive, among other factors, the level of automation of the drilling rig is also considered, which can be the key to the efficient and rapid solution of complex tasks arising during drilling.

2. Research

As is known [2], a path to the creation of automated drilling systems can be represented as the passage of the three stages (levels) of automation:

1. A system offering recommendations:
   - proposes a full range of alternative solutions and actions;
   - proposes a number of alternative solutions (minimizes their number);
   - proposes one action plan.
2. A system that makes a decision but requires approval:
   - selects a recommendation and executes it if approved;
   - provides limited time to disable the action before proceeding to its automatic execution;
   - executes the action upon obtaining approval.
3. An autonomous system:
   - executes the action automatically, provides information only on request;
   - executes the action automatically, informs about the execution of the action;

Content from this work may be used under the terms of the Creative Commons Attribution 3.0 licence. Any further distribution of this work must maintain attribution to the author(s) and the title of the work, journal citation and DOI.
Published under licence by IOP Publishing Ltd

IOP Conf. Series: Earth and Environmental Science **181** (2018) 012002 doi:10.1088/1755-1315/181/1/012002

- makes all decisions and acts autonomously.

The possibility and efficiency of drilling automation depends on the type of drive used in drilling rigs.

3. Results and discussion

One of the ways to improve the quality and increase profits of oil and gas enterprises in the field of drilling is the use of drilling rigs equipped with a top drive system (TDS), which have proven themselves in operation in difficult climatic conditions. An example of such drives, widely used both in the North of Western Siberia and in other regions of the post-Soviet space (Azerbaijan, Belarus) are TDS by Bentek.

TDS by Bentek can be used on wells with a large vertical bottom-hole deviation, as well as for wells with a deviation in the direction of drilling. They meet the requirements of the existing market (in terms of reliability and power) for a large torque load, allow reducing repair costs, provide a high level of safety. In addition, they are accompanied by an affordable and high-quality service.

Using TDS by Bentek, the following functions can be implemented:

- rotation of the drill string with frequency control while drilling, reaming and expanding the wellbore, while raising/lowering the drill string;
- braking of the drill string and its retention in the set position;
- tripping operations (build-up/dismantling of the drill string with stems and single pipes, screwing/unscrewing of drill pipes, fastening/unfastening of threaded connections of sub-assemblies and ball valves, feeding drill pipes to the wellbore/taking the swivel away from the wellbore);
- carrying out operations to run casing strings into the well.
- flushing the well and simultaneously turning the drill string;
- setting and maintaining torque and speed, their measurement and display of readings on the display of the control cabinet, remote display, control panel and the station of geological and technical studies;
- remote control.
- sealing the inside of the pipe with ball valves.

TDS by Bentek are equipped with variable frequency drive (VFD) modules. Figure 1 shows a diagram illustrating the work of the VFD.

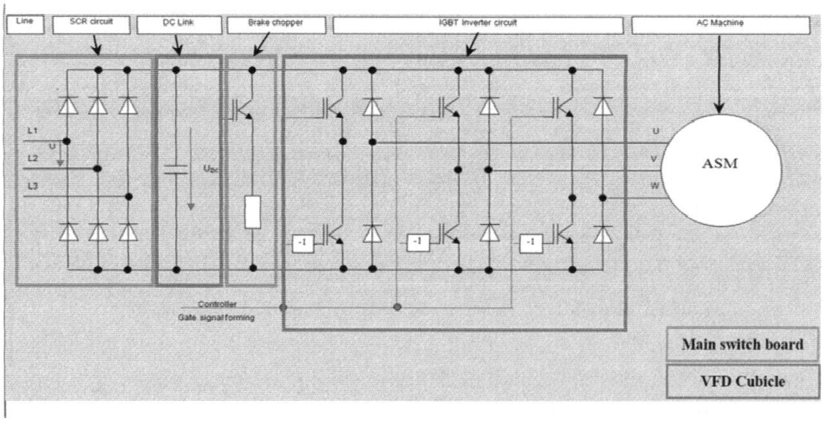

Figure 1. Diagram of functioning of VFD.

As can be seen from the diagram, regulation is performed using insulated-gate bipolar transistors (IGBT). These modules regulate the rotational speed of the electric motor rotating the drill string.

In the event of an error during the power-up procedure, or if it has already occurred and has not been acknowledged, the automation does not allow the unit to be switched on. In the event of a malfunction during the operation of the drive, the inverter will be disabled by the following algorithm:

- shutdown of the frequency drive by locking the power field IGBT and shutdown of the power circuit-breaker; if the failure conditions allow, then before the shutdown, the drive is controllably stopped;
- alarm signalling by means of a contact relay;
- displaying and saving the cause of the controller's memory fault.

In addition to local optional indicators (displays, LEDs), fault signals are also displayed on the front panel of the converter control unit and can be interrogated in the ecoWin parameterization program.

TDS by Bentek are equipped with the InfoDrill drilling control system. Let us consider in more detail the features of this system.

Pages of the InfoDrill system screen are always divided into the following sections: a section for failure messages at the top of the screen, a section for displaying parameters in the centre and a navigation section at the bottom of the screen (Figure 2).

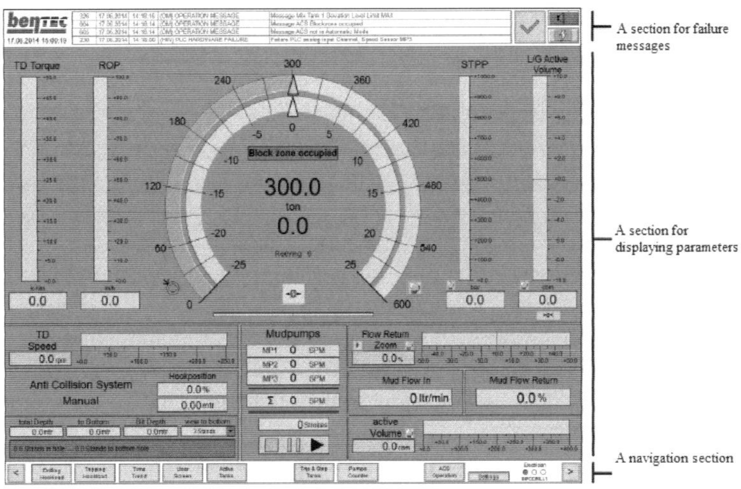

Figure 2. InfoDRILL screen.

In the section for failure messages, a table displays the actual faults (latest from the top). There you can find information about the failure number, the date/time of occurrence and a text message for easy troubleshooting.

Failure messages are coloured in the appropriate colours depending on the status which are as follows:

- incoming failure;
- confirmed failure;
- current failure.

Selecting the appropriate symbol (mouse click) initiates the following actions:

- confirmation;
- turning off the alarm signal;
- displaying the failure overview page.

The navigation section allows you to select the desired page with drilling parameters.

Two monitoring pages serve to indicate the parameters during drilling and during tripping operations (Figure 3). The management and layout of both pages are almost identical. The corresponding page is called up in the navigation section.

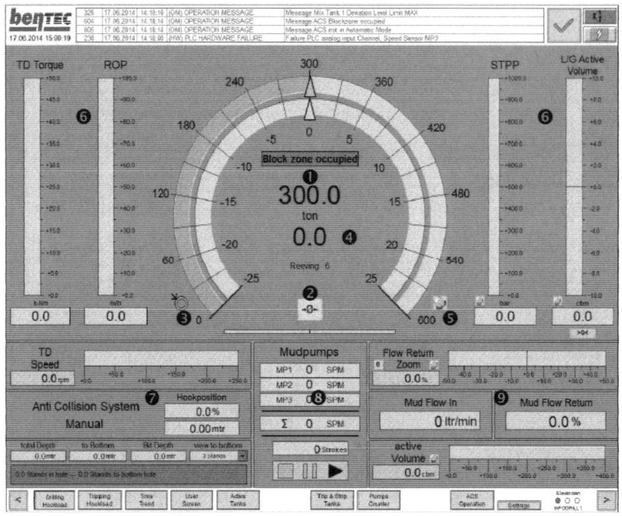

Figure 3. Drilling parameters.

All the necessary drilling process data (drilling/tripping operations) are displayed on these respective screen pages. In the middle of the screen there are circular indicators of load on the hook and on the bit, which also have digital displays in the centre showing numerical values of these parameters. An indicator in this area also shows the status of the path of the hook block movement - free/busy.

There are also additional functions for managing and tuning the above-named instruments.

To display and analyse time trends, there are three pages of InfoDrill system screens, such as time trends for drilling, tripping and user-defined ones (Figure 4). The management of all these pages is similar, except for the user-defined time trends which they can customize at their own discretion. After activating the screen, time trends in the navigation section must be selected from the drop-down menu of the trend page of interest.

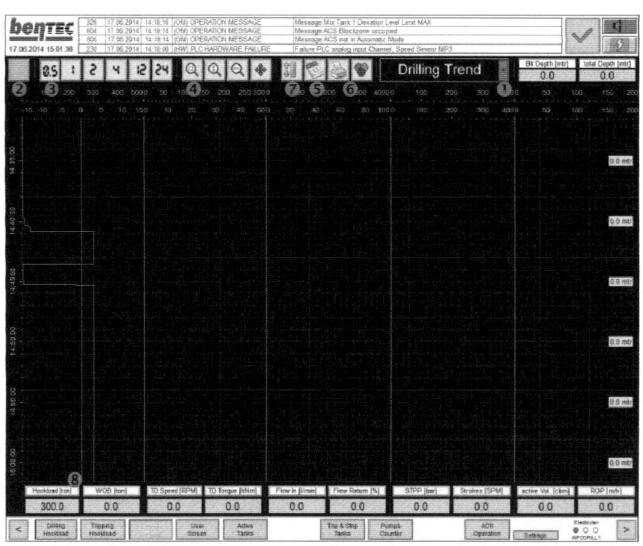

Figure 4. Time trends for drilling

IOP Conf. Series: Earth and Environmental Science **181** (2018) 012002 doi:10.1088/1755-1315/181/1/012002

If the time trends screen in the navigation section is activated, then you need to select the trend page of interest in the drop-down menu. The indicator shows the status of data transfer for time trends.

The time trends screens contain 5 columns of 2 recorders each, which display the actual data of the current drilling and tripping parameters.

To view the status of the tank system, two screens are available. The management and layout of both pages is almost identical.

The corresponding page is called up in the navigation section.

This monitoring page (Figure 5) displays active drilling mud tanks, as well as the time trend for changing its active volume.

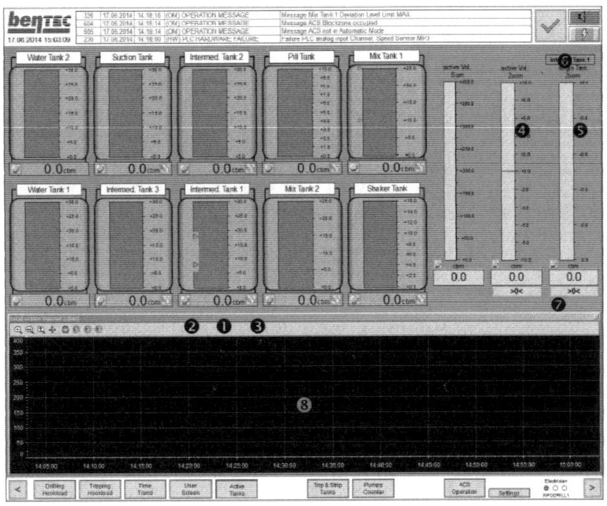

Figure 5. Active tanks.

In addition to displaying the drilling fluid level for each tank, it is also possible to set the upper and lower alarm limits. To set the level of the alarm signal, as well as the resolution of the scale using the pop-up window, you must click on the symbol of the tank of interest. The output of the drilling fluid level in the tanks beyond the limits of the set values causes an alarm. Using the buttons, this alarm can be activated or deactivated for each tank separately.

Tanks that are part of the active volume of the drilling fluid can be activated separately.

The first linear indicator on the left shows the total active volume of the tanks. The next two linear indicators show the input/output of the total flow volume and specific tanks. Just like in tanks, here you can set the upper and lower alarm limits for each line indicator.

The InfoDrill system provides a page for displaying the status of the operation of mud pumps. The corresponding page is called up in the navigation section (Figure 6).

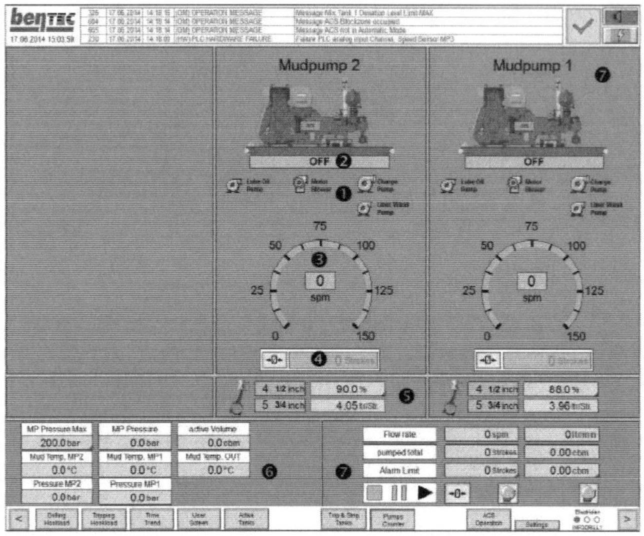

Figure 6. Mud pumps and stroke counter.

This page displays the status of mud pumps and their auxiliary equipment in a separate window for each pump. The following information is displayed for each pump separately:

- status of auxiliary drives, depending on the colour of the symbol (OFF (grey), ON (green), FAULT (red));
- status of mud pumps.

Under the graphic symbol of the mud pump is an indicator showing the mode of operation as a plain text on a coloured background. In addition, the background colour changes depending on the mode of operation:

- OFF (grey);
- ON (green);
- FAULT (red);
- Warning (yellow).

In the InfoDrill system, it is possible to set up and control the anti-collision system (ACS). The start screen is called up in the navigation section. There are two modes of operation: hand and automatic.

When working in hand mode, it is necessary to pay attention to the fact that the restriction and tracking of movements of the hook block are not activated. Calling up the automatic mode of ACS immediately initiates the system setup procedure.

When you click on the corresponding symbol, the following ACS functions are available on the screen:

- setting upper/lower work points;
- ENTER and ESCAPE to confirm or reset the entered value;
- calling up the procedure for determining the length of the brake distance;
- confirming the procedure for determining the length of the brake distance;
- cancelling the procedure for determining the length of the brake distance;
- setting tool compensation;
- crown block in emergency mode.

In the automatic mode (Figure 7), the current position of the hook block, as well as the upper and lower work points, are displayed on the linear indicator. In addition, there is a numeric display which shows all values in percent.

IOP Conf. Series: Earth and Environmental Science **181** (2018) 012002 doi:10.1088/1755-1315/181/1/012002

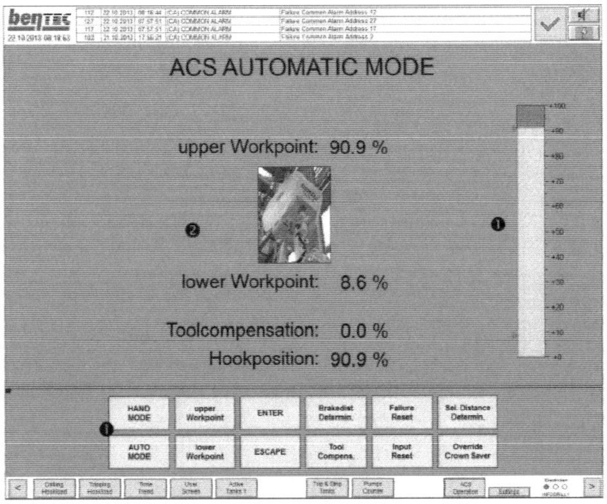

Figure 7. ACS Automatic mode.

Switching on the soft torque rotary system (STRS) is carried out by a switch on the driller's console, and its adjustment is made on the page of the InfoDrill system screen. The start screen is called up in the navigation section (Figure 8).

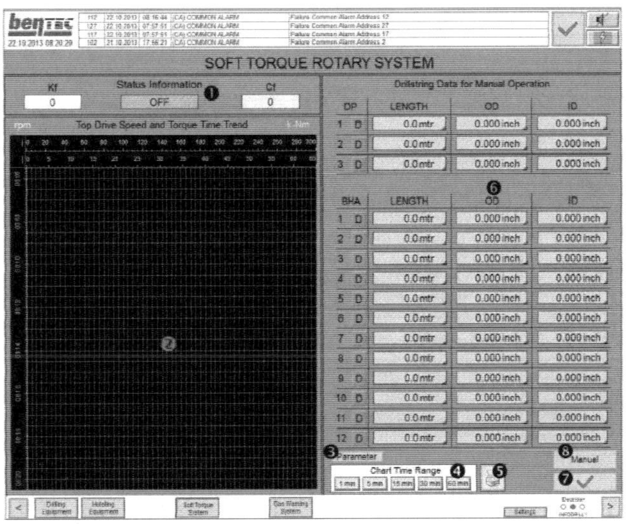

Figure 8. The STRS system with parameter calculations based on the drillstring data.

The screen is divided into sections as follows:

1. Status Information:
   - actual Kf / Cf parameters of STRS (set or calculated);
   - STRS response s (OFF/READY/ON).
2. Time trend (parameters):
   - TDS speed and torque time trend;
   - enable/disable the additional display, time trend parameters;
   - 1/5/15/30/60min time scale.

IOP Conf. Series: Earth and Environmental Science **181** (2018) 012002 doi:10.1088/1755-1315/181/1/012002

3. Printing out the screen by an external printer.
4. Entering drillstring data and Cf, Kf for calculating.

If the parameter input section for calculation is on, the values of the parameters Cf and Kf are calculated based on the length of the drill pipe, its outer and inner diameters, and the same parameters for the bottom-hole assembly (BHA). The input of the BHA parameters should be performed sequentially, beginning with the bit and up, i.e. first - BHA 1, then - BHA 2, and so on. If there are several components, the length of which is very small, then they can be neglected or if the components are the same, then they can be combined into one. When entering the parameters of the drill pipe (DP), there are three sections for entering different lengths, outer and inner diameters. When entering parameters, pay attention to the correct sequence, DP1 is first drill pipe from the bit. Once you specify the outer and inner diameters, you only need to adjust the length based on the number of stands added during the drilling process.

In the InfoDrill system, it is possible to display the status of the gas warning system. The corresponding page is called up in the navigation section (Figure 9).

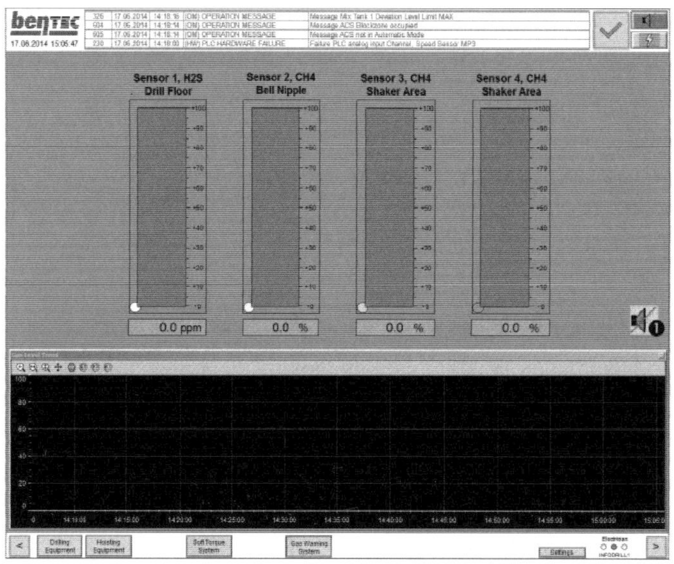

Figure 9. Gas warning system.

The status of the gas level is displayed on this page. Gas levels measured by sensors are displayed as a percentage on linear indicators, as well as a function of time on the time trend. Exceeding the limit values immediately causes an emergency message.

When switching to the third panel in the navigation section, additional Data Export (Figure 10) and Equipment Diagnostics (Figure 11) buttons appear. While on the data export page it is possible to export user-defined measurement parameters, the Equipment Diagnostics page displays the status of the components of the InfoDrill system.

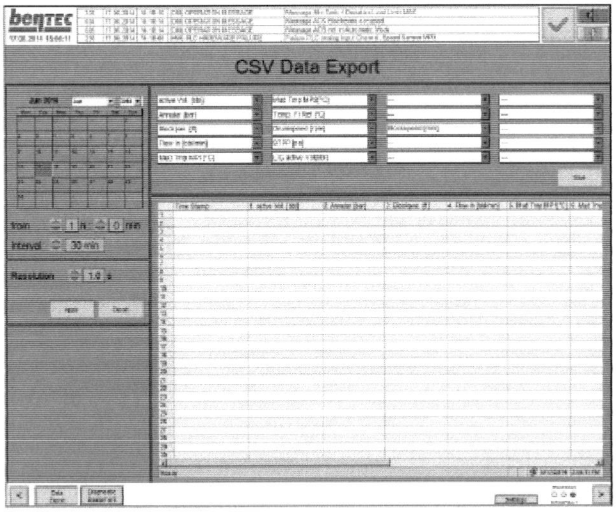

Figure 10. Data export to CSV format.

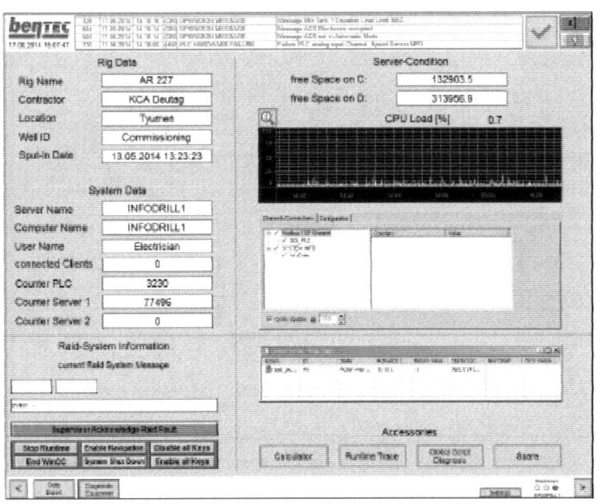

Figure 11. Diagnostics system.

4. Conclusion

The mechanized equipment of the drilling rig, proposed to date, allows replacing the same type of operations previously performed manually with the operation of technical means, minimizing the number of personnel present in the potentially hazardous area of the drilling site. However, this can hardly be called "drilling automation". Direct automation of drilling should allow the driller to penetrate productive layers more efficiently in comparison with the results of manual control, with less effort and resources, and strict observance of safety rules.

Drilling automation has to deal with a problem similar to the one that had to be overcome in the automation of other industries: many drilling specialists treat it with suspicion and see it as a threat to their professional experience.

Some side-effects of drilling automation are the reduction in the number of personnel and the ability to perform tasks with remote control.

To automate the drilling process, a system is needed that can function effectively under conditions of variability and uncertainty. Downhole and surface data introduced directly into such systems make them respond to changes, for example, of lithological characteristics, in order to maximize the efficiency of the well, increasing its production time and production performance.

References

[1] Prospects of the oil and gas industry in Russia. [Electronic resource]. URL: http://assoneft.ru/activities/press-centre/tek/4121/.
[2] Oil and gas review. Drilling automation. [Electronic resource]. URL: http://www.slb.com/~/media/Files/resources/oilfield_review/russia12/sum12/02_drilling.pdf.

IOP Conf. Series: Earth and Environmental Science **181** (2018) 012003 doi:10.1088/1755-1315/181/1/012003

Application of numerical modelling for evaluation of operating conditions of water intakes for domestic and drinking purposes in the area of oil fields operation

Yu V Bespalova[1], V K Popov[2], K I Kuzevanov[2] and N S Trofimova[1]

[1]Industrial University of Tyumen, Tyumen, Russia
[2]National Tomsk Polytechnic University, Russia

E-mail: bespalova_y@mail.ru

Abstract: The research is focused on preserving the resources of domestic and drinking water supply in the areas of operation of numerous oil fields. The basis of the study is the well-known principle of protection of groundwater from contamination during its operation, which is implemented at existing water intakes for domestic and drinking water supply through the organization of sanitary protection zones (SPZs). In the conditions of mass application, in order to calculate their boundaries for the most common analytical methods of calculation, in practice, a simplified schematization of hydrogeological conditions is often allowed, not fully considering the complex spatial structure of the filtration flows. The solution of the predicted hydrodynamic problem in the form of a spatial distribution of heads under the influence of water intake is obtained with the calculation of all balance characteristics on a numerical model. The model distribution of heads is used to solve the hydrodynamic problem in the nonstationary formulation. Recommendations for schematization of hydrogeological conditions for the purpose of justification of local hydrogeodynamic models and the study of filtration parameters in the course of experimental filtration work for the calculation of operational reserves were developed. The results of numerical modelling allow evaluating the influence of any potential source of contamination by its geographical coordinates.

1. Introduction

Hydrodynamic modelling was carried out in co-authorship with the staff of the National Tomsk Polytechnic University, within the framework of the scientific project No. 16-35-50181 with the financial support of the Russian Foundation for Basic Research. Modelling of hydrogeological conditions was carried out with the goal to estimate features of formation of groundwater reserves with reference to the substantiation of research of protection of operating groundwater intakes of domestic-drinking purposes from potential surface contamination in the areas of producing oil fields. The work is based on the results of the generalization of the sampling data of 30 reference hydrogeological wells, stock cartographic and report materials for the Middle Ob basin of the groundwater flow (the Vartovsky oil and gas bearing region). Data processing was aimed at assessing the conditions for the development of potential contamination with the aim of developing recommendations for clarifying the size of sanitary protection zones using the capabilities of numerical modelling. The main advantage of modelling lies in the possibility of taking into account such elements of the hydrogeological section as the layered occurrence of water-bearing rocks and segregated strata, the complex shape of the outer boundaries of the filtration area, the spatial

Content from this work may be used under the terms of the Creative Commons Attribution 3.0 licence. Any further distribution of this work must maintain attribution to the author(s) and the title of the work, journal citation and DOI.
Published under licence by IOP Publishing Ltd

heterogeneity of filtration parameters, the complex nature of exciting wells, etc., which cannot be taken into account in the framework of schematization of hydrogeological conditions for the purposes of analytical hydrodynamic calculations. The numerical model of the filtration area was created on the basis of the available cartographic material represented by heterogeneous maps in uncoordinated coordinate systems. The processing of the initial data was carried out in stages using geoinformation systems and eventually merged in the GMS (Groundwater Modelling System) as a set of electronic layers allowing for their joint processing, including the ability to digitize key objects.

2. Research

The tasks of hydrodynamic modelling are primarily related to the need to quantify the degree of protection of groundwater from contamination, which is formed under the influence of the geological structure and hydrogeological conditions of the water-bearing strata and can be most reliably estimated with the most complete consideration of the influence of many interrelated factors on the basis of water balance equations. From practical points of view, it is of interest to identify the spatial structure of the filtration flows, which can be used as a basis for assessing the shape and size of sanitary protection zones for both active water intakes and those being designed.

3. Results and discussion

The solution of the predicted geofiltration problem in a stationary setting is represented by the calculated field of heads, which is formed under the influence of the total impact of natural (infiltration recharge, constant head in the river channel) and artificial (operation of individual water intakes) boundary conditions.

Based on the predicted field of pressure on the regional model of the operational water-bearing complex of Paleogene deposits, the maximum dimensions of the capture zone of the filtration flow of water intake wells have been obtained in a stationary setting (for an unlimited lifetime). Shown on the map (Figure 1), they provide an opportunity to assess the potential conditions of contamination of producing aquifers, taking into account the spatial location of potential contamination sources. The flow path diagram allows localizing those areas where surface or deep contamination can get into the catchment zone of the water intake well and indicate the areas within which foci of contamination cannot impair the operation of production wells (Figure 1) under any circumstances.

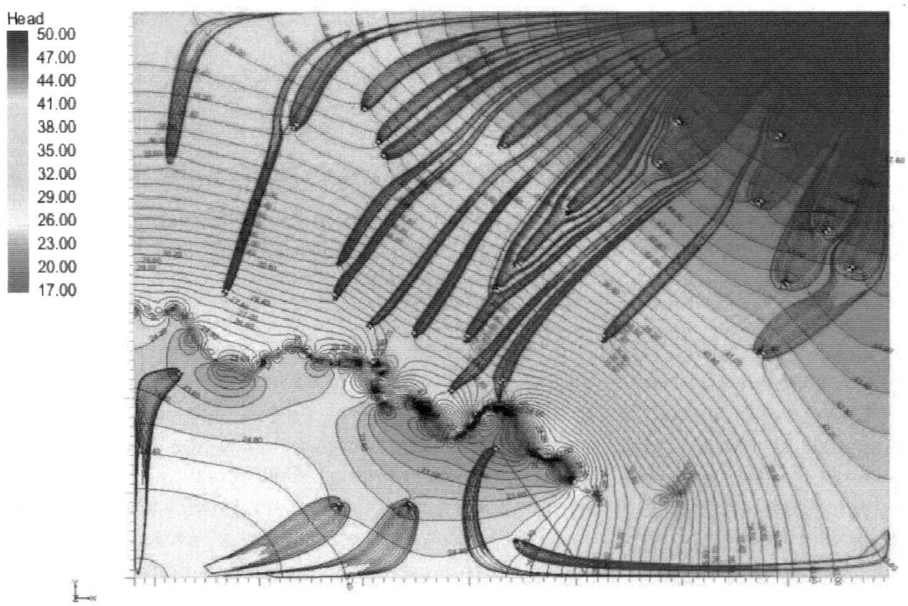

IOP Conf. Series: Earth and Environmental Science **181** (2018) 012003 doi:10.1088/1755-1315/181/1/012003

Keys:

 - flow paths;  - heads and their marks;

Figure 1. Scheme of the flow paths development for an unlimited lifetime (limiting dimensions of the filtration flow zones covered by production wells are shown).

The above scheme allows assessing the danger of any potential source of contamination, according to its geographical coordinates. If the location of the pollution source is outside the areas occupied by flow paths, it cannot impair the operating conditions of any water intake under any conditions. Conversely, if a potential source of contamination falls within the boundaries of the field of flow paths, contamination of water intake is inevitable, but the question of the period of time during which this can occur remains open. The results of numerical modelling provide a reasonable answer to this question. To do this, it is sufficient to analyse the time during which filtration takes place in the direction of the flow paths of the filtration flow. An example of such an analysis is shown in Figure 2. From the point of view of practical evaluation of the structure of the filtration flow, the limiting dimensions of the filtration flow zones covered by production wells should be limited. For the sake of demonstration, in Figure 3 the time for the filtration flow to reach the wells is reduced to 2000 years. In this case, the size of the water-capture area in the projection to the surface is on average 6 km.

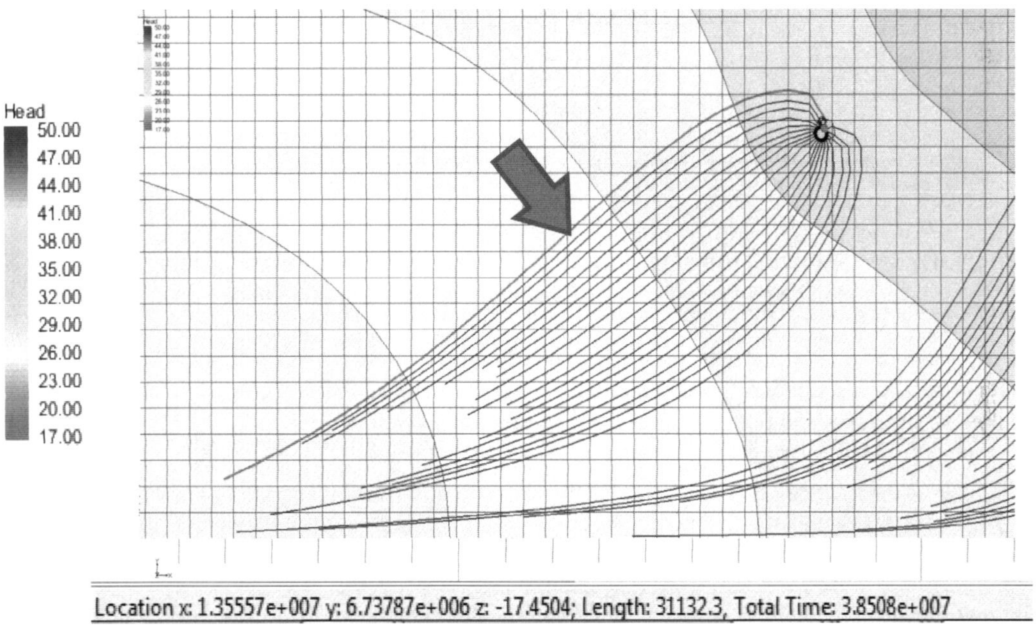

Figure 2. Parameters of the selected flow path (indicated by an arrow): location, length (~ 31 km) and the time of motion of a fluid particle along the indicated trajectory (3.85 × 10^7 days or 105 thousand years).

IOP Conf. Series: Earth and Environmental Science **181** (2018) 012003 doi:10.1088/1755-1315/181/1/012003

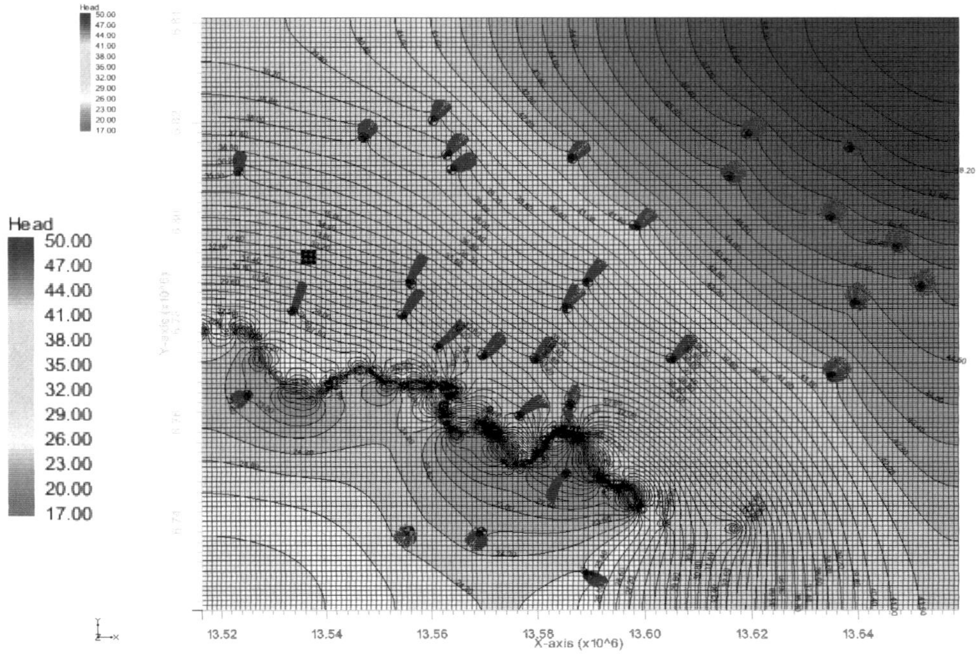

Keys:

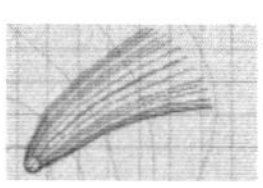 - flow paths; 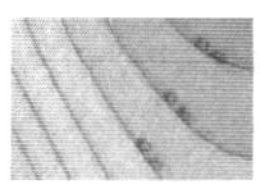 - heads and their marks;

Figure 3. Dimensions of the filtration flow zones covered by production wells being formed for 2000 years (average size of the water intake zone is 6000 m).

Restriction of the estimated time by the duration of operation of water intakes of 10,000 days, which is taken as the standard for calculating the operational reserves of groundwater, sharply reduces the dimensions of the filtration flow zone covered by production wells.

An analysis of the structure of filtration flows in the vicinity of water intake wells shows that during their operation during the service life the dimensions of the filtration flow coverage zone practically do not exceed the boundaries of the calculated cell of the regional numerical model. Consequently, in the formation of artificial filtration flows under the influence of water intake, the priority remains for vertical overflows, which does not contradict the schematization of the hydrogeological conditions adopted for analytical calculations of operational reserves. The dimensions of the filtration flow zones covered by production wells correspond to the boundaries of the third belt of the sanitary protection zone. An analysis of their spatial location allows us to conclude that the mass exploitation of the groundwater of Paleogene deposits does not lead to the unification of pressure sinks. The water intakes according to the nature of their impact on groundwater can be considered as single dispersed sources of disturbance of the filtration flow.

The influence of potential foci of contamination on the operation of the water intake well has been modelled. Based on the prediction field of head, the structure of probable movement of groundwater near a single water intake well 7-857 is analysed. The location of probable potential contamination sources is shown in Figure 4.

Figures 5 and 6 show the planned structure of the filtration flows in the zone of production well No. 7-857. For the northern focus of potential contamination, a quantitative characteristic of the filtration flow along four distinct flow paths is given. The length of the flow paths: minimum - 2753 m, maximum - 3856 m; average - 3278 m. The time of groundwater movement: minimum - 16622300

IOP Conf. Series: Earth and Environmental Science **181** (2018) 012003 doi:10.1088/1755-1315/181/1/012003

days or 45540 years; maximum - 24953800 days or 68367 years; average - 20840700 days or 57098 years.

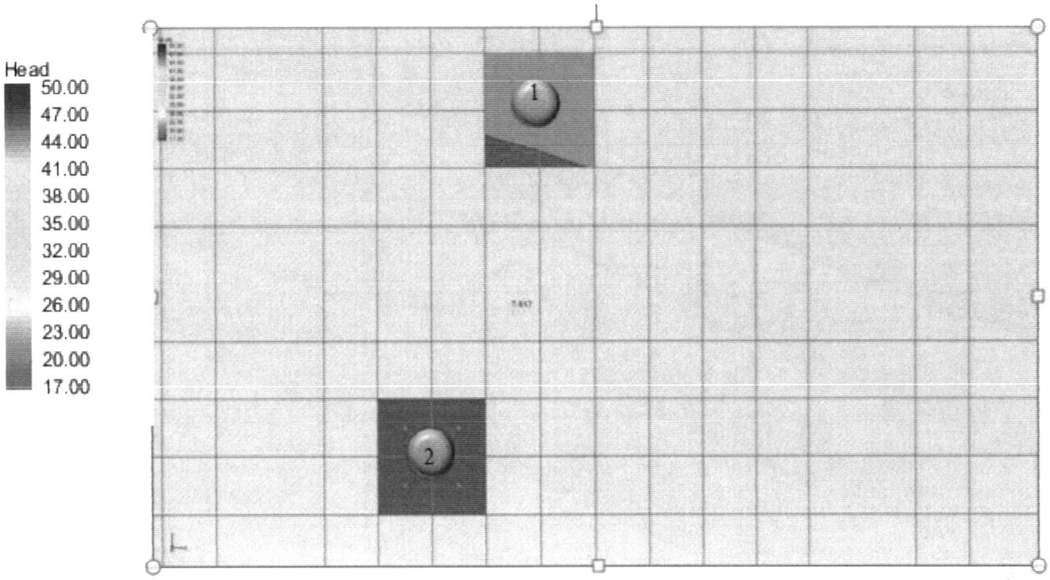

Key:

- northern focus of potential contamination (4 grid cells) and its number;

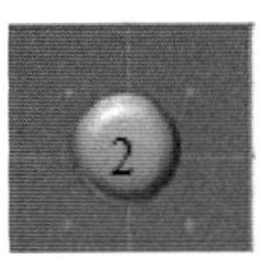

- southern focus of potential contamination (4 grid cells) and its number;

Figure 4. Location of probable surface potential contamination sites in the vicinity of production well No. 7-857: 1, 2 - conditional numbers of potential contamination foci, background shading corresponds to the prediction field of heads.

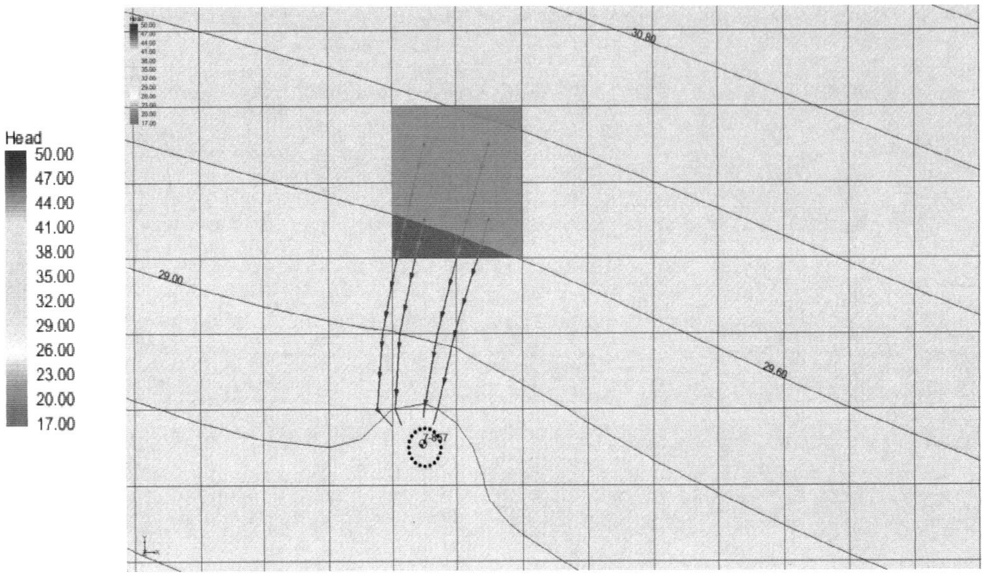

Key:

- focus of potential contamination;

- production well and its number;

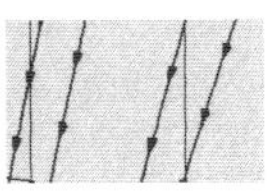
- flow paths, direction is indicated by arrows;

Figure 5. The filtration flow in the zone of production well No. 7-857 ((directed along the flow paths from the centres of the cells of the northern focus of contamination to the well), the background shading corresponds to the prediction field of heads.

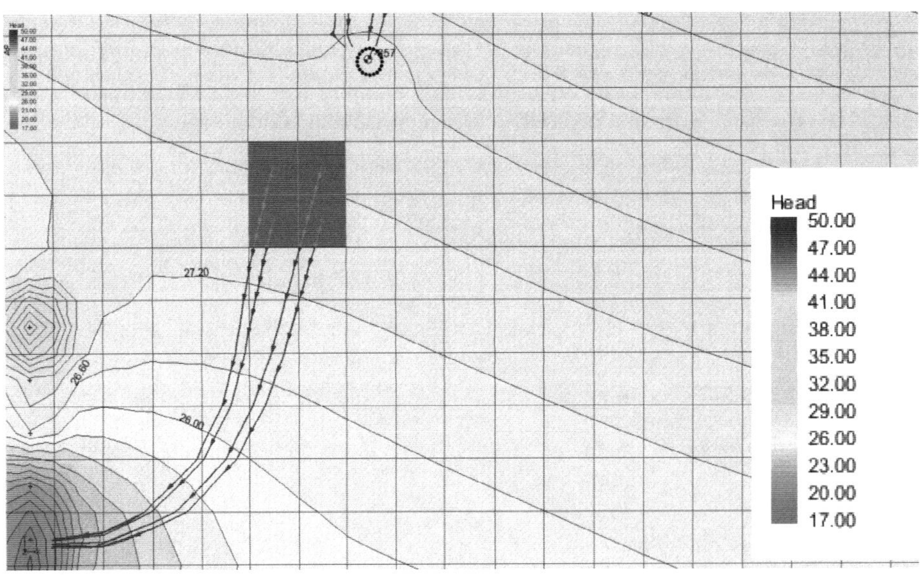

Figure 6. The filtration flow in the zone of production well No. 7-857 (directed along the flow paths from the centres of the cells of the southern focus of contamination to the river), the background shading corresponds to the prediction field of heads (for other notations, see Figure 5).

4. Conclusion

The influence of potential sources of contamination depends to a large extent on the structure of filtration flows, which helps to identify numerical modelling. The regional model shows that with the intensive exploitation of groundwater in Paleogene deposits, existing water intakes by the nature of their impact on groundwater can be considered as single dispersed sources of disturbance of the filtration flow. When using modelling, the calculated zones of sanitary protection reduce the dimensions in the plan by taking into account the vertical component of flow paths.

The influence of the location of a potential source of contamination on the quality of groundwater in an operational aquifer under the influence of the direction of the filtration flow for a local water intake site is shown. For a single water intake, represented by well 7-857, the southern focus of potential contamination cannot threaten the quality of groundwater.

In this case, the source of substandard solutions is located below the well along the flow of groundwater, but it can influence the composition and quality of the surface water of a river. On the contrary, the halo of potential contamination of the northern outbreak, located at a distance of about 2.5 km to the north, can be almost completely intercepted by well 7-857. However, the filtration time in these directions is multiples of the criteria for the calculated parameters of the third belt of the sanitary protection zone of water intakes from persistent chemical contamination. Therefore, the location of potential sources of contamination shown in Figures 5 and 6 cannot have a significant effect on the change in the composition of groundwater operated by well 7-857.

Thus, the application of numerical modelling can significantly increase the reliability of predictive estimates of hydrodynamic predictions associated with practical calculations of the size of sanitary protection zones for both existing and planned water intakes, especially in areas of active economic activity.

The investigations were supported by The Russian Foundation for Basic Research (Project No 16-35-50181).

References

[1] Kuzevanov K I, Savichev O G and Reshet'ko M V 2012 *Mathematical Modeling of Processes in the Components of Nature* (Tomsk: TPU Press) p 144

[2] Wang H F and Anderson M P 1982 *Introduction to Groundwater Modeling-Finite Difference and Finite Element Methods* (San Francosco: W.H. Freeman)

[3] Young H L et al 1986 Nothern Midwest Regional Aquifer System Study (*In Regional aquifer-system analyses program of the U.S. Geological Survey Circular 1002*)

IOP Conf. Series: Earth and Environmental Science **181** (2018) 012004 doi:10.1088/1755-1315/181/1/012004

Creating a geological model of the field taking into account information on horizontal wells and the analysis of inter-well reservoir connectivity

N N Bozhenyuk, V A Belkina and A V Strekalov

Industrial University of Tyumen, 38 Volodarskogo st., Tyumen, 625000, Russia

E-mail: bojenuk_nn@mail.ru

Abstract: The deposits of the Vikulov suite of the Krasnoleninsky arch are highly productive, characterized by a complex geological structure. Herewith, a large number of deposits are drilled with a rare grid of wells, and often extremely uneven, which results in a significant lack of empirical information for constructing an adequate three-dimensional geological or hydrodynamic model.

1. Introduction

In administrative terms, according to [2], the modeled field is located in the northern part of the Krasnoleninsk oil-and-gas bearing region (OGBR) at the border of the central part of the Krasnoleninsk oil-and-gas bearing region. It was discovered in 1992 after testing the first well, where the industrial oil-bearing capacity of the Vikulov suite was established. In tectonic terms, the site in question is located on the West Siberian Plate, within the Uvat-Khanty-Mansiysky middle massif of the Riphean geosynclinal complex, in the northwest of the Krasnoleninsky brow formed during the Baikal folding stage, which was reworked by Hercynian tectonogenesis. This site is the structure of the deutero-orogenic stage of development in the form of brows - ridge faults of the basement.

The deposits of the Vikulov suite are of a marine genesis. The rocks accumulated at the expense of tidal energy and are represented by a different-order rhythmic interlayering of sandstones, siltstones and clay varieties. Sedimentation of the Vikulov suite was carried out in the central part of the inner shallow isolated basin. Deposition of the lower part of the suite accumulated at the stage of the regressive development cycle, which was replaced by the accumulation of the upper part of the formation into a transgressive.

At the time of the creation of the geological model, 34 wells were drilled, including 3 exploration, 13 prospecting and 18 operational (including 8 horizontal) wells, and they are located extremely uneven. During the creation of the model, the drilling of the field by several drilling teams continued in two sections, the data on these wells were constantly added to the project and were taken into account in the construction of the model. On the whole area of the field, seismic works of the CDPM-3D were carried out.

Objective. The creation of a methodology for constructing 3D geological models that takes into account the unevenness of measurements, the criteria for reservoir connectivity, and data on horizontal wells that significantly increase the accuracy of the geological model, and, as a consequence, significantly reduce the number of iterations in the process of creating a hydrodynamic model.

Content from this work may be used under the terms of the Creative Commons Attribution 3.0 licence. Any further distribution of this work must maintain attribution to the author(s) and the title of the work, journal citation and DOI.
Published under licence by IOP Publishing Ltd

Research methods. Methods of lithological analysis of the composition of rocks, separation of sediment accumulation cycles by lithological characteristics into cyclites, detailed correlation, adaptation of the results of variogram analysis on drilled sections to un-drilled ones, geosteering along horizontal wells, criteria of reservoir connectivity, construction of three-dimensional geological and hydrodynamic models.

Results. Based on the analysis of the core material, three cyclites in the VK_1 and VK_2 formations of the studied deposit were identified, which allowed a detailed correlation with the allocation of three packs in each of these layers. The use of variogram parameters obtained in areas with a higher density of measurements and the availability of seismic data, the involvement of data on horizontal trunks made it possible to clarify the structural framework and three-dimensional geological models of lithology, reservoir properties and oil saturation. The above-mentioned models showed good consistency with external data. Increasing accuracy of geological models significantly reduced the computational complexity of constructing a hydrodynamic model.

2. Research

2.1. Detailed correlation

To create a 3D DM, seismic studies, GIS data, cores descriptions and studies, reservoir testing results, etc. were analyzed. One of the first tasks before the correlation of sediments is creating a sedimentological model of the reservoir, which allows to determine the conditions and processes of sediment accumulation and to identify the main facial elements of the section. In the VK_1 reservoir, according to the analysis of the core material, three sedimentation cycles have been identified, in which the sequence of the transition of the sediments of the distal parts of the delta front to sediments of the delta is reflected.

The general correlation of the object was carried out with the identification of the main marking horizons (roof and bottom of the VK_1 formation, roof and bottom of the VK_2 formation), and then a detailed one with the allocation of packs (Figure 1).

Drawing the boundaries is based on the assumption of geostatistics, the main modeling tool for this work, due to the small amount of wellbore data relative to the study area, the stationarity and representativeness of data within a single layer.

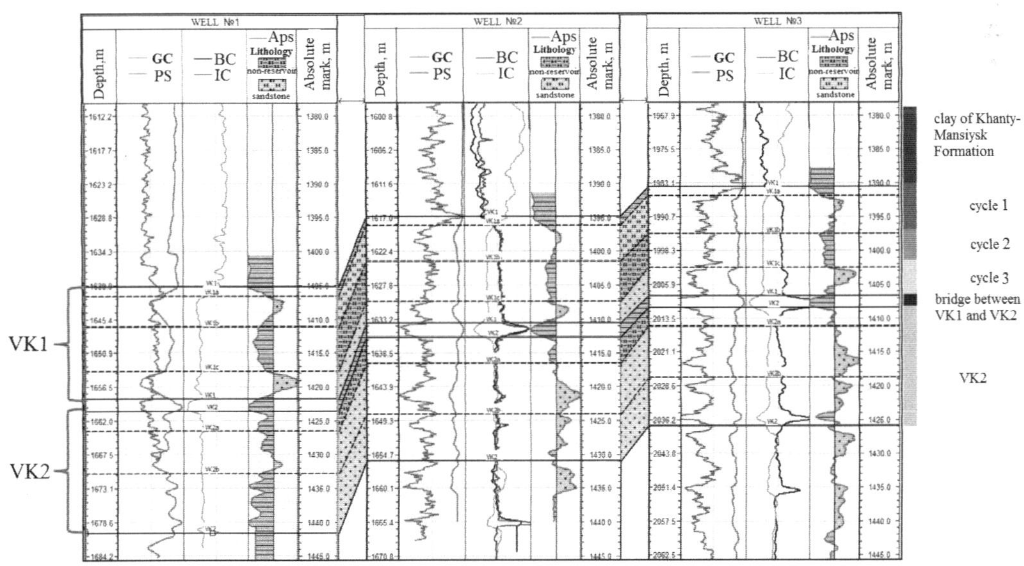

Figure 1. Detailed scheme of the correlation of VK1 and VK2 layers of the Vikulov suite.

2.2. Construction of structural framework

In view of the insufficient knowledge of the area under consideration due to the low degree and, what is also important, the significant unevenness of the well network in the field, the construction of the structural framework of the model was carried out taking into account the uncertainty analysis of the data. Uniformity of geological heterogeneity throughout the reservoir makes it possible to increase the reliability of the formation forecast in the inter-wellbore space, to increase the accuracy of the inventory calculation in areas poorly studied by drilling and more reasonably approach the design of a project document for the further drilling of the field.

The construction of a structural map is carried out in several stages:

1. To perform a variogram analysis of the drilled section of the deposit, the VK1 horizon map obtained by interpretation of seismic materials is converted into a cube of absolute marks, in the presence of areas not characterized by seismic data, and the interpolation is carried out. Then, to analyze the structural features of the drilled part and to construct a semivariogram from the obtained cube, the absolute marks in the attributes of the wells are taken.
2. Next, a cube of distances is calculated, in which the value in each cell is the distance to the nearest well. Further calculations are carried out based on the assumption that the further the site from the site with the actual information and the deeper from the day surface is, the higher the variability is. In accordance with this assumption, the dependence of the variability on the remoteness of the points with the actual information and depth is obtained.
3. Then, multi-variant (stochastic) calculations of the structural map of the roof are carried out, taking into account the selected grades according to the variogram analysis and variance variability, which give a set of realizations with minimal differences in the drilled zone, but with different structure behavior outside the drilled part. Ultimately, according to the probabilistic approach, implementations that fall within the range P50 +/- 5% (where P is the probability) are selected and averaged over them (Figure 2).

Thus, in the undeveloped areas of the field, it was possible to refine the geometry of the structural surface using the surface properties of the drilled area. Subsequently, this led to a change in the oil-drainage boundaries of the field and a decrease in the geological reserves of the deposit.

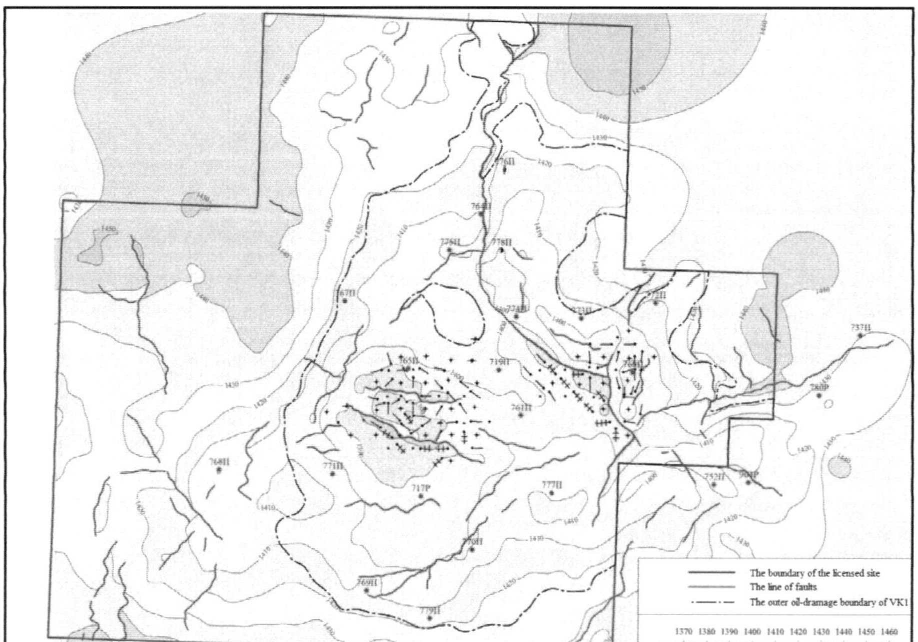

Figure 2. Structural map on the formation top of VK_1 (author's model).

IOP Conf. Series: Earth and Environmental Science **181** (2018) 012004 doi:10.1088/1755-1315/181/1/012004

2.3. Lithological-facial modeling

To reproduce the adequate variability of rock layers, the construction of the lithology cube was carried out in a combined way in two stages. At the first stage, a trend cube of lithology was made in a deterministic way, at the second stage - the final cube of lithology by one of the stochastic methods - by the indicator method. In turn, the algorithm for creating a trend cube of lithology consists of several steps: first, maps of the total and effective thicknesses on the well data are created, which are calculated by extrapolation, taking into account the variability of the drilled part into the undrilled one. On these maps, a map of the coefficient of sand content is calculated. The next stage is the creation of a geological-statistic section and a trend cube of lithology along the GSS and a map of sand content. Further, a variogram analysis is carried out, according to which zonal anisotropy is established on the object under study, which is very typical for bioturbated sediments and sediments with a highly variable genesis that cannot be described only by geometric anisotropy [4]. According to the results of variogram analysis, ranks ellipsoids are built.

Then, the cube of lithology is modeled by the indicator method taking into account the 3D trend and the ellipsoids of ranks obtained in the previous step [5]. Simulation of the "dense" lithotype is carried out on the same principle and is interpolated within the "Paleo sandstone", which is determined by its origin. A feature of constructing an ellipsoid of ranks of a "dense" lithotype is that this lithological difference is often spread sporadically.

Due to the fact that the field is characterized by a complex geological structure, small effective oil saturated thicknesses (h_{ef}) with low reservoir properties (porosity and permeability coefficients– K_{por}, K_{per}), the development of this field is effective only with the predominant horizontal drilling. Due to the small amount of actual wellbore information (the number of directional wells on the area of the field is clearly insufficient for studying the detailed geological structure) and limited vertical resolution of seismic surveys (>10 m), structural constructions have a high degree of uncertainty. To clarify the structural framework of the model, it is necessary to involve information obtained from horizontal wells (HW) throughout the horizontal section, which provides additional information on the slope of the structural surface in the inter-well space.

To do this, we propose to use the technique of "geosteering", which is used when tracking HW drilling in real time, but to clarify the structural surface, this technology is used after drilling the HW [6-9]. The technique is based on the creation of synthetic diagrams of the GR method of one of the nearest wells in the most similar section and comparison with the actual logging data. A comparison of actual and synthetic logs allows us to predict the change in the structural surface in the inter-wellbore space. The coincidence of the corresponding synthetic and actual diagrams means that the current model adequately shows the geological cross-section, if the diagrams differ significantly, then the structural surface correction is necessary. Thus, a set of points of the corrected structural surface in the inter-well space is obtained, at which the synthetic and actual diagrams coincide, the additional points obtained in this way are used in reconstructing the structural model.

This technique was first tested on a simulated field to refine the structural surface of the VK_1 roof according to all horizontal wells data. An example of adjusting the structural framework of the An example of adjusting the structural framework of the VK1 formation model for one horizontal well by comparing the synthetic and actual GR diagrams is shown in Figure 3.

IOP Conf. Series: Earth and Environmental Science **181** (2018) 012004 doi:10.1088/1755-1315/181/1/012004

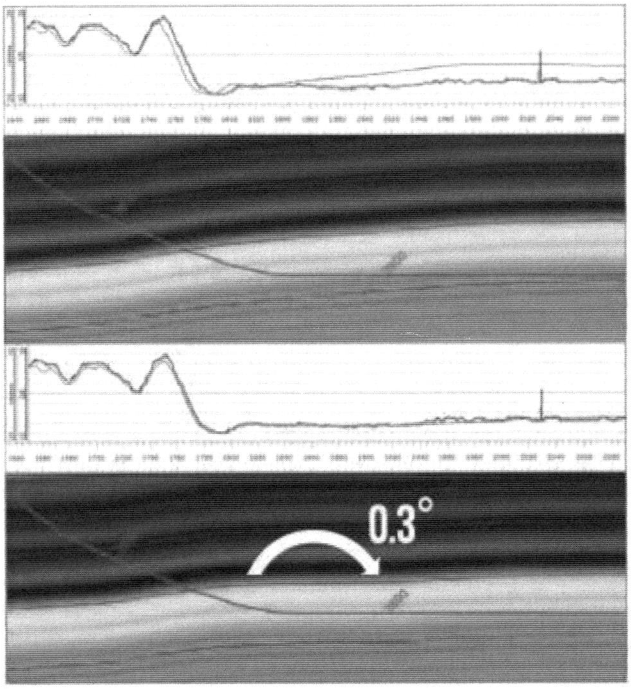

Figure 3. Comparison of the synthetic and actual GR diagrams (gray line - synthetic GR, green - actual GR).

This figure shows how the change in the angle of inclination of the structural surface made it possible to improve the convergence of the GR and ER diagrams on a given interval and obtain a more accurate model of the structure (Figures 4-5).

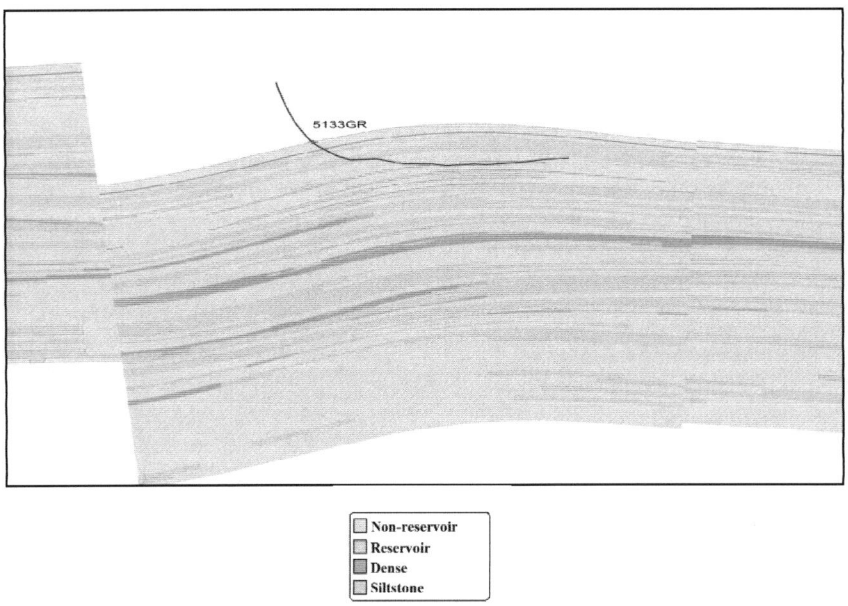

Figure 4. Structural model and lithology cube on the VK object before carrying out "geosteering" (only the entry point of the HW into the reservoir was taken into account).

IOP Conf. Series: Earth and Environmental Science **181** (2018) 012004 doi:10.1088/1755-1315/181/1/012004

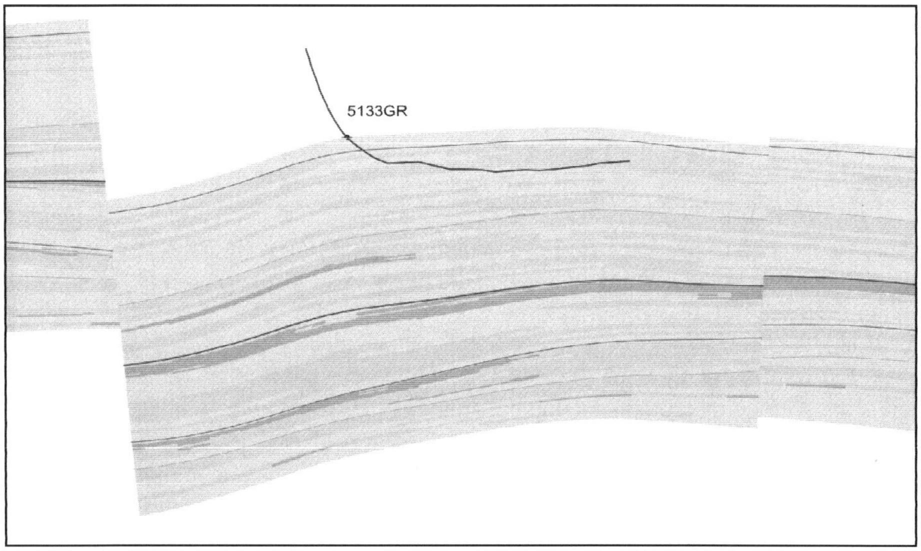

Figure 5. Corrected structural model taking with account of changes in the formation along the entire horizontal wellbore.

Thus, the use of this technology allows a more correct reproduction of the structure in the inter-well space in the areas of horizontal wells. The location of the horizontal trunks corresponds to the actual data and in this particular case is located in the reservoir of the geological model, which was previously impossible to achieve by changing the altitude of the well, only a part of the trunk was adapted, and not the whole horizontal section.

2.4. Analysis of the reservoir connectivity for hydrodynamic modeling

A characteristic feature of geological modeling is the lack of capacity to control the nature of the spatial change in geological properties, the assessment of reservoir connectivity, inhomogeneity of properties, although these parameters have a significant impact on development indices. When constructing a geological model, the conformity of the model and actual histograms of the distribution of the lithological parameter, K_{por}, statistical characteristics is monitored, while the distribution in the inter-well space is not controlled almost. However, the spatial variation of these values often does not affect the total amount of reserves, it is possible to build an unlimited number of geological models that have identical frequency histograms and insignificantly different values of statistical characteristics, but with practically identical values of reserves. For the reasons described, in this paper, to evaluate the uncertainty of data in the inter-wellbore space, it is proposed to use the joint reserve estimation along with the reservoir connectivity criterion. The above-mentioned criterion allows further clarification of the spatial variation of the reservoirs and the reservoir properties.

The approach described below allows taking into account the dynamic characteristics of the object when choosing the implementation of the geological model. Dynamic characteristics are calculated using the concept of flow paths and evaluate the hydrodynamic connection of the reservoir when calculating the flow between wells [10-21]. This method allows one to choose different models (implementations), not only in terms of the volume of hydrocarbons, but also in terms of fluid filtration characteristics in the reservoir. To be more precise, simplified hydrodynamic calculations are sufficient only to evaluate the hydrodynamic connectivity of the reservoir, and to determine the development indices, it is necessary to perform a full hydrodynamic simulation.

This procedure is carried out by means of Workflow, the sensitivity analysis of the model is carried out when the geological properties of the formation are changed, the reservoir connectivity is analyzed taking into account the current flow model and the selection of adequate models for further hydrodynamic calculations. This algorithm is incorporated as a plug-in to the program complex Petrel.

In this paper, a real example of the practical application of a plug-in is provided on one of the fields in Western Siberia.

In the geological model, the following parameters varied:

- Frac1 – the fraction of the reservoir taking into account the dense lithological difference in the reservoir volume (%), frac11 – the fraction of clay in the reservoir volume; RX1, RY1, RZ1 (m) – grades of facies in the modeling of lithological differences: clay and reservoir in the directions OX, OY and OZ respectively,
- Frac2 – the fraction of the reservoir in the volume of Frac1 (%), frac22 – fraction of dense lithologic difference in volume Frac1; RX2, RY2, RZ2 (m) – grades of facies in the modeling of reservoir and dense reservoir,
- RX3, RY3, RZ3 (m) – ranks of porosity coefficient change.

A total of 420 implementations of the geologic model of the reservoir were calculated + current lines calculations. Among the computed implementations, the probability range P10 +/-10%, P50 +/-10%, P90+/-10% is distinguished by the number of geological reserves (pore volume) and by the collector connectivity coefficient (Figure 6 - rectangular areas of red, green and blue colours). From these areas, options are selected based on two criteria: deposits (pore volume) and connectivity with the same probability values: P10, P50, P90. As a result, in the clustering process, 9 models were selected for further calculation using a hydrodynamic simulator.

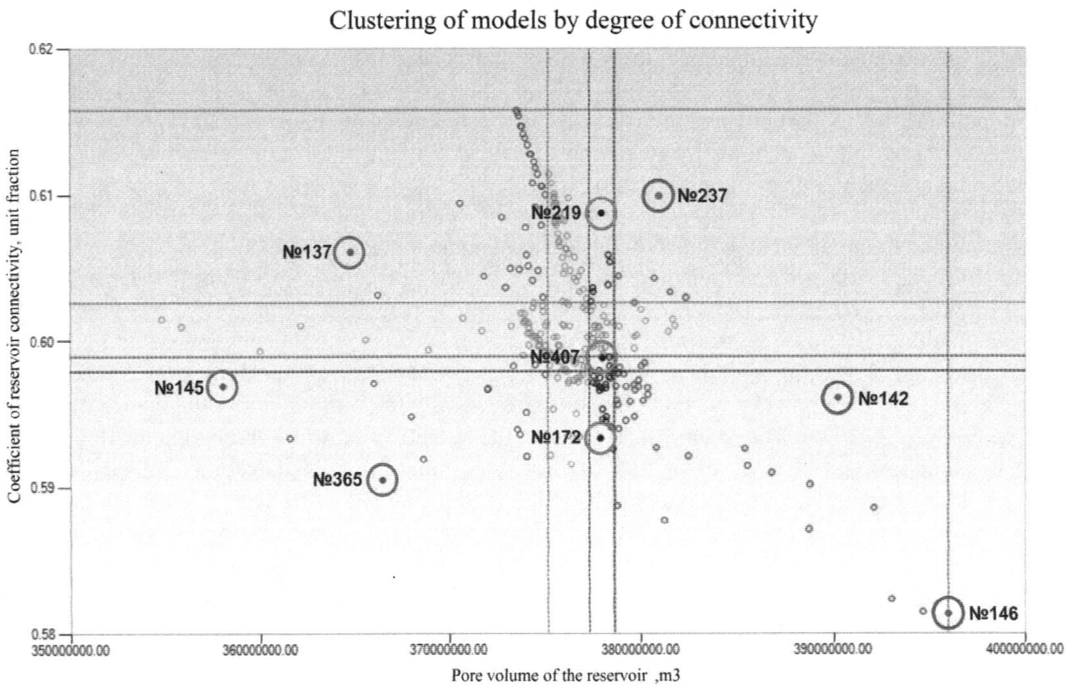

Figure 6. Declustering of models by connectivity and deposits (pore volume).

According to the results of hydrodynamic calculations of nine implementations of the geological model, a fairly good initial convergence of model calculations with historical data was obtained, depending on the degree of reservoir connectivity (Figure 7). The implementation 219 which is characterized by high connectivity and the size of deposits P50 has the minimum error (Table 1), and the variants (237, 172, 407) characterized by different connectivity with the size of deposits ~ P50 also have minimal errors. Proceeding from the above, the implementation 219 is recommended to be used for further fine-tuning of the history of exploitation.

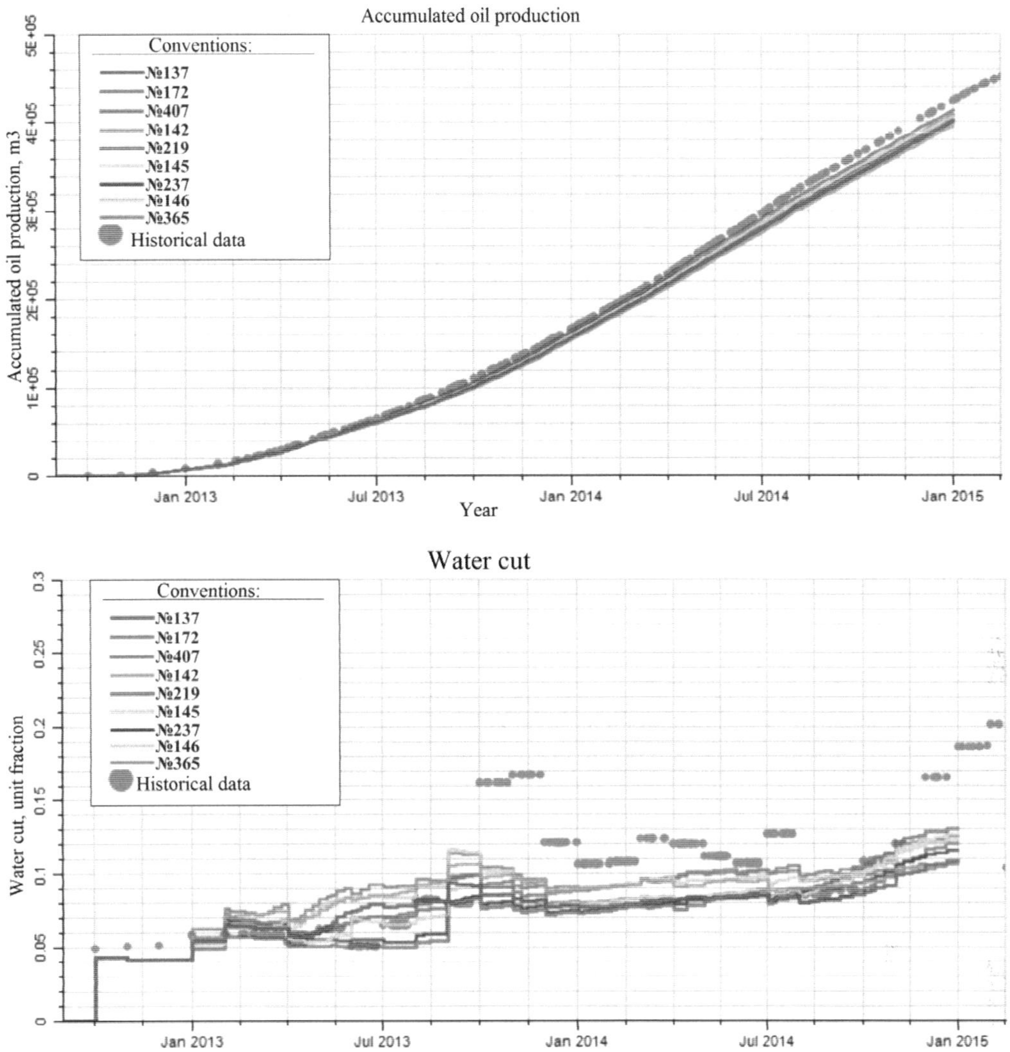

Figure 7. Results of hydrodynamic calculations for 9 geological models.

Table 1. Absolute errors of geological and reservoir data for nine representative geomodels.

№ of variant	Oil production rate, m^3	Accumulated oil production, m^3	Water cut, %	Bottom-hole pressure, bar	Water injectio n rate, m^3	Water production rate, m^3	Sum error
137	202	41733	4.00	1296	557	138	43930
145	187	36264	4.03	1265	564	138	38422
365	178	37815	3.66	1354	570	122	40041
219	186	32621	4.34	1266	555	152	34784
407	184	35703	4.09	1278	555	137	37861
172	184	35440	3.95	1300	552	123	37603
237	192	34706	4.19	1264	578	145	36891
142	205	46302	3.93	1303	569	128	48515
146	212	45984	4.30	1300	558	137	48195

After the adaptation of the selected hydrodynamic model, substantiation of production wells is carried out. According to the data of 26 drilled directional wells, an assessment of the quality of geological models has been carried out, so the average deviation of the absolute marks of the roof of the reservoir according to the model from the actual results according to the author's model is 1.3 m, and the previous one - 3.9 m, the average deviation of effective oil-saturated thicknesses according to the author's model is 0.9 m, which indicates a fairly good predictive ability of the author's model.

3. Conclusion

In general, it follows from the foregoing that the use of the proposed approach makes it possible to more efficiently and consistently assess uncertainties of different nature, use a wide range of data and criteria for the creation of geological models, which significantly increases their accuracy, and therefore reduces risks and improves technological indicators of exploitation. And what is more important, the time to adapt the hydrodynamic model is reduced by 3-5 times.

Thus, a methodology for constructing a structural model has been developed, taking into account the uncertainty of the input data. This technique is especially relevant for fields with a low degree of study and/or with a significant unevenness in the observational network. The methodology for constructing a geological model also allows taking into account reservoir connectivity in the inter-wellbore space and data on horizontal wells. The final choice of the most probable and adequate model for hydrodynamic modeling is based on the total error of a number of geological and field parameters.

References

[1] Karagodin Yu N 2008 Oil and gas of the Vikulov suite *Science in Siberia* (Publishing house of the SB RAS) 41 pp 10-3

[2] Atlas 2004 *Geology and Petroleum Potential of the Khanty-Mansijsk Autonomous District* (Khanty-Mansiysk) p 148

[3] 2014 *Technological scheme of experimental program of the oil field exploitation Surgut Research and Design Institute* (SurgutNIPIneft) p 337

[4] Aleksandrov V M, Kazanskaya D A and Belkina V A 2015 *Features of the geological structure of tempestites in the sediments of the Vikulov suite Oil and Gas* (Tyumen) 5 pp 10-5

[5] Belozerov B V, Butorin A V, Gerasimenko P N, Zhuravleva E V and Faizov R Z 2015 *Practical Advice on 3D Geological Modeling* (St Petersburg: Gazpromneft Research and Development Center) p 354

[6] Kuvaev I, Uvarov I and Pairazyan K 2016 Gesteering, Modern approaches to the optimization of horizontal drilling *Oil & Gas J. Russia* 6 pp 48-52

[7] Vorobyov V S and Petrov A N 2017 The use of horizontal wells in the construction of geological models *Geology, Geophysics and Development of Oil and Gas Fields* 12 pp 24-32.

[8] Kulchitsky V V 2015 Geosteering of horizontal wells and georeactors in the fields of Western Siberia *Oil, Gas, Innovations* 3 pp 8-12

[9] Kulchitsky V V, Grigashkin G A, Larionov A S and Schebatov A V 2008 *Geosteering of Wells* (Moscow: MAKS Press) p 312

[10] Gene M Narahara, John J Spokes, David D Brennan 2004 Incorporating Uncertainties in well-count optimization with experimental design for the Dcepwator Agbami field *SPC 91012, Presented at the SPE Annual Technical Conf. and Exhibition, Houston, Texas, USA*

[11] Idrobo Eduardo A, Choudhary Manoj K and Datta Gupta 2000 A swept volume calculations and ranking ot geostatistical reservoir models using stream-line simulation *SPE 62557*

[12] James R Gilman, Hai-Zui Meng, Michael J Uland 2002 Statistical ranking of stochastic geomodels using streamline simulation. A field application *SPE 77374*

[13] Saad N and Maroongroge V 1996 Ranking geostatistical models using tracer production data *SPE 35494* pp 131-42

[14] 2016 *FrontSim Technical Description* Version 1
[15] Garmash V A 2014 The use of dynamic characteristics in the selection of representative realizations of the geological model *Oil Industry* 12 pp 110-1
[16] Datta-Gupta and King 2007 Streamline simulation: theory and practice *SPE*
[17] Shook and Mitchell 2009 A robust measure of heterogeneity for ranking earth models: The F-PHI curve and Dynamic Lorentz Coefficient *SPE 124625*
[18] Inge Brandsaeter et al 2005 Reservoir modeling and simulation of the Lajas Formation outcrops (Argentina) to constrain tidal reservoirs of the Halten Terrace (Norway) *Petroleum Geoscience* 11 pp 37-46
[19] Harun Ates et al 2004 Ranking and upscaling of geostatistical models by use of streamline simulation: a field case study *SPE 81497*
[20] Utilization of the GRM (Geological Representative Models) to integrate different types of uncertainties in the decision making process, Petrobras, *SPE 120924*
[21] James Gilman et al 2002 Statistical ranking of stochastic geomodels using streamline simulation: a field application *SPE 77374*

IOP Conf. Series: Earth and Environmental Science **181** (2018) 012005 doi:10.1088/1755-1315/181/1/012005

Calculation of the operating mode of the mixing refrigerator

O P Deriugina and A O Larchenko

Industrial University of Tyumen, 38 Volodarskogo st., Tyumen, 625000, Russia

E-mail: olga-7266@mail.ru

Abstract. This article considers the calculation of a mixing refrigerator (contact refrigerator) taking into account the absorption of gaseous components. Calculation of operating mode and definition of optimal operating conditions of the machine that provides the given final gas temperature is carried out.

1. Introduction

The surface heat exchangers are widely used in the industry. At the same time, contact refrigerators and condensers are used mainly in those cases where it is possible to mix hot and cold coolants. Similar machines can be used if the technological scheme provides the subsequent separation of the cold coolant from the main flow. However, contact refrigerators can be used in cases where it is necessary to quickly cool the gas flow, a flow containing solid particles (dust, coke particles) or it is required to absorb a number of gaseous components. This type of machines has a number of advantages: low cost, high heat exchange surface, simple construction, low hydraulic resistance. So, when the regeneration gas of the reforming or dehydrogenation catalyst is cooled, it is necessary to quickly reduce the temperature from 450 ... 500 ° C to about 60 ... 70 ° C [1] without significant resistance to a gas flow. The regeneration gas or waste gas contains harmful components: chlorine, hydrogen chloride, sulfur dioxide and other. Consequently, a desirable aspect during cooling will be partial absorption of harmful substances. Unfortunately, the calculation of contact refrigerators cannot always be carried out accurately. As a result, we can see both insufficient cooling of the gas (which entails a decrease in the efficiency of the trapping of harmful substances in subsequent machines) and an excess of temperature in the machine that can lead to an emergency. The methods of calculating contact refrigerators that are used for reducing the temperature of waste gases often do not allow foreseeing possible modes of operation of the apparatus (including critical thermal loads) [2]. Also in a number of sources approximate calculation methods are given that do not have sufficient flexibility and completeness and that allow a number of approximations and assumptions [3].

In this paper the calculation of the operation modes of mixing refrigerator of spray type including absorption processes is given. The method of obtaining the curves of the operating modes of the apparatus and finding the optimal operating conditions are considered.

2. Research objective

The aim of this paper is to calculate a spraying type mixing refrigerator made to reduce the temperature of the waste gas in the regeneration processes of the reforming and dehydrogenation catalyst proceeding continuously - Continuous Catalyst Regeneration (CCR). Data on the composition of the waste gas are

Content from this work may be used under the terms of the Creative Commons Attribution 3.0 licence. Any further distribution of this work must maintain attribution to the author(s) and the title of the work, journal citation and DOI.
Published under licence by IOP Publishing Ltd

IOP Conf. Series: Earth and Environmental Science **181** (2018) 012005 doi:10.1088/1755-1315/181/1/012005

taken from the patent literature, for example [4]. When solving the problem, the following conditions are accepted:

1. The absorption of all components contained in the gas, depending on their solubility in the cold coolant is possible
2. The neutralization reactions of the components are described by a number of chemical reactions proceeding mainly in the straight direction.

Assumptions are also made:

1. The cold coolant is first heated to the final temperature and then part of it is vaporized in order to reach the given final temperature of the gas;
2. The thermal effect of the chemisorption reactions is taken into account integrally, regardless of the height of the apparatus.

Thus, in the following version of the calculation of the contact refrigerator, assumptions having an insignificant error are accepted.

3. Results and discussion

The initial data for the calculation are given in Table 1. At the same time, the waste gas consumption is assumed equal to 1000 kg /h with a probable increase to 1100 kg / h (project safety factor) [1]. As a cold coolant, a soda-alkaline solution is adopted, which ensures a more complete absorption of harmful substances compared to water. The remaining salts that are in the solution are the products of the reactions of soda and alkali with the components of the waste gas. As mixing refrigerators have low hydraulic resistance (usually not more than 1 kPa), the pressure change in the system can be neglected.

Table 1. Composition and characteristics of coolants

Waste gas (hot coolant)		Soda-alkaline solution (cold coolant)	
Parameter	Value	Parameter	Value
1	2	3	4
Temperature, °C	470	Temperature, °C	50…60
Absolute pressure, MPa	0.101325	Absolute pressure, MPa	0,101325
Flow compositions			
Composition, % mass.:		Composition, % mass.:	
N_2	63.38	H_2O	90.42
H_2O	13.18	Na_2CO_3 и $NaHCO_3$	4.99
CO_2	19.07	$NaCl$	2.00
O_2	3.30	Na_2SO_3 и Na_2SO_4	0.30
HCl	0.62	$NaClO$	0.30
SO_2	0.10	$NaOH$	2.00
Cl_2	0.35	-	-
Total:	100.00	Total:	100.00

In order to simplify the processing of information, all the physical properties of the components are described by the general formula (1):

$$A_t = a_5 \cdot t^5 + a_4 \cdot t^4 + a_3 \cdot t^3 + a_2 \cdot t^2 + a_1 \cdot t + a_0, \quad (1)$$

where $A\,t$ – the defined property of substance or mixture

$a_5, a_4, a_3, a_2, a_1, a_0$ – coefficients of equation (are not given in the article).

The properties of mixtures are calculated by formulas *DIPPR*. Table 2 shows the equations of chemical reactions between the components of the solution and the absorbed substances.

Table 2. Chemical reactions equation

Number of reaction	Equation of summary reaction
1	$CO_2 + NaOH = NaHCO_3$
2	$CO_2 + Na_2CO_3 + H_2O = 2NaHCO_3$
3	$2HCl + Na_2CO_3 + NaOH = 2NaCl + NaHCO_3 + H_2O$
4	$HCl + NaOH = NaCl + H_2O$
5	$Cl_2 + SO_2 + 4NaOH = Na_2SO_4 + 2NaCl + 2H_2O$
6	$Cl_2 + 2NaOH + Na_2SO_3 = Na_2SO_4 + 2NaCl + H_2O$
7	$3Cl_2 + 6NaOH = 5NaCl + NaClO_3 + 3H_2O$
8	$SO_2 + 2NaOH = Na_2SO_3 + H_2O$

Chemisorption reactions taking into account two stages - physical absorption and the actual chemical reaction - are given in Table 3 with the indication of the thermal effect per 1 kmol of the absorbed substance. It is necessary to note that in Table 3 the coefficients for equation (1) are given, with the help of which it is possible to calculate the thermal effect of the reaction with a given chemisorption temperature. Absorption of inert components (nitrogen, oxygen) is not considered as the share of absorbing these gases does not exceed 0.01%.

An approximate thermodynamic calculation gives (with rounding up to 0.05) the values of selectivities for the reactions: for the first 0.75; for the second 0.25; for the third 0,75; for the fourth 0.25; for the fifth in sulfur dioxide, 0.50; for the sixth (for the remainder of chlorine) 0.50; for the seventh (for the remainder of chlorine) 0.50; for the eighth 0.50.

The thermal calculation of the mixing refrigerator is based on the solution of the thermal balance equation (2):

Table 3. Chemisorption reactions with the indication of the thermal effect

Reaction	Coefficients of the polynomial for finding enthalpy, kJ/mol				
	a_4	a_3	a_2	a_1	a_0
Absorption CO_2	$-3.117 \cdot 10^{-8}$	$9.256 \cdot 10^{-6}$	$-1.065 \cdot 10^{-3}$	$1.982 \cdot 10^{-2}$	-24.71
Absorption HCl	$-2.924 \cdot 10^{-8}$	$8.645 \cdot 10^{-6}$	$-8.859 \cdot 10^{-4}$	$1.228 \cdot 10^{-2}$	-86.28
Absorption SO_2	$1.708 \cdot 10^{-12}$	$6.984 \cdot 10^{-11}$	$5.050 \cdot 10^{-4}$	$2.463 \cdot 10^{-2}$	-32.67
Absorption Cl_2	$6.961 \cdot 10^{-8}$	$-2.028 \cdot 10^{-5}$	$2.405 \cdot 10^{-3}$	$-9.603 \cdot 10^{-2}$	-22.33
Reaction1	-	-	-	$-1.858 \cdot 10^{-2}$	-43.95
Reaction 2	$8.130 \cdot 10^{-7}$	$-2.004 \cdot 10^{-4}$	$1.720 \cdot 10^{-2}$	$-8.553 \cdot 10^{-2}$	5.849
Reaction 3	$-7.331 \cdot 10^{-7}$	$1.767 \cdot 10^{-4}$	$-1.470 \cdot 10^{-2}$	$4.133 \cdot 10^{-2}$	-46.83
Reaction 4	$-7.737 \cdot 10^{-7}$	$1.886 \cdot 10^{-4}$	$-1.597 \cdot 10^{-2}$	$5.435 \cdot 10^{-2}$	-48.35
Reaction 5	$-1.741 \cdot 10^{-6}$	$4.334 \cdot 10^{-4}$	$-3.895 \cdot 10^{-2}$	1.197	-551.20
Reaction 6	$-8.937 \cdot 10^{-7}$	$2.236 \cdot 10^{-4}$	$-2.011 \cdot 10^{-2}$	$7.551 \cdot 10^{-2}$	-410.10
Reaction 7	$-2.693 \cdot 10^{-6}$	$6.740 \cdot 10^{-4}$	$-6.029 \cdot 10^{-2}$	2.476	-356.50
Reaction 8	$-8.470 \cdot 10^{-7}$	$2.098 \cdot 10^{-4}$	$-1.884 \cdot 10^{-2}$	$4.411 \cdot 10^{-1}$	-141.00

$$G_g \cdot \left(I_g^{t_{g1}} - I_g^{t_w}\right) + G_{wv} \cdot \left(I_{wv}^{t_{t1}} - I_{wv}^{t_w}\right) + Q_{chem} = G_l \cdot \left(I_l^{t_w} - I_l^{t_{l1}} + e \cdot r_l^{t_w}\right), \quad (2)$$

where G_g, G_{wv}, G_l– consumption of dry gas, water vapor and neutralization solution, respectively, kg/h;

$I_g^{t_{g1}}$ and $I_g^{t_w}$ – enthalpy of dry gas at initial gas temperature t_{g1} and at a final temperature (the temperature of a wet thermometer at which thermal equilibrium is reached) t_w respectively, kJ/kg;

$I_{wv}^{t_{t1}}$ and $I_{wv}^{t_w}$ - enthalpy of water vapor contained in the gas at temperatures t_{g1} and t_w respectively, kJ/kg;

$I_l^{t_w}$ and $I_l^{t_{l1}}$ – the enthalpy of the neutralization solution (cooling liquid) at the initial temperature of the liquid t_{l1} and t_w, respectively, kJ/kg;

Q_{chem} – total heat released during chemisorption, kJ/h;

e –mass fraction of liquid evaporated water (liquid);

$r_l^{t_w}$ – specific heat of evaporation of the liquid at a temperature t_w, kJ/kg.

Calculation is described above.

1. Set the temperature of wet thermometer t_w, °C and the initial flow rate of the neutralization solution G_l, kg/h.
2. Calculate the thermal effect of the reactions taking place in the solution (including the thermal effect of absorption) upon the absorption of carbon dioxide in an amount $F'_{CO_2} = F_{CO_2} - F''_{CO_2}$. For the analysis of the proceeding reactions, it is assumed that the reactions fully proceed. In this case, the data of Tables 2 and 3 are used. Find the thermal absorption effect. Calculate the total heat flux during chemisorption Q_{chem}, kJ/h.
3. Find the value of the left side of the equation (2).
4. With the temperature t_w define the pressure of saturated water vapor in accordance with the formula (3):

$$P_{sat} = 1{,}12 \cdot 10^{-3} \cdot t^4 - 6{,}43 \cdot 10^{-2} \cdot t^3 + 5{,}508 \cdot t^2 - \\ -22 \cdot t + 808{,}6, Pa. \quad (3)$$

5. Find the mass content of water vapor y_{H_2O} in a gas mixture at the final temperature in accordance with the formula (4):

$$y_{H_2O} = \frac{\frac{P_{sat}}{P_{abs}} \cdot M_{H_2O}}{\frac{P_{sat}}{P_{abs}} \cdot M_{H_2O} + \left(1 - \frac{P_{sat}}{P_{abs}}\right) \cdot M_g}, \quad (4)$$

where M_g – average molar mass of dry gas which is found in accordance with the formula (5), kg/Kmol:

$$M_g = \sum y'_i \cdot M_i, \quad (5)$$

where y'_i - mole fraction of the i-component;

M_i–molar mass of the i-component.

6. The mass flow rate of water vapor, converted into the gas is found in accordance with the formula (6):

$$G''_{wv} = \frac{y_{H_2O}}{1 - y_{H_2O}} \cdot \left(G''_{N_2} + G''_{CO_2} + G''_{O_2} + G''_{HCl} + G''_{SO_2} + G''_{Cl_2}\right)_{g-2}, \quad (6)$$

where G''_{N_2}, G''_{CO_2}, G''_{O_2}, G''_{HCl}, G''_{SO_2}, G''_{Cl_2}- mass flow rates of the relevant components in the output flow, kg/h.

7. Determine the degree of evaporation of water vapor in accordance with the formula (7):

$$\varepsilon = \frac{G''_{wv} - G_{wv}}{G_l}. \quad (7)$$

8. Choose a value of G_l such that the right side of equation (2) becomes equal to the left side.

IOP Conf. Series: Earth and Environmental Science **181** (2018) 012005 doi:10.1088/1755-1315/181/1/012005

Based on the calculations carried out by the authors of this article, it could be argued that when cooling the gas in the mixing refrigerator, no more than 20 MJ/h of heat will be released during the absorption of the above components. In further calculations in this article, exactly this thermal effect of chemisorption will be used as the maximum. In this case, a margin is provided for the degree of extraction of components of the order of 5 ... 7%. The reader can independently calculate the heat effect of chemisorption in order to have result that is more accurate.

In the case of design calculations, the adopted value of the thermal effect is justified by the margin on the machine operation stability.

To find the operating modes of the apparatus (the set of points for plotting the curves), the above calculation is repeated several times for different initial temperatures of the cold coolant and the required final temperatures of the cooled gas. Below is a calculation of one of the options of the operation of the machine

In the first approximation, we take the temperature of the wet thermometer equal to $t_w = 65$ °C. This temperature is higher than the initial temperature of the liquid which is equal to $t_{l1} = 50$ °C. The initial gad temperature $t_{g1} = 470$ °C. The gas is cooled to the temperature $t_w = 65$ °C.

The heat flow of chemisorption is assumed to be equal to $Q_{chem} = 20000 \frac{kJ}{h}$.

Enthalpy for dry gas under the given conditions for the initial temperature $I_g^{t_{g1}} = 523{,}88 \frac{kJ}{kg}$ and for the final $I_g^{t_w} = 65{,}51 \frac{kJ}{kg}$.

Enthalpy for water vapor under the given conditions for the initial temperature $I_{wv}^{t_{t1}} = 3471{,}07 \frac{kJ}{kg}$ and for the final $I_{wv}^{t_M} = 2617{,}96 \frac{kJ}{kg}$.

Enthalpy for neutralization solution under the given conditions for the initial temperature $I_l^{t_{l1}} = 248{,}65 \frac{kJ}{kg}$ and for the final $I_l^{t_w} = 190{,}31 \frac{kJ}{kg}$.

The value of the left side of the heat balance equation (2) at a known flowrate of the gas flow (1000 kg/h is assumed):

$$G_g \cdot \left(I_g^{t_{g1}} - I_g^{t_w}\right) + G_{wv} \cdot \left(I_{wv}^{t_{t1}} - I_{wv}^{t_w}\right) + Q_{chem} =$$

$$= 868{,}32 \cdot (523{,}88 - 65{,}51) + 131{,}77 \cdot (3471{,}07 - 2617{,}96) = 530429 \frac{kJ}{h}.$$

We find the pressure of saturated water vapor at the temperature of a wet thermometer in accordance with the formula (3):

$$P_{sat} = 1{,}12 \cdot 10^{-3} \cdot 65^4 - 6{,}43 \cdot 10^{-2} \cdot 65^3 + 5{,}508 \cdot 65^2 -$$

$$-22 \cdot 65 + 808{,}6 = 24988{,}04\ Pa.$$

Then the water vapor content in the outgoing gas at absolute pressure $P_{abs} = 101325\ Pa$ and average molar mass of dry gas $M_g = 30{,}7618 \frac{kg}{kmol}$ is equal in accordance with the formula (4):

$$y_{H_2O} = \frac{\frac{24988{,}04}{101325} \cdot 18{,}0153}{\frac{24988{,}04}{101325} \cdot 18{,}0153 + \left(1 - \frac{24988{,}04}{101325}\right) \cdot 30{,}7618} = 0{,}1609.$$

As the flowrates of components in the gas flow do not change (except for the flowrate of water vapor), the mass flowrate of water vapor in the gas phase in the outgoing gas flow is found as follows:

$$G''_{wv} = \frac{y_{H_2O}}{1 - y_{H_2O}} \cdot G_g = \frac{0{,}1609}{1 - 0{,}1609} \cdot 862{,}32 = 166{,}46 \frac{kg}{h}.$$

We take the flowrate of neutralization solution equal to $G_l = 7697{,}35\frac{kg}{h}$

We find the degree of evaporation of the liquid in accordance with the formula (7):

$$\varepsilon = \frac{166{,}46 - 131{,}77}{7697{,}35} = 0{,}0045.$$

We calculate the right side of the heat balance equation (2), knowing the specific heat of evaporation at the temperature of a wet thermometer $r_l^{t_w} = 2345{,}73\frac{kJ}{kg}$:

$$G_l \cdot \left(I_l^{t_w} - I_l^{t_{l1}} + e \cdot r_l^{t_w}\right) =$$

$$= 7697{,}35 \cdot (248{,}65 - 190{,}31 + 0{,}0045 \cdot 2345{,}73) = 530429\frac{kJ}{h}.$$

In order to determine the optimal operating conditions of the apparatus, the analysis of the apparatus is carried out with the repetition of the calculation for different modes: different initial liquid temperatures (we take 50, 55, and 60 ° C), and also different final temperatures of the cooled gas (in the range of 52 to 76 ° C). The results of the analysis are given in Table 4.

It should also be noted, that in order to obtain a better understanding of the operation of the apparatus, the step between the temperatures of the cold coolant must be made less.

Table 4. Results of the analysis of the operation of the mixing refrigerator

Final gas temperature t_w, °C	Removable heat kW	Liquid flow rate at an initial temperature t_{l1}, kg/h		
		50 °C	55 °C	60 °C
52	151.4	86449	-	-
54	150.8	41596	-	-
56	150.2	26533	158905	-
58	149.5	18906	50324	-
60	148.9	14242	28432	-
62	148.3	11047	18903	66041
64	147.7	8679	13476	30266
66	147.0	6815	9894	18107
68	146.4	5271	7285	11818
70	145.8	3936	5238	7843
72	145.1	2732	3529	4990
74	144.5	1603	2022	2739
76	143.9	504	622	816

The obtained data are graphically shown in Figure 1.

It is clearly seen that the curves of the flowrate of the liquid at different initial temperatures have a bend which meet a transition to a low-efficiency mode of heat removal when moving from right to left or when the final temperature of the gas decreases. It can also be seen that as the initial temperature of the liquid decreases, the minimum achievable final temperature of the gas decreases.

On the basis of the data obtained, it can be concluded that effective cooling corresponds to a flat curve located in the area of higher final gas temperatures. In our case, at a liquid temperature of 50 °C of such area, the final gas temperature will be above 55-56 °C.

The curves of the operation of the machine have the following boundary conditions:

1. When the final temperature of the cooled gas approaches the final temperature of the cold coolant, the irrigation flow increases significantly, as a result of which the moisture contained in the initial gas starts to condense (the opposite effect is observed – the cold coolant starts evaporating);

2. With a small amount of irrigation, all the coolant coming into the mixing refrigerator completely evaporates; the gas temperature does not decrease to the given temperature.

For other initial conditions (composition and flow parameters), similar curves characterizing the operation of the mixing refrigerator can be obtained.

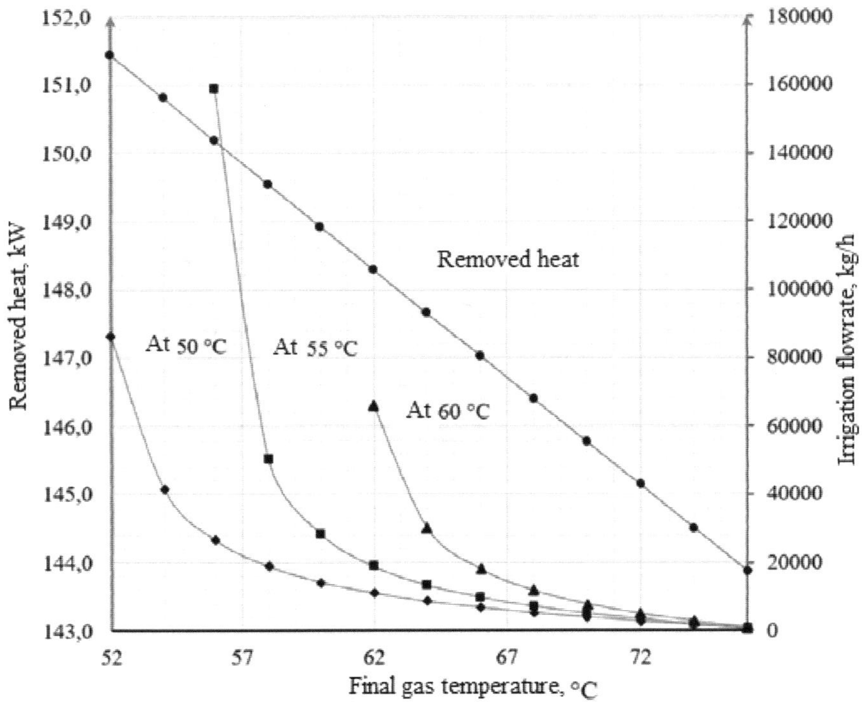

Figure 1. Characteristic of operation of the mixing refrigerator.

4. Conclusion

In this paper, a method of calculating a contact refrigerator of a spraying type taking into account the proceeding chemisorption is given. On the basis of the derived equations, the curves characterizing the operation of the apparatus (load irrigation, depending on the initial and final temperatures of coolants) are constructed. An approximate method for estimating the operation's parameters of the apparatus providing cooling of the gas to a given temperature is proposed. At a higher initial temperature of the cold coolant, the irrigation flowrate is reduced. In order for the final temperature of the gas to be close to the final temperature of the cold coolant, it is required to increase the irrigation flowrate. The optimum irrigation flow is at the lower flat curves until the inflection points clearly visible in Figure 1. Further investigations can be directed to the refined calculation with the use of a differential calculation of the height of the apparatus as well as taking into account the effects of mass transfer.

References

[1] Larchenko A O and Deryugina O P 2016 Cleaning of industrial waste gases from chlorine and its compounds *Energy Saving and Innov. Technologies in the Fuel and Energy Complex: Mater. of the Intern. Sc.-Techn. Conf. IUT* ed Khalin A N (Tyumen) pp 171-5

[2] Birger M I 1983 *Handbook on Dust and Ash Collecting* ed Rusanova A A (Moscow: Energoatomizdat) 312 p

[3] Boehm R F 2014 *Direct Contact Heat Transfer* (Las Vegas: Univ. of Nevada) 42 p

[4] Selclen A Patent 105879562 China IPC B01D 53/02 *Reforming Catalyst Regeneration Discharged Air Dechlorination Method* № 201410468016.5; filed 09.05.2014; date of patent 24.08.2016 8 p

Processing of high-viscosity oils by complex electromagnetic treatment

V O Dovbysh, S A Leontyev, L I Maksimov and V A Ogay

Industrial University of Tyumen, Tyumen, Russia

E-mail: Vadim932@mail.ru

Abstract: At present, a promising method for improving the rheological properties of high-viscosity oils is their electromagnetic treatment. This is confirmed by a number of scientific and theoretical works of modern researchers. The positive effect of the impact is due to the presence of dispersion areas in the electromagnetic fields of certain frequencies. In order to improve the energy efficiency of the preparation of high-viscosity oils and develop an innovative installation, studies were carried out. A complex treatment process was used: a high-frequency electromagnetic field; microwave electromagnetic field; a static magnetic field, and then the change in the rheological characteristics of the oil was recorded.

1. Introduction

The high-viscosity oil of the Russian Federation (more than 30 mPa-s) is concentrated in the West Siberian (54%), Volga-Ural (26%) and Timano-Pechora (17%) oil and gas provinces, its geological reserves are estimated at 7.2-11.0 billion tons. [1]. High-viscosity, ultra-high viscosity (200-10,000) [mPa·s] oil and solid bitumen are characterized by a high content of arenes, resinous-asphaltene substances, high concentration of metals and sulfur compounds, high density, cocking behavior, which leads to high production costs, oil-trunk pipeline and low-profit oil refining. In 2014, the Russian Federation produced about 5.4 million tons of oil from deposits containing ultra-high viscosity oil and solid bitumen, which accounts for 18% of the total production of hard-to-recover hydrocarbons in Russia. But measures of state support and economic stimulation of involvement in the development of hard-to-recover hydrocarbon reserves are predictably leading to an increase in oil production from hard-to-recover reserves in Russia from the current 30 million tons per year to 80 million tons per year in 2035, which is about 16% of the total oil production in the country.

For example, when calculating the extraction tax rate for ultra-high viscosity oil, a special coefficient is provided, resulting in a rate of about 35% of the standard, for deposits of solid bitumen (with a viscosity of more than 10,000 [mPa·s]), the extraction tax rate is zero [2]. It is also indicative that the duty in the Russian Federation for the export of high-viscosity oil in 2016 is indicative: in January to $ 8.7 from $ 10.9 per ton in December 2015, in February to $ 5.9.

These trends will undoubtedly affect the transportation volumes of hard-to-recover reserves, including high-viscosity and ultra-high-viscosity oils. Predominantly in the territory of the Russian Federation, transportation of high-viscosity oil is carried out by pipelines. On the basis of the analysis it was found out that today the most common technologies for transportation of high-viscosity oils are: pumping in mixture with a low-viscosity diluent, heat treatment and addition of depressors. The main objective of each method is to create favorable conditions for transporting oil through the pipeline, among which the main one is to reduce the dynamic viscosity. Each of the described methods has

Content from this work may be used under the terms of the Creative Commons Attribution 3.0 licence. Any further distribution of this work must maintain attribution to the author(s) and the title of the work, journal citation and DOI.
Published under licence by IOP Publishing Ltd

IOP Conf. Series: Earth and Environmental Science **181** (2018) 012006 doi:10.1088/1755-1315/181/1/012006

significant drawbacks: high energy costs; use of the transported product as fuel; corrosion of pipes; high operating costs; The absence of universal additives and the need for their development for a particular oil product [3, 4].

2. Formulation of the problem

To date, a number of developments are being explored in large oil and gas companies and scientific and technical communities to identify the effectiveness of electromagnetic treatment methods to improve the rheological characteristics of high-viscosity oils. Special attention is paid to the following methods of treatment:

- static magnetic field;
- high-frequency electromagnetic field;
- microwave electromagnetic;
- magnetic vibration activation.

Each of these methods has its advantages in processing high-viscosity oils before conventional methods. However, the method is not always economically feasible for operation in industrial conditions because of the insufficiently developed scientific and practical basis for research in this area.

The research is aimed at solving the problem of reducing the viscosity of oils. A characteristic feature of this type of oil is the high content of resins, paraffins and asphaltenes, which are in need of creating special conditions. It is necessary to heat oil to a certain temperature during preparation.

The object of the study is high-viscosity oil under the influence of complex processing by a high-frequency magnetic field with a frequency of oscillations of 140 MHz, a microwave electromagnetic field with a frequency of oscillations of 2.4 GHz, and the action of a static magnetic field based on neodymium magnets with an intensity of 8 A/m. Characteristics of the component composition of high-viscosity oil are presented in Table 1.

Table 1. - Characteristics of the component composition of high-viscosity oil

No.	Dipole molecule	Percentage in total volume, %
1	Water	0.5 - 1
2	Asphaltenes	3.2
3	Paraffins	15.6
4	Resins	23

The main tasks of the scientific research:

- to substantiate the choice of the optimum frequency of high-frequency electromagnetic radiation and microwave electromagnetic radiation;
- to experimentally investigate changes in the dynamic viscosity of oil during treatment by each selected method of electromagnetic treatment separately and in complex;

3. Results and discussion

There are many works related to the study of the effect on the high-viscosity oil of the electromagnetic field. A method for treating high-viscosity oil by a magnetic inductor is widely used, which results in a decrease in asphaltene deposits and corrosive activity in wells and pipelines. Also, a positive effect was achieved using the method of oil product separation and melting of rod wax in wells and pipelines based on microwave radiation. [5, 6].

Analyzing the trends in the development of pipeline transport of high-viscosity oils, taking into account the basic requirements for high-viscosity oil preparation systems for transportation via the main pipeline, the main features of its component composition were identified, a solution based on the combined effect of electromagnetic fields and a static magnetic field was formulated. The method consists in a high-frequency action of an electromagnetic field of 140 MHz frequency, microwave interference with a frequency of 2.4 GHz, and exposure to a static magnetic field produced by neodymium magnets. The frequency spectrum was chosen on the basis of the positive experience of work of this direction of other researchers and preliminary estimation. A complex solution allows each component to locally affect a certain component of a high viscosity oil product [7].

Experiments were carried out in the laboratory on the basis of the Industrial University of Tyumen. Five identical samples of high-viscosity oil with a component composition: 23.5% paraffins, 3.64% resins, 1.4% asphaltenes, 1% water, at an initial oil temperature of 25 °C, were individually exposed to each of the above fields and sequentially complexed (a constant magnetic field (MF), a high-frequency electromagnetic field (HFMF) and a microwave electromagnetic field (MEF)). The graphs (Figures 1, 2, 3, 4) show the results of the action, which reflect the decrease in the dynamic viscosity of the oil.

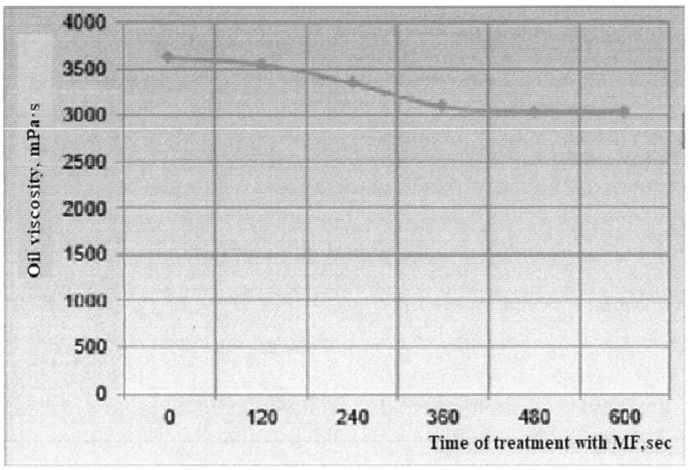

Figure 1. Dependence of the change in dynamic viscosity on the effect of a static magnetic field, initial oil temperature 25 °C.

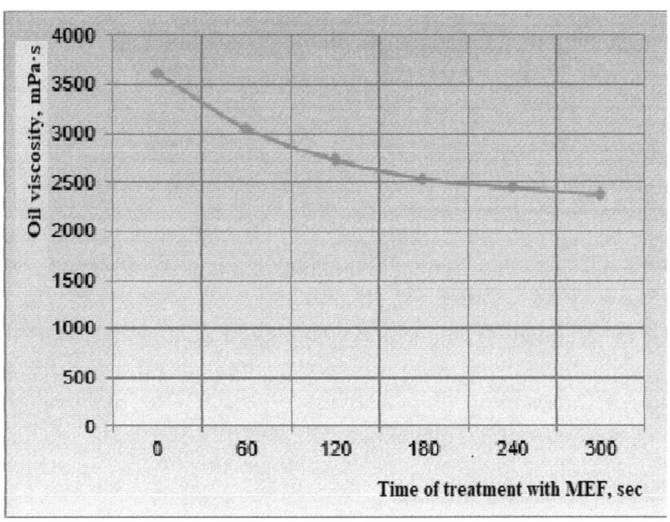

Figure 2. Dependence of the change in the dynamic viscosity of oil on the effect of microwave radiation, the initial oil temperature of 25 °C.

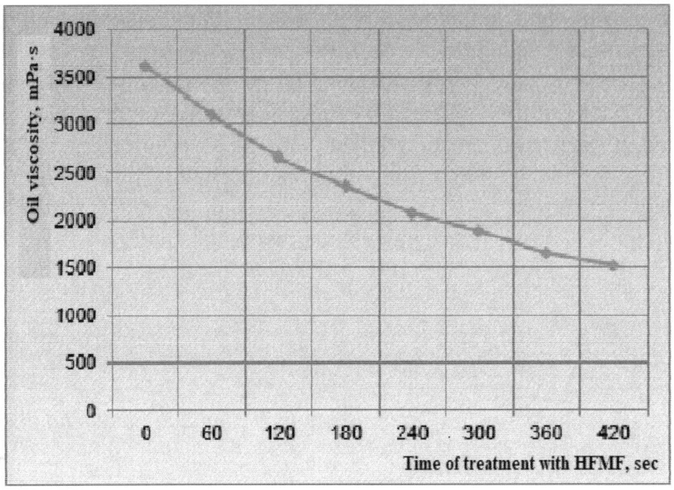

Figure 3. Dependence of the change in the dynamic viscosity of oil on the influence of high-frequency radiation, the initial oil temperature of 25 °C.

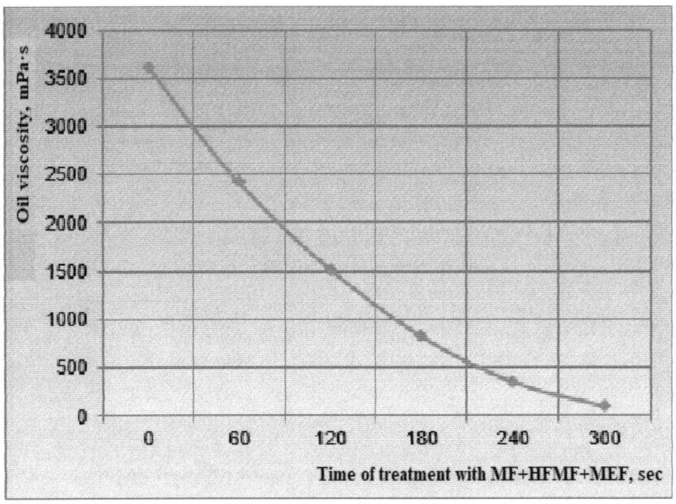

Figure 4. The dependence of the change in the dynamic viscosity of oil on the complex effect, the initial oil temperature of 25 °C.

The positive effect of treatment is due to the presence of dispersion areas in the electromagnetic fields of certain frequencies. In order to increase the efficiency of transportation of highly viscous oils and to develop a unit for transporting high-viscosity oils along the main oil pipeline, studies were conducted on the effect of the complex treatment type on the change in rheological characteristics.

The degree of influence of the electromagnetic field depends on the frequency of the field and the dielectric properties of the components of high-viscosity oil, such as the relative permittivity ε 'and the tangent of the dielectric loss angle tgδ.

The water molecules that are present in the transported oil product in small quantities have resonant peaks tgδ at a frequency of 2.4 GHz, when the prototype oil images are affected by an electromagnetic field of the appropriate frequency, the energy of the microwave waves is absorbed by the water molecules and converted into energy of thermal motion, which is transferred to all components of the oil product.

IOP Conf. Series: Earth and Environmental Science **181** (2018) 012006 doi:10.1088/1755-1315/181/1/012006

To assess the influence of the microwave electromagnetic field on high-viscosity oil, the problem of determining the degree of field influence on an emulsion drop was considered. This approach is relevant for the concentration of water molecules in oil not more than 1%, provided that the droplets do not behave like dipoles. Under these conditions, it is necessary to consider the process of thermal conductivity through a system of equations in oil and water in a spherical coordinate system (r, θ, φ), where φ and θ are the azimuth and zenith angles, respectively.

Since the thickness of the armor shell is Δ << r0 (r0 is the radius of the drop), the polarization in it can be neglected. A drop of water is placed in an external uniform microwave field. The center of the drop is the center of the coordinate system. The axis of rotation of the drop passes parallel to the field, so there is no dependence with the coordinate φ. There is also symmetry with respect to the angles θ = 0 and π/2, the drop consists of 4 equal quadrants. So, solving the problem, it is necessary to investigate the 1-st quadrant in the region from θ = 0 to π/2. Taking into account the above-mentioned conditions, the problem is formulated by the following system of heat equation:

$$\frac{\partial T_i}{\partial t} = \frac{1}{r^2 c_i \rho_i}\frac{\partial}{\partial r}\left(\lambda_i r^2 \frac{\partial T_i}{\partial r}\right) + \frac{1}{r^2 \sin\theta \cdot c_i \rho_i}\frac{\partial}{\partial \theta}\left(\lambda_i \sin\theta \frac{\partial T_i}{\partial \theta}\right) - U_{ir}\frac{\partial T_i}{\partial r} - \frac{U_{i\theta}}{r}\frac{\partial T_i}{\partial \theta} + \frac{q_i}{c_i \rho_i} \quad (1)$$

where $i = 1,2$ indices related to water and oil, respectively;
T – area temperature, ^{0}C
λ – thermal conductivity, W/(m·K)
c – specific heat, J/(kg•K)
ρ – density of medium; kg/m^3
Ur and Uθ – components of the velocity vector of motion both inside and outside the drop (oil, water), initiated by the electric stress tensor,
q – thermal sources, J.

Calculation of distribution of thermal sources is made solutions of an electrodynamic task

$$q_i^e = \frac{\omega \varepsilon_0 \varepsilon_i' \, tg\delta_i}{2} E_i^2 \quad (2)$$

where E – electric field strength, V/m
ω – rate of phase change, Hz
ε_0 – electric constant

The boundary and initial conditions are based on the conductor heat transfer of the environment and the drop:

$$T_1(r,\theta,0) = T_2(r,\theta,0) = T_0 = const, \quad (3)$$

$$\lambda_1 \frac{\partial T_1(r_0,\theta,t)}{\partial r} = \lambda_2 \frac{\partial T_2(r_0,\theta,t)}{\partial r}, \quad (4)$$

$$T_1(r_0,\theta,t) = T_2(r_0,\theta,t), \quad (5)$$

$$\frac{\partial T_1(0,\theta,t)}{\partial r} = 0, \quad T_2(l,\theta,t) = T_0, \quad (6)$$

$$\frac{\partial T_1(r,0,t)}{\partial \theta} = 0, \quad \frac{\partial T_2\left(r,\frac{\pi}{2},t\right)}{\partial \theta} \quad (7)$$

where l – boundary of the region of thermal action.

The system of equations (1) with conditions (2) - (6) is solved by a finite-difference method according to the implicit scheme. At the end of the calculations, the temperature fields were determined both outside the drop and inside it.

If we talk about the degree of influence of a high-frequency electromagnetic field on high-viscosity oil, it should be taken into account that the resonant peaks of tgδ paraffins and resins, which are present in significant amounts in the component composition of oil, appear at a frequency of 140 MHz. Therefore, when the experimental field of oil is exposed to an electromagnetic field of 140 MHz frequency, intensive absorption of energy by molecules of paraffins and resins takes place, which leads to a rapid decrease in the viscosity of the oil product. In this case, the formula for sources of heat release is preserved analogous to the process of action of the microwave field:

$$q_i^e = \frac{\omega \varepsilon_0 \varepsilon_i' \, tg\delta_i}{2} E_i^2 \tag{8}$$

The process of absorption of energy of the RF field is described by the equation of heat conduction for given external heat sources:

$$\rho c_T \frac{\partial T}{\partial t} = \frac{1}{r}\frac{\partial}{\partial r}\left(\lambda r \frac{\partial T}{\partial r}\right) + \frac{\partial}{\partial z}\left(\lambda \frac{\partial T}{\partial z}\right) + Q(r,z,t) \tag{9}$$

where T – ambient temperature, ^{0}C
λ – thermal conductivity, W/(m·K)
c_T – specific heat, J/(kg•K)
ρ – density of medium, kg/m^3
Q – power density, W/m^3.

Decrease in viscosity of high-viscosity oil in the treatment of high-frequency and microwave fields is caused by the absorption of energy fields according to the process described above and is accompanied by an increase in the temperature of the oil product. The process description is reduced to the following equation:

$$\mu = \mu \exp\left(-\gamma (T - T_0)\right) \tag{10}$$

It should also be taken into account that, in contrast to the thermal methods of impact on oil environments, electromagnetic fields, in addition to the thermal effect, also exhibit "nonthermal" fractures, manifested at the nano- and micro-levels.

The effect of a constant magnetic field on the decrease in the viscosity of oil is due to a significant increase in the number of paraffin crystallization centers, and is accompanied by the transformation of a coarsely dispersed suspension of paraffin crystals into a finely dispersed volumetric stable suspension. These processes help reduce the viscosity of oil.

It should be noted that during the simultaneous processing by a constant magnetic field, high-frequency and microwave fields in the course of experiments, a synergetic effect is observed and a decrease in the dynamic viscosity of the oil product occurs at a high rate. The process is determined by the combination of all the above-mentioned phenomena of individual influences, which contribute to the intensification of the absorption of field energy by oil.

4. Conclusion

The results of the study can be used in the development of a design relating to devices for preparation for transportation of high viscosity paraffins and tar oils, the design can be used in the oil industry for field and main preparation for transportation of high viscosity oil by reducing its viscosity and increasing the relaxation time of its rheological properties.

The work performed in the future is aimed at solving the task of reducing energy costs in preparation for transportation of high-viscous oils along main pipelines in severe natural and climatic conditions under negative temperature conditions. In particular, the results of the work can be used on sections of the pipeline similar to the area “Zapolyar'ye – Purpe”.

An important factor for the development of the work is compliance with both the policy of import substitution policy of the Russian Federation and the program of innovative development of PJSC "Transneft" for the period until 2020.

References

[1] Sidorov I V 2014 *Substantiation of the development of high-viscosity oil deposits by horizontal wells* Dis. Cand. Tech. Sc. (Moscow) p 131
[2] Klubkov S 2015 Stimulating the development of TRIZ will help support the level of oil production in Russia *Oil&Gas J. Russia* 5 pp 6-11.
[3] Hangs F E 1966 More insulated, heated - oil pipelines are prospect *Oil & Gas J.* **64** 40 pp 117–22
[4] Lebedko A G 2010 The economic justification of geoecological risks in the oil and gas industry *Engineering Bulletin of the Don* 4 URL: ivdon.ru/ru/magazine/archive/n4y2010/252/
[5] Balakirev V A, Sotnikov G V, Tkach Yu V and Yatsenko T Yu 1998 HF method of eliminating paraffin plugs in oil well equipment and oil pipelines *Electromagnetic Phenomena* **1** 4
[6] Masanobu Wasaka and Hisaharu Hayshi 1996 Magnetic field effect on the Hydrogen Abstraction Reactions Triplet Benzophenone with Thiophenon in Nonviscous Homogeneous Solutions *J. Phys. Chem.* 100
[7] Bugautdinov N Ya 2007 *Scientific bases and technologies of physical fields influence on hydrate - paraffin deposits in oil wells* Abstract. Dis. D.Sc. (Ufa)
[8] Laptev A B, Inyushin N V, Kashtanova L E, Mugtabarov F K, Khaydarov R F, Khalitov D M and Shaydakov VV 2001 *Apparatus for magnetic treatment of liquids* (Moscow: Nedra) p 145
[9] Viryasov A N, Gostinin I A and Semenova M A 2013 Transport innovations of the Tuapse refinery *Engineering Bulletin of the Don* 1 URL: ivdon.ru/ru/magazine/archive/n2y2014/2381/
[10] Program of innovative development of PJSC "Transneft" for the period until 2020 Internet resource: http://innovation.gov.ru/sites/default/files/documents/2016/5360/556.pdf

Development of installation for the preparation of biosorbents from waste petrochemical industry

V O Dovbysh, A A Kuznetsova, L I Maksimov and V A Ogay

Industrial University of Tyumen, Tyumen, Russia

E-mail: Vadim932@mail.ru

Abstract: The article considers the problem of environmental pollution by oil spills. A solution is proposed in the form of a developed device and a method for its implementation. Its technical characteristics for the preparation of biosorbents from wastes from the petrochemical industry, in particular surplus activated sludge after heat treatment, are presented. The parameters that are necessary for the effective preparation of the biosorbent are given. Parameters are obtained by experiment.

1. Introduction

Beginning in 2011, Russia has seen steady growth in oil production volumes and at the same time the amount of oil pollution is increasing, so the amount of breaching in oil pipelines has increased by 31% over the past five years. [1]

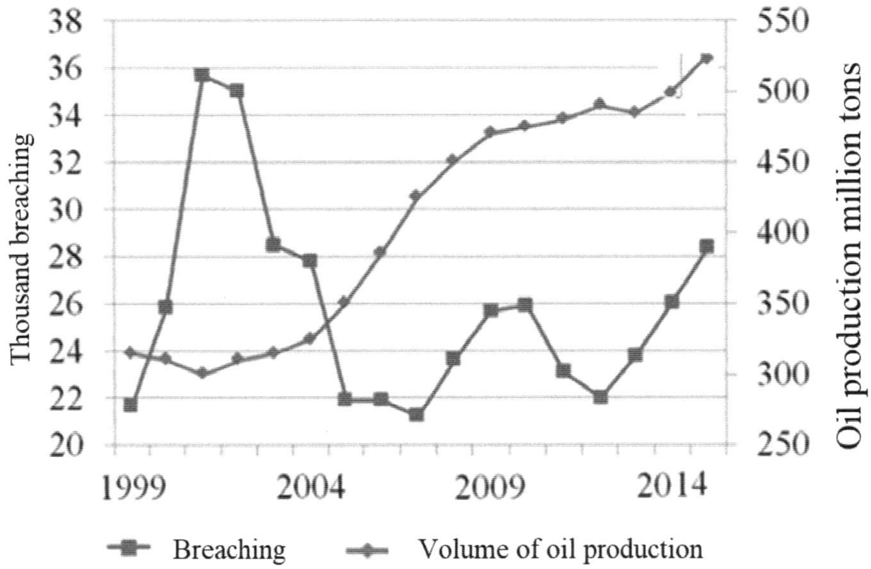

Figure 1. Amount of pipeline breaching and oil production.

Taking into account the current dynamics, the production and technical base and the program for the development of the fuel and energy complex, the scale of pollution will only increase.

Content from this work may be used under the terms of the Creative Commons Attribution 3.0 licence. Any further distribution of this work must maintain attribution to the author(s) and the title of the work, journal citation and DOI.
Published under licence by IOP Publishing Ltd

IOP Conf. Series: Earth and Environmental Science **181** (2018) 012007 doi:10.1088/1755-1315/181/1/012007

The importance of minimizing the damage inflicted is reflected in the state policy, namely: "Socio-economic development of the Arctic zone of the Russian Federation for the period until 2020" No. 336 of 21.04.2014; Federal target program "Elimination of accumulated environmental damage" for 2014 - 2025"; The program of import substitution in the direction "Perspective technologies of management and environmental protection of water resources". So, the state annually allocates about 17 billion rubles through targeted state programs. Only in 2015, oil companies allocated more than 11 billion rubles for land reclamation.

The greatest demand is observed in biosorbents, which are a complex of different types of sorbents and oil-oxidizing microorganisms immobilized on it. [2]

The analysis of the effectiveness of the currently used biosorbents shows the presence of a number of shortcomings. The most significant shortcomings are high cost and insufficient effectiveness of sorbent carriers. Investigating the technological scheme for the production of biosorbents, processes were identified modifying which it would be possible to improve the adsorption properties of the bases used for the biosorbent and the efficiency of the ligament of oil-oxidizing bacteria and the base.

As a result of the analysis of the process of biosorbents production, the necessary parameters of the effective sorbent carrier and the possibility of utilization of waste from the petrochemical complex, it was found that the most optimal is the carbon substrate from the heat treated activated sludge that is taken to a specialized storage facility. By developing special equipment, it will be possible to reduce the labor and financial costs of enterprises and increase the efficiency of refining. [3,4]

2. Formulation of the problem

In oil companies, there has been a shift from mechanized oil spill cleanup to the use of sorbents and biosorbents. To date, the market for biosorbents in Russia is about $450 million, with an annual growth rate of 6%. Using the SWOT-analysis of the market, the following problems were identified: labor intensity of preparation, a high proportion of expensive imported equipment and the need for recycling of wastage from the petrochemical complex. Since the market of sorbents is represented by foreign companies by more than 70%, the policy of import substitution is of great importance.

The main tasks of the research are:

- Develop a schematic diagram and a method for preparing a carbon-based biosorbent;
- Obtain the boundary parameters for the preparation of the sorbent;
- Construct a 3-D model of a biosorbent preparation plant.

3. Results and discussion

To achieve these goals, a device for the preparation of biosorbents from wastes from the petrochemical industry is proposed.

The installation is a vertical cylindrical technological equipment intended for the production of biosorbents by creating the necessary physical conditions for the successful immobilization of the microorganism strain on a carbon base. Installation consists of waste petrochemical enterprises. Installation has a mechanical and pneumatic mixing system. The installation regulates such parameters as: stirring speed, temperature, pH (acidity of the medium), pressure and humidity. This is done by means of the presence of the following elements in the installation: a circulation circuit for temperature control; mobile diffusers which create circulation of air masses throughout the volume and, at high speeds of air masses, it provides active mixing with blowing and drying; stirring mechanical device; outlet for loading nutrient media and strains of microorganisms and unloading the finished material; a vacuum pump to provide pressure parameters.

The process of preparation of biosorbent in the installation occurs at a temperature regime of 18 to 60 degrees Celsius, under conditions of excess pressure. The humidity range is from 20 to 95%. Air flow rate is up to 15 m^3/min.

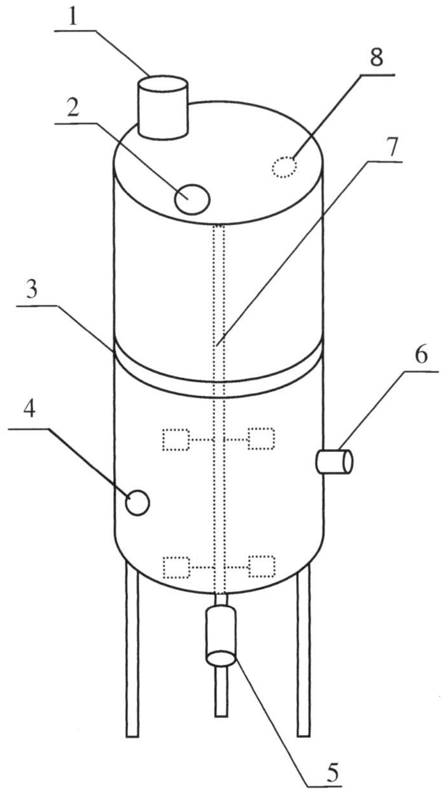

Figure 2. Sketch of the installation for the preparation of biosorbents from waste petrochemical industry. 1- access hatch, 2-sight window, 3-section circulation circuit, 4-hatch of bacteria, 5- water disposal system, 6-sampler, 7-agitating mechanical device, 8-mobile diffuser.

Method of preparation of biosorbent. Sludge with low adsorptive properties received as a result of low-temperature separation is loaded into the reception hatch of installation. Then sludge is exposed to activation, i.e. crushing and hashing at the expense of the mixing mechanical device and establishment of temperature condition for couples of a section circulating contour with the subsequent drying by the mobile diffuser. As a result, necessary structural parameters, namely the size of a macropores which is suitable for a further immobilization, turn out since the basis has high adsorptive properties. After stopping the supply of steam and drying, certain parameters are set: humidity 40% -50%, temperature 150-180 °C, speed of the mixing mechanical device 150-200 rpm for 20-30 minutes.

Then, the mixture is loaded into installation which consists of various chemical compounds which contains the strains of microorganisms relating to the sorts Pseudomonas sp., Rhodococcus sp., Mortierella. The pH of the medium is set to 5.8-6.0; temperature 26±5 °C; the speed of the mechanical stirrer in the range of 80-100 rpm. Then process of hashing by means of the device mixing mechanical and/or giving a stream of air by the mobile diffuser is carried out. Then via the reception hatch commodity oil with a density up to 910 kg/m^3 immobilization moves in a small amount for an additional. After thorough mixing, the drying process begins, at which the mixing process stops. The temperature, humidity and pressure parameters change and dry warm air flows through the mobile diffuser. Diffusers, moving in a small interval, contribute to the drying process. The liquid that settles at the bottom is disposed of using a water disposal system.

At the end, the resulting substance, called biosorbent, is removed from the installation and packaged.

The installation for the preparation of biosorbents from the waste petrochemical industry, the vertical type has a cylindrical shape and works as a fermenter, inoculator and adsorber simultaneously.

The installation performs sequential actions in a single system, where there is no need to extract or redistribute the biosorbent. [5,6]

During the use of the offered Installation key difference of a way of preparation of a biosorbent is the possibility of use of silt after heat treatment as an effective basis for a biosorbent.

During the implementation of the research, the obtained biosorbent has a component composition with a component ratio of dry matter, % (Table 1), has characteristics (Table 2) and can be reproduced under the conditions of preparation (Table 3).

Table 1

Surplus activated sludge	89-93
Suspension of microorganisms	7-11

Table 2

Humidity	15-19%
Macropore volume	0.551 – 0.538 cc/g
Water capacity	0.62 cc/g
Oil capacity	0.51 g/g

Table 3

Humidity range, %	15-85
Temperature range, °C	+15...+250
Pressure ange, atm	0.7 to 2.5
Range of turns at mechanical hashing, rpm	0-700
Range volume air supply at pneumatic mixing, l/min	0-100
Range of pore size of the carbon base biosorbent, micrometer	0.2 -0.5

Design features of the installation. The installation case is a cylindrical container consisting of two cylinders of identical shape made of stainless steel 08H18N10 or 10H17N13M2T, differing in structural dimensions such that a small cylinder is placed in a large cylinder and forms a gap into the inner cavity of which the thermal insulation is placed. Each cylinder consists of a solid sheet of metal with structural dimensions: length - 1350 mm, width - 8 mm, height - 1600 mm. A cylinder is created by welding a sheet into the shape of a cylinder with a welding top cover.

In the tank occurs a process of fermentation, inoculation and drying.

Dimensions of the installation: diameter of the cylindrical base - 420 mm, height - 1500 mm

Insulation layer: glass fibre mineral insulation «ursa geo», parameters:

- flammability class; non-combustible
- noise figure: 56 dB;
- material density: 15 kg/m^3;
- heat conduction: 0.036 W/m ^{0}C;
- heat-transfer resistance: 1.39 m^{2}K/W;
- operating temperature: from -60 to +220 ^{0}C

The system for regulating the temperature in the inner cavity of the main tank is represented in the form of a pipeline made of stainless steel. The pipeline is located along the perimeter of the main tank and has a closed cycle of electrical type with the possibility of controlled heating and heat exchange (cooling system), parameter:

- pipeline diameter – 20 mm
- wall thickness – 3 mm
- heat carrier – liquid, steam
- range of temperatures of the heat carrier: 12-300 ^{0}C

System of mechanical hashing. It is carried out by the bladed mixer connected to a shaft and the electric motor (power: 3-5 kW) with a range of turns up to 700 rpm.

Pneumatic mixing system. It is carried out by means of the directed air stream from the diffusers defining the direction. Volume giving: up to 100 liters of air a minute by means of the pneumatic membrane pump. In addition coal sorption with function of aeration for purification of air and creation of the clean environment contains the filter.

The diffuser is adjustable, a mobility corner (the angle of change of the direction) in each part is up to 15°.

The system of liquid discharge is located in the lower part of the camera and presented in the valve form with automatic discharge with the vacuum engine. In the camera, the formed condensate flows down and the remains from processes of fermentation and an inoculation. Discharge volume: 2 liters.

The sight window is located in the opening of the top cover with double thick glass. It has a diameter of 50 mm and a glass thickness of 7 mm.

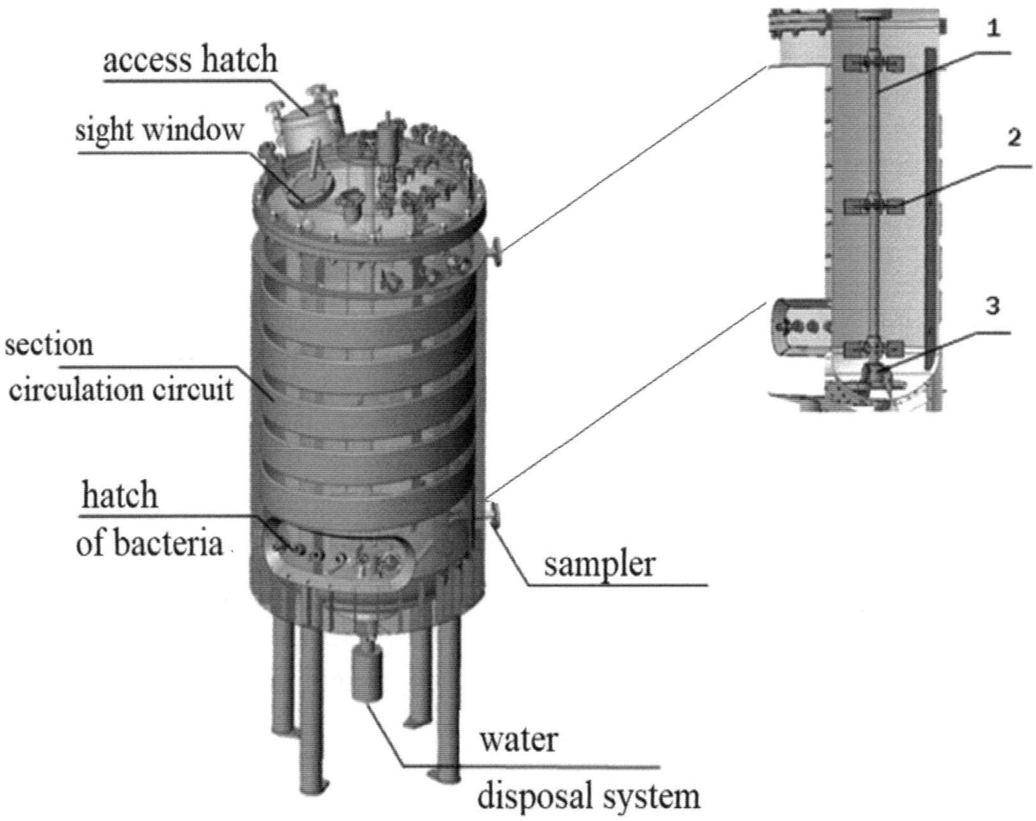

Figure 3. - 3D model installation for the preparation of biosorbents from waste petrochemical industry.
1 - Shaft of agitating mechanical device; 2 - Blade of agitating mechanical device; 3 - operating head of agitating mechanical device.

A comparison was made with analogs.

Analogs among technological installations are:

1 - Fermenter LP351, made in Russia, the company Bioengineering AG.

2 - Fermenter F-250, made in Russia, the company NPG "Priority".

Analogues among patents are:
1 - Installation for receiving a biosorbent RU 2012110076;
2 - Method of producing a sorbent for cleaning of firm surfaces of oil and liquid oil products RU 2396112;
3 - Method of producing a sorbent RU 2362619.
The main a shortcoming in listed analogs:
1 - Absence of combined systems of mixing (mechanical and air flow), reducing the efficiency and quality of mixing and fixing oil-oxidizing microorganisms;
2 - Absence of a combined thermodynamic parameters control system;
3 - Absence of speed regulation system and angular range of mechanical agitator;
4 – Absence of a system for the complex preparation of biosorbents;
5 – Absence of a system for increasing the adsorption properties of carbon sorbents.
Also assessment of costs of purchase of the equipment has been made by production of one installation for preparation of a biosorbent (Table 4).

Table 4 - Assessment of costs of purchase equipment for production of one installation for preparation of a biosorbent.

Costs of purchase equipment Designation	Cost, rub.
Mechanical mixing system (mechanical agitating, rod, electric drive)	23 000
Metal case from stainless steel	31 000
Thermal insulation	7 800
Vacuum pump	6 600
The control system (pressure sensors – 1 pieces, temperature sensor – 4 pieces, the humidity sensor – 2 pieces)	22 000
Metal framework	9 600
Thermal control system	11 000
System of pneumatic mixing	7 400
System of discharge of liquid	7000

Thus, total expenses make 125 400 rubles. At prime cost of the equipment in 125 400 rubles the estimated cost of production will be 520 000 rubles.

4. Conclusion

In the course of the studies, based on theoretical calculations, laboratory studies, analysis of scientific and industrial literature, the following results were obtained:

- A schematic diagram and a method for preparing a carbon-based biosorbent have been developed;
- Parameters for the preparation of biosorbents;
- 3-D model of installation on preparation of a biosorbent.

The developed installation can be used in the field of environmental protection, in particular for production of the biosorbent made from waste of petrochemical production. The biosorbent is intended for elimination of emergency oil spills and oil products.

Thus, the installation is intended for creation of a carbon-based biosorbent.

The installation has to surpass the existing analogs both in a cost index, and efficiency indicators: the adsorptive properties, efficiency of absorption at low temperatures and degree of immobilization of the petrooxidizing bacteria.

Functional purpose: Ensuring technical, physical and chemical parameters for effective preparation of a biosorbent on the basis of waste of a petrochemical complex (thermally processed excess activted sludge).

The developed complex installation will solve the following problems:

1 – low degree of immobilization of the existing bases of biosorbents for the petrooxidizing microorganisms;

2 – unsatisfactory efficiency of bioremediation of the existing biosorbents;

3 –high cost and complexity in operation of the existing technological conveyors on preparation of biosorbents;

4 – presence of large volume of secondary waste of a petrochemical complex which are disposed of on specialized grounds, polluting the environment;

5 –high cost of services for placement of secondary waste on specialized grounds;

6 - large size of the redeemed areas under specialized grounds.

The main direction of further research is the approbation of the obtained biosorbent in commercial conditions in order to identify possible shortcomings of the obtained biosorbent and, if necessary, subsequent design modifications of the installation parameters.

References

[1] Khaustov A P and Redina M M 2002 *Economics of Nature Management: Diagnostics and Reporting of Enterprises* (Moscow: Publishing House of the PFUR) p 216

[2] A large encyclopedia of oil and gas. [Electronic resource]: URL: http://www.ngpedia.ru/ (reference date: 25.10.2013)

[3] Kamenshikov F A and Bogomolny E I 2003 *Oil Sorbents* (Izhevsk, Moscow: Institute of Computer Science) p 268

[4] Loginov O N 2000 *Biotechnological methods of cleaning the environment from technogenic pollutants* (Ufa: Reaktiv) p 100

[5] Shamraev A V and Shorina T S 2009 Influence of oil and oil products on various components of the environment (Bulletin of the OSU) 6 pp 642-3

[6] Ilarionov S A 2004 *Ecological aspects of the restoration of oil-contaminated soils* (Ekaterinburg: URORAN) p 194

IOP Conf. Series: Earth and Environmental Science **181** (2018) 012008 doi:10.1088/1755-1315/181/1/012008

Comparative analysis of oil properties from Western Siberia oil fields

G I Egorova, N I Loseva and A N Egorov

Industrial University of Tyumen, Tobolsk branch, Tobolsk, Russia

E-mail: egorovagi@list.ru

Abstract. The article provides a comparative analysis of oil properties from Western Siberia fields. Experimental data on fractional composition, oil density, sulphur content, chloride concentration, kinematic viscosity, water content are shown, which proves the quality of oil produced, which is important for both producers and consumers of oil, including Russian and foreign enterprises.

1. Introduction

The strategy of Russia's development in the field of power engineering provides for an increase in the volume of oil refining up to 220-225 million tons per year. A significant part of the obtained oil products is planned to be exported, including to Western Europe. However, the continual tightening of environmental and quality requirements of the European Union to consumed petroleum products may lead to a reduction in export opportunities for the Russian oil refining industry.

Due to this, the task of ensuring a world-class level of quality of products is becoming increasingly relevant for domestic refineries. The complexity of its solution is largely determined by the quality of raw materials coming in for processing. Consequently, determination of the quality of oil extracted from various deposits on the territory of the country acquires great significance both for producers and consumers of oil [1].

Of course, oil is a mutual conjugate solution of hydrocarbons and heteroatomic organic compounds. Oil as a dispersed system is a complex, multicomponent, inter-soluble mixture of gaseous, liquid and solid hydrocarbons of different chemical structure with the number of carbon atoms up to 100 or more with an admixture of heteroorganic compounds of sulphur, nitrogen, oxygen and some metals. It's not a secret that the chemical composition of oil from various fields is very diverse. The elemental composition of oil is the least variable: 82.5 - 87% carbon; 11.5 - 14.5% hydrogen; 0.05 - 0.35, rarely up to 0.7%, oxygen; up to 1.8% nitrogen and up to 5.3%, rarely up to 10%, sulphur. In oil, metals are also found (Ca, Mg, Fe, Al, Si, V, Ni, Na, etc.). Oil consists mainly of hydrocarbons of four groups: paraffin (alkanes), naphthenic (cycloalkanes), aromatic (arenes) and hybrid - paraffin-naphthenic-aromatic. Oxygen, sulphur and nitrogen are present in the form of oxygen, sulphur and nitrogen compounds. The relative content of hydrocarbon groups in oil fractions is very different. The predominance of hydrocarbons gives different properties to oil, which inevitably affects the processing methods and the fields of application of petroleum products. Paraffin hydrocarbons - alkanes C_nH_{2n+2} - are the basis of the group components of oil, the total content of which is 25-35% of the mass (not counting the dissolved gases). Alkanes of the normal structure and isoalkanes, mostly monomethyl-substituted, with different positions of the methyl group in the chain

Content from this work may be used under the terms of the Creative Commons Attribution 3.0 licence. Any further distribution of this work must maintain attribution to the author(s) and the title of the work, journal citation and DOI.
Published under licence by IOP Publishing Ltd

IOP Conf. Series: Earth and Environmental Science **181** (2018) 012008 doi:10.1088/1755-1315/181/1/012008

are most widely represented. With increasing molecular weight of oil fractions, the content of alkanes in them decreases. Naphthene hydrocarbons - cycloalkanes (cyclanes) - are included in all fractions of oil, except gases. On average, in oils of various types, they amount to 25-80% of the mass. The distribution of naphthenic hydrocarbons by oil fractions is very diverse. Their content usually grows as the fractions become heavier. The distribution of cycloalkanes by types of structures is determined by the chemical composition of oil and the temperature limits of the fractions. Aromatic hydrocarbons - arenes with the empirical formula C_nH_{2n-6} - are contained in oil, less than 10-15% by weight, and are represented by homologues of benzene and derivatives of polycyclic arenes. In hybrid hydrocarbon molecules, structural elements of all types exist: mono- and polycyclic arenes, mono- and polycyclic five- or six-ringed cyclanes and alkanes of a normal and branched structure. They can be divided into three types: 1) alkane-cyclane; 2) alkane-arene and 3) alkane-cyclane-arene. Hybrid hydrocarbons with mono- or bicyclic arenes with long alkyl chains can be part of paraffins and ceresins. The third type of hybrid hydrocarbons is most common among hydrocarbons of high-molecular part of oil [2]. Abroad, when determining the qualitative indicators of oil, density and distillation models of quality are applied. In the density model, the quality of oil and, consequently, its cost parameters are associated with the density and sulphur content. The distillation model of oil quality and its cost is associated with the potential of light oil fractions. The attempt to bring the quality of domestic oil to world standards led to the fact that in 1989 in our country for the first time in additions to GOST 9965 "Oil for oil refineries. Technical conditions" density and mass content of sulphur were proposed as the main indicators characterizing the consumer properties of oil [3]. Today, the following physicochemical properties of oil are indicated: the yield of fractions at temperatures up to 200, 300 and 350 degrees; oil density ρ; mass fraction of sulphur S; concentration of chloride salts. In TU 39-1623-93 "Russian oil supplied for export", according to the listed physicochemical properties, oil is divided into four types (Table 1).

Table 1. Classification of oil supplied for export.

Indicator	Norm for type			
	1	2	3	4
1. Density at 20 °C, kg/m^3, not more than	≤ 850	≤ 870	≤ 890	≤ 895
2. Fraction output, volumetric %, not less than				
at a temperature up to 200°C	≥ 25	≥ 21	≥ 21	≥ 19
at a temperature up to 300°C	≥ 45	≥ 43	≥ 41	≥ 35
at a temperature up to 350°C	≥ 55	≥ 53	≥ 50	≥ 48
3. Mass fraction of sulphur, %, not more than	≤ 0.6	≤ 1.8	≤ 2.5	≤ 3.5

2. Research

In the course of the study, the quality of oil in the following fields of Western Siberia was determined: Ust-Balykskoye, Ombinskoye, Vostochno-Surgutskoye, Yuzhno-Surgutskoye, Fainskoye. We experimentally determined the following quality indicators: fractional composition, density of oil, mass fraction of sulphur, concentration of chloride salts, kinematic viscosity, water content. Fractional composition is the defining characteristic when establishing the field of application of petroleum products. The limits guarantee the quality of products with the corresponding characteristics of volatility. The essence of the method is the distillation of 100 cm3 of the test sample in an automatic apparatus and the chromatographic determination of the indicators under conditions corresponding to the nature of the product and conducting constant observations of the thermometer readings and condensate volumes. The density of oil and oil products was determined by a pycnometer. The method is based on the determination of the relative density - the ratio of the mass of the test product to the mass of water taken in the same volume and at the same temperature. Since a mass of 1 cm^3 of water is taken as a unit of mass at a temperature of 4 °C, the density expressed in g/cm^3 will be numerically equal to the density with respect to water at a temperature of 4 °C. The essence of the method for determining the mass fraction of sulphur was that the test sample was placed in a beam of rays emitted by an X-ray source. The characteristics of the excitation energy from the X-ray radiation were measured and the obtained signal of the pulse counter was compared with the counter signals obtained by testing the pre-prepared calibration samples. The main method for determining the concentration of chloride salts was the extraction of chloride salts from oil by water and indicator or potentiometric titration into aqueous extracts. Determination of the kinematic viscosity consisted in measuring the time elapsed (seconds) by a certain volume of the test fluid under the influence of gravity at a constant temperature by a calibrated glass viscometer. The essence of the method for determining the water content consists in heating a sample of oil with a water-insoluble solvent (for example, with Separol demulsifier) and measuring the volume of condensed water.

3. Results and discussion

The chemical experiment was carried out in several stages. The task of the first stage of the experiment was to test the fractional composition. The regulators of the device were set in the positions corresponding to the sample and the test conditions for this group. The device automatically records the initial boiling point, the end point and recorded the distillation curve in the evaporation temperature coordinates - the percentage of distillation. The device controlled the time interval between the start of the test and the initial boiling point, the set distillation rate and the final heating control. After cooling the flask, the cooled liquid remaining in the flask was poured into a graduated cylinder with a division value of 0.1 cm^3 and volume was recorded. The volume was added to the percentage of distillation to obtain the total amount of distillation. An uncorrected percentage of losses was calculated. From each stated evaporation percentage, distillation losses were subtracted to calculate the appropriate percentage of distillation and the thermometer reading corresponding to this percentage of distillation was determined from the distillation curve (Table 2).

Table 2. Fractional composition.

Field	Fractional distillation of oil, volume of distillation (ml), at temperature (°C)											
	init. boil.	**100**	**120**	**150**	**160**	**180**	**200**	**220**	**240**	**260**	**280**	**300**
Ust-Balykskoye	70	3.0	7.0	10.0	12.0	14.5	17.5	21.0	24.0	28.0	31.5	35.5
Ombinskoye	58	5.5	9.5	14.5	16.0	20.5	23.0	27.5	30.5	33.5	38.0	41.0
Vostochno-Surgutskoye	50	7.5	12.0	18.0	20.5	25.0	29.0	33.0	37.0	41.0	46.0	50.5
Yuzhno-Surgutskoye	62	4.0	7.0	12.0	13.0	16.0	19.0	22.5	26.0	30.0	33.5	37.5
Fainskoye	52	6.0	9.5	15.0	17.0	21.0	24.0	28.5	33.0	37.0	41.5	45.0

Analysis of the fractional composition showed that the oil of the Yuzhno-Surgutskoye, Ust-Balykskoye fields does not meet the requirements of TU 39-1623-93. Fractional distillation of oil from the remaining fields shows quality results. The second stage of the experiment included determination of density. A pycnometer was selected based on the properties of the test product. The latter was weighed with an error of not more than 0.0005 g, if the capacity of the pycnometer was more than 25 cm^3, and with an error of not more than 0.0002, if the capacity of the pycnometer was less than 25 cm^3. The pycnometer, with the "water number" set, was filled with the test product at a temperature of 18 - 20 °C. The pycnometer was closed with a stopper, immersed in a thermostat at a temperature of 20 °C and kept until the level of the test product changed (at least 30 minutes). The density was calculated by the formula:

$$\rho_t = \frac{(m_t - m_0) \cdot \rho_c}{(m_c - m_0)} + C,$$

where ρ_t - sample density at the determination temperature, kg/m^3; ρ_c - water density at the sample oil temperature, kg/m^3; m_0 - mass of the pycnometer in the air, g; m_c - mass of the pycnometer with water in the air at the sample oil temperature, g; m_t - mass of the pycnometer with the sample in the air at the test temperature, g; C - correction for air pressure, kg/m^3 (Table 3).

Table 3. Oil density.

Field	Density of drained oil by pycnometer, kg/m^3
Ust-Balykskoye	876.6
Yuzhno-Surgutskoye	866.2
Ombinskoye	868.2
Vostochno-Surgutskoye	836.4
Fainskoye	846.6

The density of drained oil by a pycnometer, kg/m^3, showed the quality results of the Vostochno-Surgutskoye and Fainskoye fields.

The third stage of the experiment was connected with the determination of the sulphur concentration. 3/4 of the volume of the cuvette was filled with the test sample. Viscous samples were heated to ensure their fluidity; air bubbles were absent in the space between the cuvette window and the surface of the sample. Each sample was measured according to the recommended counting time for a certain concentration range. The sulphur concentration in the sample was calculated automatically from the calibration curve (Table 4).

Table 4. Mass fraction of sulphur.

Field	Sulphur content of oil, %
Ust-Balykskoye	1.75
Yuzhno-Surgutskoye	1.74
Ombinskoye	1.14
Vostochno-Surgutskoye	1.16
Fainskoye	1.22

The sulphur content of oil corresponds to the norms of TU 39-1623-93 for all fields.

The fourth stage of the experiment was connected with the determination of the concentration of chloride salts by their potentiometric titration in aqueous extract. The sample of the analysed oil was stirred for 10 minutes, shaking in a bottle filled 2/3 of its capacity, then the sample of the analysed oil was transferred to a separatory funnel with a stirrer. The remainder of oil was removed from the pipette walls with toluene. 100 cm^3 of hot distilled water were added to the sample of the analysed oil and the chloride salts were extracted by stirring for 10 minutes. If an oil emulsion formed during the extraction of chloride salts, then a 2% demulsifier solution was added to destroy it. During potentiometric titration of aqueous extracts of chloride salts, it was evaporated to a volume of 15 cm^3 in a 150-cm^3 beaker, then transferred to a titration beaker. Then it was cooled to room temperature, 7 cm^3 of acetone was added, acidified with 6 mol/dm^3 sulfuric acid solution and titrated. The mass concentration of chloride salts (*X*1) in milligrams of sodium chloride per 1 dm^3 of oil was calculated by the formula:

$$X_1 = \frac{(V_1 - V_2) \cdot T \cdot 1000 \cdot A}{V_3},$$

where V_1 - volume of 0.005 mol/dm^3 of a solution of mercuric nitrate or 0.01 mol/dm^3 of silver nitrate during potentiometric titration consumed for titration of aqueous extract, cm^3; V_2 - volume of 0.005 mol/dm^3 of a solution of mercuric nitrate or 0.01 mol/dm^3 of silver nitrate during potentiometric titration consumed for titration of solution in the control experiment (without an oil sample), cm^3; V_3 - volume of oil taken for analysis, cm^3; T - titre of 0.005 mol/dm^3 of a solution of mercuric nitrate or 0.01 mol/dm^3 of silver nitrate during potentiometric titration, in milligrams of sodium chloride per 1 cm^3 of solution; 1000 - coefficient for recalculating the mass concentration of chloride salts in 1 dm^3 of oil; A - coefficient expressing the ratio of the volume to which the aqueous extract of the analysed oil was diluted to the volume of the solution taken from the volumetric flask for titration (when titrating the entire aqueous extract of the coefficient A =1). The mass fraction of chloride salts in oil in percentage of sodium chloride (*X*2) was calculated by the formula:

$$X_2 = \frac{X_1 \cdot 100}{BC\rho},$$

where X_1 - mass concentration of chloride salts in oil in milligrams of sodium chloride per 1 dm^3 of oil; B and C - conversion factors for cubic decimetres into cubic centimetres (1000) and grams into milligrams (1000); ρ - density of the analysed oil, g/cm^3 (Table 5).

Table 5. Concentration of chloride salts.

Field	Chloride salts in emulsion, mg/dm^3
Ust-Balykskoye	114.2
Ombinskoye	214.6
Vostochno-Surgutskoye	21.7
Yuzhno-Surgutskoye	31.8
Fainskoye	14.5

The maximum content of salts was determined in the Ust-Balykskoye and Ombinskoye oil fields, which does not meet the requirements of TU 39-1623-93.

The fifth stage was based on the method of determining the water content. The test included heating the flask to boiling so that the rate of condensation of the distillate into the receiver was 2 to 5 drops per second. The distillation was stopped as soon as the volume of water in the receiver-trap took the temperature of the air in the room. The volume of water was recorded, with an accuracy up to one upper division of the part of the receiver-trap occupied with water. The mass (X) or volume ($X1$) fraction of water was calculated by the formulas:

$$X = \frac{V_0}{m} \cdot 100, \qquad X_1 = \frac{V_0}{V} \cdot 100,$$

where V_0 – volume of water in the receiver, cm^3; m – sample mass, g; V – sample volume, cm^3 (Table 6).

Table 6. Water content.

Field	Water content in oil, % vol.
Ust-Balykskoye	0.70
Yuzhno-Surgutskoye	0.20
Ombinskoye	0.50
Vostochno-Surgutskoye	0.10
Fainskoye	0.01

The water content was discovered in abundance at the Ust-Balykskoye field.

At the stage of determining the kinematic viscosity, the selected oil sample was heated in a container for 1 hour (at 60 °C). The sample was stirred and shaken. After the flowability was reached, the sample was poured into a glass flask with a capacity of 100 cm^3 in the amount to fill the two viscometers. It was heated for 30 minutes in a bath with boiling water. Two viscometers were filled. If the sample was heat treated before the test, a pre-heated filter was used to prevent coagulation. The time was measured to within 0.1 s. This time was required for the sample to pass from the first mark of the viscometer to the second [4]. The kinematic viscosity v, mm^2/s, was calculated by the formula:

$$v = C \cdot t,$$

where C — calibration constant of viscometer, mm^2/sec^2; t — arithmetic mean value of the flow time, sec. The dynamic viscosity η, MPa·s, was calculated from the kinematic viscosity:

$$\eta = v \cdot \rho \cdot 10^{-3},$$

where ρ — density at the same temperature at which the kinematic viscosity was determined, kg/m^3; v — kinematic viscosity, mm^2/s. The results of the tests are given in Table 7.

Table 7. Kinematic viscosity.

Field	Kinematic viscosity of oil, mm²/sec	
	20 °C	50 °C
Ust-Balykskoye	30.61	10.19
Yuzhno-Surgutskoye	17.80	7.09
Ombinskoye	18.03	7.05
Vostochno-Surgutskoye	5.51	2.86
Fainskoye	8.02	3.81

The kinematic viscosity of oil from the Vostochno-Surgutskoye and Fainskoye fields corresponds to the norms of TU 39-1623-93.

4. Conclusion

On the basis of the results of the oil studies, the following conclusion can be drawn: from all the fields studied, oil from Vostochno-Surgutskoye and Fainskoye meets all the requirements of TU 39-1623-93 for export raw materials, while most of the investigated oil does not meet these standards for a number of parameters. A comparative analysis of the quality of oil by its physico-chemical characteristics underlies the development and implementation of the development strategy of Russia in the field of fuel and energy complex.

References

[1] Egorov A N 2016 *Oil and Gas Processing Industries. Technology. Security. Ecological compatibility* (Tyumen: IUT) 150 p
[2] TU 39-1623-93 *Russian Oil, Supplied for Export* Introduced 01.02.93 IPTER Belarus Academy of Sciences 11 p
[3] Polishchuk Yu M 2012 Analysis of oil quality *Oil Industry* 1 p 68
[4] GOST 33-2000 *Interstate Standard Petroleum Products. Determination of Kinematic Viscosity and Calculation of Dynamic Viscosity* 24 p

IOP Conf. Series: Earth and Environmental Science **181** (2018) 012009 doi:10.1088/1755-1315/181/1/012009

Waste refrigerant fluid of the process of n-butane dehydrogenation as a source of new commercial products: resource-ecological aspect

A N Egorov and G I Egorova

Industrial University of Tyumen, Tyumen, Russia

E-mail: egorovandre@mail.ru

Abstract: The article deals with the technology of disposal of waste refrigerant fluid for the process of one-stage dehydrogenation of n-butane by oxidative thermal destruction with the production of bituminous varnish as a commercial product, which allows solving the problems of neutralizing and processing the waste of petrochemical industries and bottoms with the aim of saving resources and raw materials.

1. Introduction

The analysis of the prospects for the development of world oil and gas chemistry and oil refining shows a significant shift in priority problems towards a comprehensive assessment of the environmental friendliness and enterprises safety as well as possible consequences of their impact on the environment, life and health of the population. The objects of oil and gas processing, like many enterprises of other industries, are sources of the inevitable, potential, until now practically quantitatively negligible risk to the population and the environment. The fact should be noted, that on the one hand, oil refineries are the main driving factor in the development of all branches of industry, transport, municipal engineering and agriculture, the basis for raising labor productivity and the human wellbeing. On the other hand, there is an indisputable fact that the environmental friendliness and safety of enterprises are on the forefront of the development of the Russian Federation. Environmental friendliness and safety of enterprises are directly linked to the solution of the problem of waste disposal. Let us reveal some key points.

First, the problem of waste becomes extremely acute due to the depletion of natural resources, degradation of terrestrial ecosystems, deterioration of the quality of the natural environment caused by the growth of the anthropogenic load on the nature, including waste from the oil and gas industry. Secondly, the colossal amount of waste from the oil and gas industry and the related flow of man-made substances into the environment lead to the formation of new geological, biological and geochemical parameters of the environment, creating a threat to life on the Earth. Thirdly, the problem of disposal of oil and gas waste is reflected in normative documents of federal and regional significance. So in the decree of the Government of the Tyumen Region on the management of production and consumption wastes for the period up to 2020 (the regional target investment program), priorities are noted for reducing the negative impact of production waste on the environment, public health and innovative methods of waste processing.

Of course, an aggravation of the problem of waste became one of the consequences of the rapid development of industry, the scientific and technological revolution. Scientists, sociologists and

Content from this work may be used under the terms of the Creative Commons Attribution 3.0 licence. Any further distribution of this work must maintain attribution to the author(s) and the title of the work, journal citation and DOI.
Published under licence by IOP Publishing Ltd

environmentalists derived a direct relationship between welfare growth and a decrease in the amount of waste from the population and industry [1]. Waste is something that cannot be used in this production. Waste volumes are determined based on this level of technology development. If initially the problem solution was seen primarily in the destruction of wastes – landfill, dumping into the seas and oceans and incineration, then with increasing environmental pollution, other, environmentally more acceptable measures for eliminating waste came to the fore – sorting and reusing, that is, resources recycling.

In the world production it must be strived for a situation that takes place in the biosphere: a complete recycling of matter – the secondary use of all atoms. The implementation of recycling with minimal costs (energy and others) is a grandiose technical task which requires changing all equipment and technology. At present, using up-to-date equipment and technologies, less than 10% of raw materials from the earth's interior and from the surface of the planet are converted into finished products, and 90% goes to wastes which pollute the biosphere. Large-scale technological researches and development carried out in developed countries are associated with a reorientation to a waste-free lifestyle or products yielding a minimum of waste, developing methods for waste collecting and sorting which ensure the economic efficiency of their recycling.

Non-waste production can be characterized in every possible way by the disposal of wastes generated in direct technological processes. Let us note also that fact, that non-waste technology is a priority of a new direction in a science – "green chemistry".

Analysis of domestic and foreign materials shows that non-waste technology as an element and component of green chemistry can develop in four main directions: the creation of various types of wastewater free technological systems based on existing, implemented and promising methods of purification; development and implementation of a system of modern methods for processing industrial and consumer waste; organization of fundamentally new processes for obtaining traditional types of products allowing to exclude waste; development and creation of territorial industrial complexes with a closed structure of material flows of raw materials and wastes within this coplexes, having a minimum of emissions.

The content analysis of author's scientific research, monographic and patent papers showed that practically all types of production and consumption waste can be used in the national economy as a secondary raw material for obtaining many types of technical and consumer goods. The reality and technical feasibility of using wastes was proved, for example, by the practice of many domestic enterprises of various industries. What is the problem of the waste neutralizing of petrochemical industries and bottoms? Solid waste from petrochemical enterprises includes a variety of substances of organic origin. During mechanical treatment of industrial wastewater slurries and pasty precipitates are formed, which must be treated for disposal or abandonment. Dehydrogenated sludge and slurries, depending on their composition, are used in industry for a limited amount, and in most cases, they are exported to the dump. When toxic substances get into the sediments, they are subjected to thermal neutralization or disposal [2]. Bottoms and various substandard liquid products are liquid waste, in addition to wastewater. Bottoms are liquid products formed in technological processes during evaporation, extraction, rectification and filtration. Their use is practically impossible because of the high content of toxic organic and mineral substances. For this purpose, they must be neutralized.

2. Research

For the correct choice of the neutralization method, the physical and chemical properties of the components of their composition are taken into account.

These mentioned wastes cannot be neutralized by mechanical or chemical methods because of the complex composition and variety of chemicals belonging to different classes of compounds. Thermal methods for neutralizing bottoms are the most common in our country and abroad. They are selected depending on the required energy resources (steam, fuel, compressed air or electricity) and catalyzers.

In the one-stage dehydrogenation of n-butane, the cooling of the contact gas is carried out by the so-called circulating refrigerant oil. According to the design documentation, a high-boiling mixture of aromatic hydrocarbons C_9–C_{15} should be used as the refrigerant oil. But, due to the shortage and high cost of this fraction, a fraction of diesel fuel is currently used as the circulating refrigerant oil for cooling the contact gas of n-butane, butenes and isoamylenes dehydrogenation. In the work process, this fraction is saturated with resins, aromatic hydrocarbons and products of partial cracking of hydrocarbons, while losing its properties. In this regard, periodically part of the waste fraction is withdrawn from the system with replacement by a fresh fraction. The waste refrigerant oil taken out of the system is shipped to consumers as fuel oil (mazut) for boiler plants.

This production waste is a dark, high-solidified liquid having a density from 0.98 to 1.10 g/cm^3, a viscosity above 250 mm^2/s and a flash point above 80 °C.

From the standpoint of saving raw materials, energy and resource saving, a technology for the qualified salvaging of this by-product was developed. Physicochemical analysis showed that the waste refrigerating oil is identical to the high-boiling bottoms of oil refining, for example, fuel oil. Therefore, the optimal method of processing this waste is one of oxidative destruction and the production of bitumen and further the bituminous varnish on its basis.

3. Results and discussion

Let us reveal the features of the experimental part. The chemical experiment was carried out in several stages. The objective of the first stage of the experiment was the production of bituminous resin. Experiments on obtaining bituminous resin by oxidative thermal destruction were carried out in a flask with a reflux refrigerant condenser and air supply under a bed of raw materials (bubbling). The process temperature was set at 250 to 300 °C. The reaction time was controlled by the selection of the product with the determination of its properties. The yield of bituminous resin was 50% pts. wt., under optimal process conditions (reaction time 4 hours, temperature 280 °C, air consumption 2 l/min for 300 cm^3 of the loaded refrigerant oil). Bituminous resin meets the requirements of GOST 21822 - 76 (grade B), namely: appearance of the film; softening point for the ring and the ball, °C; depth of penetration of the needle at 25 °C, 0.1 mm; mass fraction of substances insoluble in toluene,%; flash point, °C; solubility in linseed oil. Prior to all determinations, a sample of bituminous resin in an amount of 1 kg was dehydrated and freed from mechanical impurities. For dehydration, samples were heated for 30 minutes in a drying oven at a temperature of no higher than 180 °C and periodically stirred. The samples were heated to 60 °C and passed through a funnel with a layer of freshly calcined sodium chloride. To remove mechanical impurities, the dehydrated resin was passed through a metal sieve with a No. 07 mesh (GOST 6613-53). By chemical composition, bituminous varnish resin is a high-molecular organic compound obtained as a result of polymerization and polycondensation reactions in the presence of air oxygen of aromatic, paraffin-naphthenic hydrocarbons, resins and asphaltenes.

The mechanism of transformations is radical chain polymerization, the initiator of transformations is molecular oxygen. In the initial period of the polymerization reaction, free radicals are formed when the initial hydrocarbon reacts with the oxygen dissolved in it. The formed radical attaches an oxygen molecule to itself and is converted to a peroxide radical (RO_2), which then breaks the hydrogen atom from the hydrocarbon molecule and forms a hydroperoxide and a new free radical (R) which ensures the further growth of the chain. The general scheme of the radical-chain mechanism includes five basic steps.

(0) $RH - R \rightarrow R - R + H^\cdot$ chain nucleation
(1) $R^\cdot + O_2 \longrightarrow RO_2^\cdot$ chain continuation
(2) $RO_2^\cdot + RH \rightarrow ROOH + R^\cdot$ chain continuation
(3) $ROOH \rightarrow RO^\cdot + OH^\cdot$ degenerate branching
(4) $R^\cdot + R^\cdot \rightarrow R - R$ chain termination
(5) $RO_2^\cdot + R^\cdot \longrightarrow ROOR$ chain termination

The chain is broken as a result of recombination of radicals. In the process of oxidation, hydroperoxide accumulates, the molecules of which decompose relatively slowly into radicals. This leads to an increase in the rate of formation of free radicals [3].

As part of our study, bituminous resin performs a dual role. On the one hand, this is the final product, in demand in the fuel and energy industries and construction, and on the other hand, we regard bituminous resin as an intermediate raw material for the production of a new product - bituminous varnish. In the course of the experiment, a yield of condensate was 50%.

Therefore, the main task of the second stage was the development of a technological chain for the production of bituminous varnish. The composition of the technological chain of production of bituminous varnish includes the following stages: synthesis and study of the parameters of bituminous varnish resin, mixing of varnish resin with solvent, filtration of varnish. The properties of the oxidized product obtained were compared with the requirements of GOST for bituminous varnish (bituminous varnish BT-577 GOST 5631-79). The physicochemical properties of the initial materials and the products obtained are shown in Table 1.

Table 1. Characteristics of raw materials, substances and manufacturing products.

Product	GOST or technical conditions	Quality indicators	Norm	Application area
1. Waste refrigerant oil (raw material)	According to the production procedure	1. ρ_4^{20} 2. Viscosity, mm²/s	≥1.05 ≥250	Raw materials
2. Varnish resin (product)	According to the project	Softening temperature	≥100 °C	Intermediate raw materials
3. Solvent: oil xylene, white spirit (reagents)	GOST 14710-78 GOST 3134-78	– –	– –	Thinner for varnish
4. Light hydrocarbons (condensate, product)	According to the project	1. Density, g/cm³ 2. Initial boiling point °C 3.T.c. End boiling point °C 4. Flash point °C	0.850…0.950 ≥100 ≤250 ≥30	Component of gasoline, diesel fuel
5. Siccative (reagent)	Technical conditions TU 20511.559-84	–	–	Film drying catalyzer
6. Bituminous varnish (received commodity product)	GOST 5631-79	1. Appearance 2. Conditional viscosity, 3. The mass proportion of non-volatile substances 4. Drying time of degree 3, h.	Black liquid 18-35 37 24	Finished product

The main objective of the third stage was the development of methods, preparation formulas, research of the commodity characteristics of bituminous varnish. The composition of the formulation of bituminous varnish, obtained by a periodic method, was: varnish resin - 50%; solvent (xylene, white spirit) - 45%; siccative - 5%.

The following indicators of commercial bituminous varnish were obtained: drying time of one layer at 20 °C, - 24 hours; conditional viscosity by viscometer VZ-4 at 20 °C, - 38 seconds; mass fraction of non-volatile substances - 37%; adhesion strength with metal - 0.2 MPa; elasticity of the film in bending, 0.9 mm; stability of the film with a static effect of water - 48 hours.

The main objective of the fourth stage of the experiment was to study the chemical composition of distillate-condensate. On the basis of the "Crystal 5000" chromatograph, after purification from unwanted components and mixing with useful additives, it was shown that in accordance with ASTM D 6729 (the "Chromatec DHA" software application), the obtained distillation condensate is a mixture of light hydrocarbons with a boiling point in the range from 100 to 250 °C; density of 0.850-0.950 g/cm^3.

At the same time, the content of unsaturated hydrocarbons was no more than 2%. Oxygen-containing compounds (methanol, ethanol, tert-butanol, MTBE, ETBE, TAME) were not determined in the obtained distillate-condensate, which is the key task of the next stage of the experiment. The chromatographic analysis data allow us to state that the chemical composition of distillate-condensate can be used as a component of gasoline (gasoline fraction NK-150 °C), diesel fuel. A high octane number is provided by the presence of oxygenates - hydrocarbon oxidation products.

Based on the developed methodology, we proposed a basic process flow diagram. We will reveal the features of the process of obtaining bituminous varnish (Figure 1).

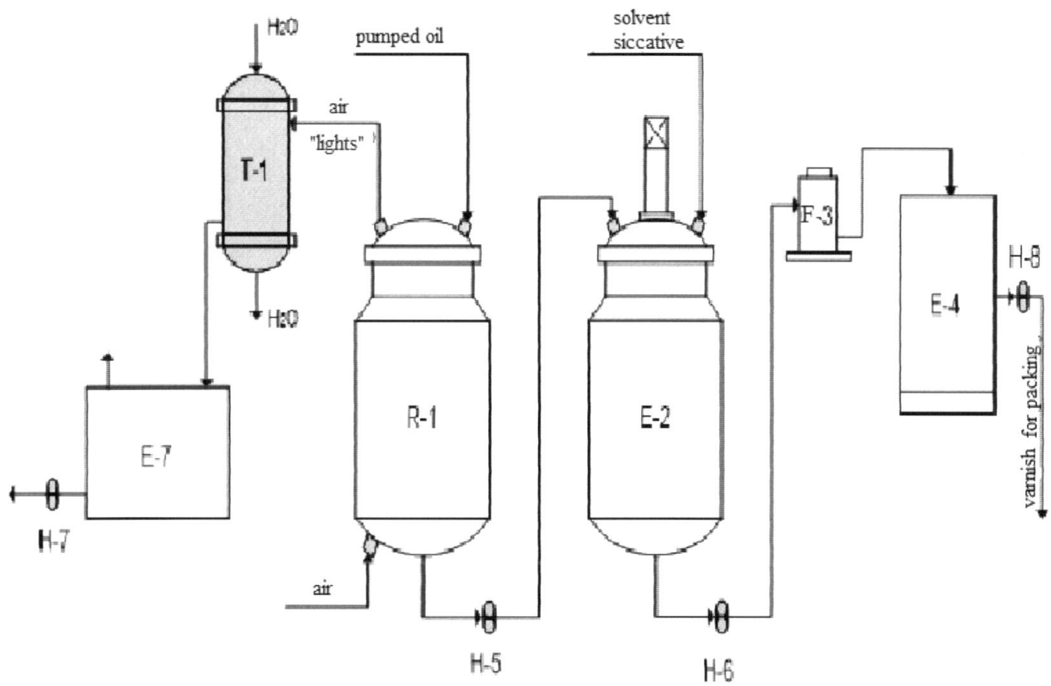

Figure 1. Principal technological workflow of production of bituminous varnish in a periodic way: T1 - heat exchanger; R1 - reactor; E2 - stirring capacity; F3 - filter "Kuno"; E4 - capacity for finished products; H5, H6, H7, H8 - pumps; E7 - capacity for collecting "light" hydrocarbons.

The waste refrigerant oil is loaded into the reactor. In the reactor, the raw material is heated and its oxidation takes place. The reaction is intensified by introducing air (up to 1500 m^3/h). A sample of the oxidized product is taken through the side of the reactor to determine the softening point. The liquid varnish resin is cooled to a temperature of 130 °C and pumped to a mixer in a pump, after which the mixer is purged with nitrogen. The solvent is poured through the counter-dispenser into the reactor and is pumped into a mixer in which the varnish resin is mixed with the remaining solvent and siccative. The obtained varnish is supplied by the pump for filtration into the cartridge filter, from which the varnish flows by gravity into the capacity of the finished product. Light hydrocarbons, which escape during oxidation through the top of the reactor, condense in the refrigerant condenser and are withdrawn for salvaging as part of light gasoline fractions. The principle technological scheme of production of bituminous varnish in a periodic way does not require expensive equipment, which is shown in Figure 1.

4. Conclusion

Thus, the raw material is a waste of the dehydrogenation process, a source of new commodity products which are in high demand in construction and engineering. Commodity qualities of the products obtained correspond to GOST and TU. It solves the problem of saving raw materials as well as energy and resource saving.

The ecological effect is related to the improvement of the ecological situation in the cities of the Tyumen Oblast on the basis of the international obligations of the Russian Federation and the environmental priorities of the Tyumen region. Utilization of heavy oil fractions will not lead to additional adverse effects and will ensure the use of new, modern technologies approved on an industrial scale. Indicators of priority of utilization of heavy oil fractions are: national (covers the territory of the Tyumen Oblast as a subject of the Russian Federation) and local (covers the enterprises of the Tobolsk industrial site).

The economic effect is obvious. Here it is appropriate to quote the words of the great scientist, our fellow countryman D.I. Mendeleev, that "burning oil is tantamount to burning accounts". The raw material effect is related to the possible use of waste as a secondary raw material for the production of bituminous resin and varnish.

The implementation of the proposed methods and technologies for the production of bituminous varnish will allow diversifying traditional production facilities, bringing them to new markets, ensuring the loading of existing production capacities, stable employment and income of the residents of the region working at these enterprises.

References

[1] Egorov A N 2016 Waste from oil and gas chemistry enterprises *Rational Ways of Recycling. Energy Saving and Innovative Technologies in the Fuel and Energy Complex Proc. of the Int. Sc.-Pr. Conf. of Students, Graduate Students, Young Scientists and Specialists* (Tyumen) **2** pp 265-8

[2] Egorov A N 2016 *Oil and gas processing industries. Technology. Security. Ecological compatibility* (Tyumen: IUT) p 150

[3] Egorova G I 2017 The spent cooling liquid of the n-butane dehydrogenation process is a source of new marketable products. Resource and ecological aspect. Status, trends and problems of development of oil and gas potential of Western Siberia *Mater. of the Int. Conf.* ed Grachev S I (Tyumen: IUT) pp 94 -106

Principles of sustainable development of new resources from waste for the oil and gas processing industry

G I Egorova and A N Egorov

Industrial University of Tyumen, Tobolsk branch, Tobolsk, Russia

E-mail: egorovagi@list.ru

Abstract: The article considers the main principles of sustainable development of new resources from waste for oil and gas processing industries. The reserves of waste-free technologies and the principles of sustainable development in the field of waste management are disclosed that provide Foresight and modern trends in the oil and gas processing industry.

1. Introduction

Today, it is difficult to imagine the life of a modern person without functionally important chemicals and materials based on them, such as: fuel, polymer materials and plastics, rubber and synthetic fabrics, paints and varnishes, metals and alloys, solvents and extractants, freons and varnishes, materials for communications, radio, television, transport, technology, household chemicals, etc. Without chemistry and chemical technology, it is impossible to imagine the life of a modern person [1]. However, using the received material substances as goods, few people think about the questions: what is the technology of obtaining this or that material? How much money is spent on obtaining a commodity product and how much waste is received at the same time? What is waste in general, what is its chemical basis and how can it be used?

Naturally, the modern world, the world community is a civilized system - open to understanding the global role of industrial and domestic waste. Therefore, the solution of the set of problems of rational utilization and processing of modern wastes of the chemical, oil and gas processing industry will allow solving not only the question of saving raw materials and energy but will also ensure an increase in the standard of living of modern people, which is especially urgent and in demand in the modern world.

The conduct of literary and patent search made it possible to analyse domestic and foreign waste-free technologies that are in demand in the oil and gas processing industry, which are developing in four main directions [2]. First of all, it is the creation of various types of drainless technological systems, based on existing, implemented and promising methods of purification. Secondly, it is the development and implementation of a system of modern methods for recycling production and consumption wastes, taking into account the fact that solid waste is a complex concentrated mixture of polluting chemicals. Thirdly, it is the organization of fundamentally new processes for obtaining traditional types of oil and gas processing products, which allow eliminating or reducing the technological stages at which the main amount of waste is formed. Fourthly, it is the creation of territorial industrial complexes (TIC) with a closed structure of material flows of raw materials and wastes inside the TIC, which have a minimum of emissions and waste. It should be noted that the large-scale technological research and developments carried out in developed countries are associated

Content from this work may be used under the terms of the Creative Commons Attribution 3.0 licence. Any further distribution of this work must maintain attribution to the author(s) and the title of the work, journal citation and DOI.
Published under licence by IOP Publishing Ltd

with a reorientation to a waste-free lifestyle or products that provide minimum waste, development of methods for collecting and sorting waste, ensuring the economic efficiency of their re-use. The problem of rational disposal is also significant for our region. As part of the development of the oil and gas processing cluster of the Tyumen Oblast (region), which will reach full capacity by 2020, the issues of effective ways of waste disposal have been resolved. "Oil and gas processing cluster - cluster without waste". This statement is not a simple slogan, but the current principle of the development of the region, which was reflected in normative documents, in the annual message of the governor to the deputies of the regional Duma [1]. The economy of the region is diversified taking into account the development priorities for the extraction and deep processing of hydrocarbon raw materials without waste: first of all, RN-Uvatneftegaz, Antipinsky oil refinery, creation of a special waste-free zone at SIBURTobolsk enterprises. A system cluster of global oil and gas chemical industries is formed from deep processing to the processes of creating commodity polymer materials [2].

Thus, sustainable development of the oil and gas processing industry is ensured by the fact that new principles are being developed as special promising targets for waste disposal, which emphasizes the relevance of our study. As part of the implementation of the key task, the goal of the study was also set: formation and development of methodological approaches for analysing the efficiency and sustainability of the development of the oil and gas processing industry without waste, as the leading sector of the economy, including the economy of the Tyumen region.

2. Research

Research objectives: identification of key trends in the long-term development of the oil and gas processing industry without waste, taking into account the features of its modern structure; systematization and analysis of the most popular trends in waste management, forecasting of the Foresight of the oil and gas processing industry, taking into account the issues of rational waste management, identification of key directions and priorities; a comparative analysis of key trends in the long-term perspective of various companies; based on the results obtained, defining global trends in the development of the oil and gas processing industry without waste, their analysis, the construction of a road map up to 2050 [2].

The problem of ensuring comprehensiveness and ecological safety during waste disposal with additional extraction of mineral and organic raw materials is considered in the aspect of development priorities and the creation of technologies of the XXI century - the technologies of "Green Chemistry". At the same time, in our understanding, three main tasks are solved ensuring:

- creation of waste-free production processes with a closed or reverse technological cycle;
- transfer of waste to environmentally safe or inert substances;
- expansion of the mineral and raw materials base due to the transformation of industrial waste into a useful commercial product.

Waste management is possible with the following principles:

- compliance with the hierarchy of waste management (priority): prevention, minimization, reuse;
- precautionary principle based on the implementation of cost-effective measures;
- principle of proximity (waste is processed, where it is formed);
- principle of producer responsibility is based on the principle "polluter pays" and establishes the physical and financial responsibility of the producer for the entire life cycle of the product.

Disposal of products is shifted to the manufacturers, which motivates them to develop products that are less hazardous to the environment and can be disassembled, reused and recycled. We are considering the principles of sustainable development of new resources from waste for the oil and gas processing industry in dynamics, from the middle of the last century. Having carried out content analysis of scientific articles and monographs in the field of waste management, we note that integrated practices have developed largely. The format of integrated practices reflects not only the development forecasts of a particular area, but also includes the content of strategic aspects of achieving the desired prospects for waste disposal in the medium and long term. This way of organizing the interaction of key principles and the plan for their implementation in the production of the oil and gas processing industry was called "Foresight" [3].

A significant point should be noted: it is necessary to carefully develop the methods of Foresight focused on waste disposal. They can be considered at the interface of methods for forecasting and waste management. At the same time, it is mandatory to include representatives of all groups of key stakeholders in the process, which makes it possible to achieve high effectiveness of programs through their general coordination and discussion at all stages of development of waste management issues. Foresight programs and waste management projects are implemented at the supranational, national, regional and corporate level, with a wide range of research areas within the framework of such basic areas as oil refining, oil and gas production.

3. Results and discussion

When solving the problem of developing new resources from waste for the oil and gas processing industry, we consider it necessary to take into account certain groups of technologies: process, basic and breakthrough. We used such a system in the proposed methodology for assessing the efficiency of waste disposal of the fuel and energy complex. We consider process technologies of waste disposal as part of industrial production; these are those that are already born with production. Basic technologies of waste recycling are those that are most in demand "now" and "here", at this point in time. Breakthrough technologies are technologies for waste recycling previously not implemented, or implemented at an insufficient level, which can qualitatively improve production conditions, thereby providing the most beneficial competitive advantages. A methodology for estimating the efficiency of waste disposal through a defined system of RES indices (Resource capacity, Ecological, Security), revealing waste from the position of resources, ecological compatibility and safety, which is necessary for a comprehensive presentation and application prospects. The methodology included monitoring of oil and gas processing companies taking into account the following criteria: availability, sufficiency (frequency of mention in open waste reports), objectivity, specificity of indicators. Considering the monitoring of the open reporting of more than 20 oil and gas processing companies, a list was drawn up. Among the set of criteria, those that most often occur in open reporting mattered.

The Foresight programs of waste management is a reference point for the future. At the current level, this priority is based on a system of real trends that have evolved in the oil and gas processing industry [4]. We will reveal the main trends of the oil and gas processing cluster in more detail.

The first trend is the rational use of APG from the production and collection of associated gas, transportation and processing of associated gas. The high level of APG utilization in the first half of 2016 was found to be 86.1% (on the example of the Tyumen region). The increase in production of natural gas and APG - at the level of 800-1000 billion m^3/year. The production of the C2+ fraction - at the level of 60-80 million tons/year. The growth rate of production in 2050 - 2.5%. The growth rate of consumption in 2050 - 3.6%. The increase in the output of primary products of NGP is projected to be 18% by 2030. From the point of view of chemical technology, natural gas and APG are the most valuable hydrocarbon raw materials along with oil, chemists and technologists know about it. Paraphrasing the words of the outstanding scientist D.I. Mendeleev, we note that burning gas is not only a significant amount of harmful emissions released into the atmosphere, but also billions of rubles lost. With the growth of oil production, associated petroleum gas production is increasing. The volume of associated gas production in 2012 amounted to 175 million m3. Utilization of petroleum gas reached 35.4% and amounted to about 62 million m3 [4]. Processing of natural gas with the receipt of commodity products is determined by the need to increase the level of socio-cultural, economic efficiency and sustainability of the economic complex of the Tyumen Oblast.

The technical and methodological literature discusses liquid and gaseous oil and gas industrial waste, characterized by specificity of origin and composition. In a generalized form, this information is presented in Figure 1.

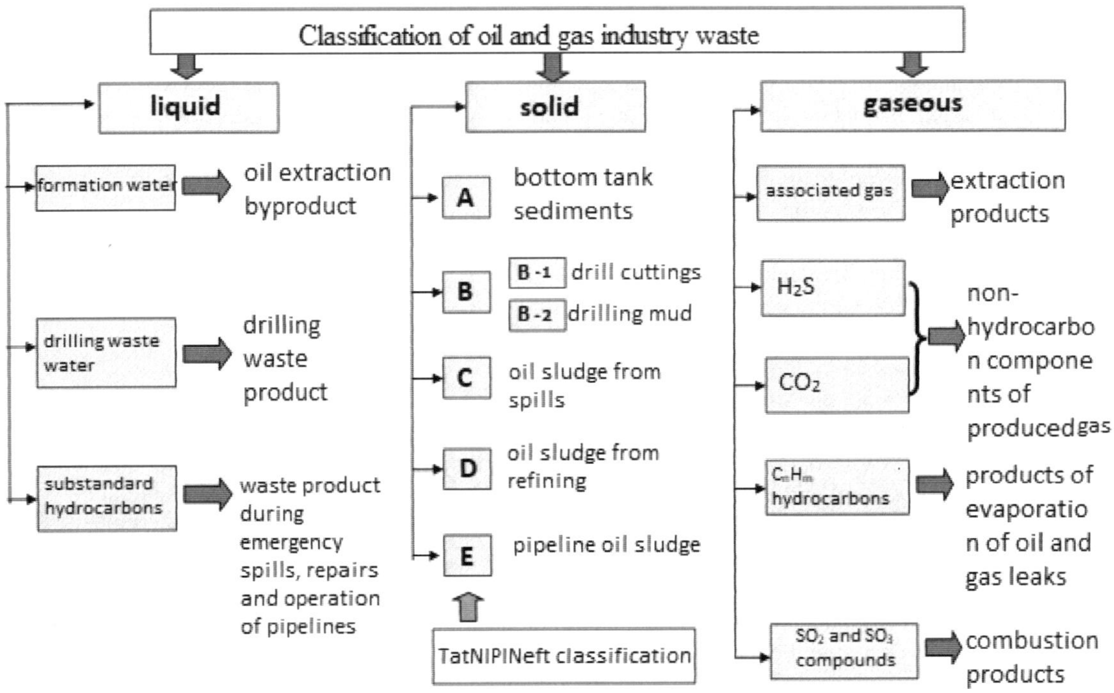

Figure 1. Products of natural gas processing.

Increasing the energy efficiency of the fields is the next priority for the oil and gas chemical cluster. In accordance with this, the first stage of the gas turbine power plant (GTPP) with a capacity of 20 MW operates at Ust-Tegusskoye field, with utilization of up to 55 million m^3 of gas per year. In 2013, the second stage of the GTES is launched, bringing the total useful capacity to 60 MW, which will further rationally dispose of 150 million m^3 of gas per year. At the Tyamkinskoye and Kalchinskoye fields, two GTPPs were built, with a capacity of 18 MW each. In this connection, the percentage of utilization of associated petroleum gas by 2017 reaches a world average of 95% [5, 6].

The project to increase the capacity of the existing central gas fractionation unit (CGFU, Tobolsk) to process a wide fraction of hydrocarbons, with an increase in the volume of NGL processing from 3.5 to 8.0 million tons, which is a comprehensive project not only for the region but also for the world industry.

The second trend is advanced processing of crude oil. The total annual capacity was 273,0 million tons, including 28 operating refineries (254 million tons), 4 GPPs (7.8 million tons), 80 mini-refineries (10.8 million tons). The share of advanced processes for primary processing is low. Deepened processes account for only 16.7% across Russia. Cracking accounts for 63%, hydrocracking - 2.9%, visbreaking - 5.4%, coking - 2.1%. Let us point out the fact that these processes need to be developed. The volume of the main output of the Tyumen cluster: gasoline (52-62%), stable gasoline (11-13%), kerosene (18-22%), diesel fuel (20%), with oil and gas losses.

The development of the oil and gas chemical cluster is connected with the activities of the Antipinsky Refinery, where 2.6 million tons of oil are processed every year, with the equipment capacity of 4 million tons. It produces diesel and motor low-sulfur fuel - 840 thousand tons; furnace fuel - 226 thousand tons; stable gasoline - 503 thousand tons; mazut - 1004 thousand tons. Today, the enterprise operates with an increase in capacity to 8 million tons per year. A wide range of refined petroleum products with a modern technological configuration and processing depth up to 95% (including secondary processes: hydrocracking, dewaxing, coke production, sulfur production according to Klaus), meeting the environmental requirements of EURO-5 standards, as well as obtaining concomitant modern products and materials in demand on the market [5]. The third trend is realization of the pyrolysis process. The current number of pyrolysis units does not allow processing existing raw materials and fully meeting the domestic demand. Construction of new pyrolysis facilities (ZapSibNeftekhim) with the launch in 2018 will increase the capacity for the production of monomers. "West Siberian complex of deep processing of hydrocarbon raw materials with a capacity of 2 million tons per year" includes a pyrolysis plant for 1.5 million tons of ethylene per year. ZapSibNeftekhim will change the face of all Russian petrochemicals. The fourth trend is production of polymers. Polymer production is a promising priority for the oil and gas chemical cluster of the Tyumen Oblast, which is being considered through the implementation of investment projects, which will make a powerful breakthrough and reach the international level in terms of volume and quality of products. Analysis of existing trends shows the possibility of additional waste production. When choosing criteria for waste selection, great attention was paid to the demand of the product on the market and the possibility of drawing the attention of ecologists, technologists, and chemists to this substance. Basic products included oil sludge, polymerization waste, atactic propylene, carbon monoxide (CO), carbon dioxide (CO2), acids, aluminum chromium catalysts of various brands, oil emulsions, spent heavy oil fractions and many others. Some of these compounds, such as oil sludge, atactic propylene, acids are already in demand, others are the subject of research as basic products, for example, in the development of waste gas processing processes - the use of carbon dioxide as a feedstock, which is a source of carbon. Hydrogen is used as a source of energy for synthesis. The highly demanded compound, isobutene, which is used in the production of rubbers, can also be obtained from ethanol. There are catalysts that make it possible to obtain tryptane from dimethyl ether and methanol, a high-energy compound that can be used as an effective fuel additive. A promising direction is the use of carbon monoxide, industrial gases to produce intermediates for the synthesis of ethanol, acetone, butanol, etc.

It is obvious that the classical approach (for example, burning, biodegradation, burial) assumes the physical destruction of a useful product contained in the waste products of enterprises in the form of organic and inorganic constituents. At the same time, there are irrecoverable losses of minerals already extracted, and the processes of ecological waste neutralization are not complete, they only partially reduce the environmental load on the technological process or the territory of the enterprise (for example, storage of spent catalysts in closed, expensive bunkers and burning of high-boiling point waste in furnaces or flares, etc.). Such an approach does not allow developing "mechanisms of stability and competitiveness of oil and gas enterprises" [5]. Let us note the fact that the problem of disposal of waste with additional extraction of mineral raw materials should be considered in the aspect of "the priorities of sustainable development of Russia, civilization and the creation of new technologies of the XXI century". Among the main factors and priorities of sustainable development in Russia are the following areas that ensure: creation of waste-free production processes with a closed or reverse technological cycle; environmental safety, which involves the transfer of all toxic waste into environmentally safe or inert substances; additional extraction of mineral resources, which implies expansion of the mineral resource base and creation of a useful commodity product. In modern conditions, the "green" economy is developing, the main purpose of which is sustainable development of civilization, and analysis of trends from the standpoint of the methodology for the efficiency of waste disposal allows us to formulate a number of key positions: to plan possible wastes; to plan synthesis methods; to transfer toxic waste to non-toxic; to take into account environmental consequences; to reduce the number of stages of the process; to neutralize catalytic systems to safe; to monitor waste in real time; to ensure minimum accidents.

4. Conclusion

Thus, the priority trends in the oil and gas refining industry provide a set of new volumes of industrial waste that can become new sources of raw materials and a combination of resources in terms of chemical composition, physical and chemical parameters. A qualitative analysis of the prospects for the development of the oil and gas processing cluster, an analytical review of existing world practices in the field of waste management within the framework of forecasting and foresight of the fuel and

IOP Conf. Series: Earth and Environmental Science **181** (2018) 012010 doi:10.1088/1755-1315/181/1/012010

energy sector development, data on existing forecasts and prospective technological directions of waste utilization are systematized. The following aspects are analyzed: modern waste structure, key trends in the development of recycling technologies; the role of local types of waste in the structure of the regional block. The role of waste as a result of globalization and the need for resource regionalization are uncovered. The need to reduce the volume of oil production is shown and the peculiarities of the growth of the gas industry are revealed. The need for reasonable integration and diversification of waste management is presented. The features of the vector of development of green chemistry are revealed with the strengthening of the influence of the ecological factor for the transition to sustainable development of the fuel and energy sector. The need for using a methodological apparatus in the field of waste management is shown, which includes the practice of forecasting, strategies and methods of foresight used in the practice of various countries. The possibilities of companies in solving the issue of waste management are considered.

References

[1] Degtyarev P A 2012 *Plan for the development of gas and petrochemical industry in Russia for the period up to 2030* Mater. of the Intern. Conf. "Gas and Oil Chemistry"- 2030 *(Moscow: Alliance-Analytics) 23 p.*

[2] Egorov A N 2016 *Oil and Gas Processing Industries. Technology. Security. Ecological compatibility* (Tyumen: IUT) 150 p

[3] Egorova G I 2012 Disposal of wastes as a social problem of improving the quality of human life *Coll. Mater. of the Intern. Sc. and Pr. Conf.* (St. Petersburg: Express) pp 46-50

[4] Egorov A N 2014 *Wastes from Petrochemical Industries* (Tyumen: Tyumen State Oil and Gas University) 126 p

[5] *The concept of long-term social and economic development of the Tyumen Oblast until 2020 and for the future until 2030* Annex to the order of the Government of the Tyumen Oblast of 05.25.2009 № 652-rp (as amended of 10.02.2016 No. 94-rp).

IOP Conf. Series: Earth and Environmental Science **181** (2018) 012011 doi:10.1088/1755-1315/181/1/012011

Influence of tectonic inversions on the process of tectonic development of the north part of the west Siberian oil and gas province

P A Gorbunov and S V Vorobyov

Industrial University of Tyumen, 38 Volodarskogo st., Tyumen, 625000, Russia

E-mail: razpas@mail.ru

Abstract: The article deals with the history of tectonic development of the territory of the northern part of Western Siberia. Particular attention is paid to the influence of alternating (inversion) tectonic movements on this process. The study presented in this article is based on paleotectonic analysis performed by the isopachian triangle method. As a result of the work carried out, the existence of tectonic inversions was established, which had a significant effect on the distribution of sedimentary strata and the tectonic structure of the northern part of the West Siberian sedimentary basin.

1. Introduction

The history of tectonic development in the northern part of Western Siberia is characterized as a complex process, influenced by a large number of different factors [1], [2], [3], [4], [5]. The most important of these, according to the authors, is the influence of alternating (inversion) tectonic movements.

Tectonic inversion is a change in the direction of tectonic movements within a single territory during geological time. As an example, one can cite a phenomenon when the uplift of a certain territory is replaced by a relative immersion, which is reflected in the plan as the transformation of the anticlinal forms of the paleo-relief into synclinal ones, and vice versa [6], [7].

Some researchers argue [8], [9] that zones of inversion tectonic development are favorable for the formation of oil and gas deposits. The identification of such zones, respectively, corresponds to the tasks of searching for new clusters of hydrocarbons. The relevance of the study, therefore, is to identify new potential areas of oil and gas accumulation, in order to maintain the resource base of hydrocarbons within the northern part of the West Siberian oil and gas province.

2. Research

In the course of the conducted research, the task was to identify the existence of tectonic inversions within the northern part of Western Siberia and to trace the change in their range during geological time. Identification of successive changes in the forms of the paleo-relief during the geological time associated with inversion tectonic movements is possible during paleotectonic analysis, which plays an important role in the structure of the study. At this stage of the work the authors set the task of qualitative evaluation of inversion tectonic movements, which is carried out by means of a complex interpretation of the results of paleotectonic analysis.

Content from this work may be used under the terms of the Creative Commons Attribution 3.0 licence. Any further distribution of this work must maintain attribution to the author(s) and the title of the work, journal citation and DOI.
Published under licence by IOP Publishing Ltd

3. Results and discussion

3.1 A method for constructing paleotectonic maps

For the qualitative evaluation of alternating tectonic movements by the authors, according to the method of V.B. Neumann [10], a set of paleotectonic maps (isopachite maps) was compiled (Figure 1), describing the distribution of sedimentary strata in the Jurassic-Cretaceous development period of the northern part of the West Siberian oil and gas province.

The basis for constructing paleotectonic maps was the structural surfaces of the reflecting horizons A, B, M, M/ and G. Horizon A corresponds to the position in the wave field of the sole of the Mesozoic-Cenozoic sedimentary front. The other reflecting horizons correspond to the roofs of the Upper Jurassic, Neocomian, Aptian and Alb-Cenomanian sedimentary complexes, respectively.

According to the V.B. Neumann's methodology of paleotectonic studies, the difference between the maps A-B, B-M, M-M/ and M/ -G characterizes the distribution of sediments accumulated at the end of the Jurassic, Neocomian, Aptian and Alb-Cenomanian times, respectively.

In order to characterize the processes of changing the direction of tectonic movements, for each of the paleotectonic maps, the authors identified the main (largest) tectonic elements characterizing the paleo-relief of the investigated territory at the corresponding moment of geological time. When distinguishing these structures, the features of modern tectonic zoning of the Mesozoic-Cenozoic sedimentary cover of Western Siberia [11] were taken into account, as well as regional peculiarities of changes in the thickness of sedimentary complexes. When carrying out paleotectonic zoning, the authors tried to preserve a certain continuity with the names of modern tectonic structures.

3.2 Analysis and interpretation of the results of paleotectonic constructions.

In the Jurassic time (Figure 1, a), over-Taymyr, over- Paykhoy-Yamal and over-Enisey paleomonoclysis and Yamal paleomegalithic bank were the most hypsometrically expressed in the Paleoplane positive structures. It should be noted that these tectonic elements experienced the predominant relative uplift during the entire Mesozoic-Cenozoic period. The thickness of the Jurassic sediments formed within these structures varies from 200 to 1600 m. The intermediate position at the considered time was occupied by the Yamal paleomegalithic bank and the Nadym-Variegan paleomonoclysis within the limits of which the thickness of the Jurassic sedimentary strata reached 1600-1800 m.

At this time, the individual elements of one of the most contrasting modern positive tectonic structures occurred - the Messoyahsky Paleomegalithic bank.

The largest negative structures at the end of the Jurassic period were Yamal-Gydan and Pur-Tazovo paleomegasineclises, as well as the West Yamal paleobasin. The above structures experienced the most intensive diving, as a result of which, the most powerful (1600-4000 m) thickness of Jurassic sediments formed.

Thus, in the Jurassic, downward tectonic movements covered the central and north-eastern parts of the study area, while the peripheral areas of the sedimentary basin and its southern part experienced relative uplift.

IOP Conf. Series: Earth and Environmental Science **181** (2018) 012011 doi:10.1088/1755-1315/181/1/012011

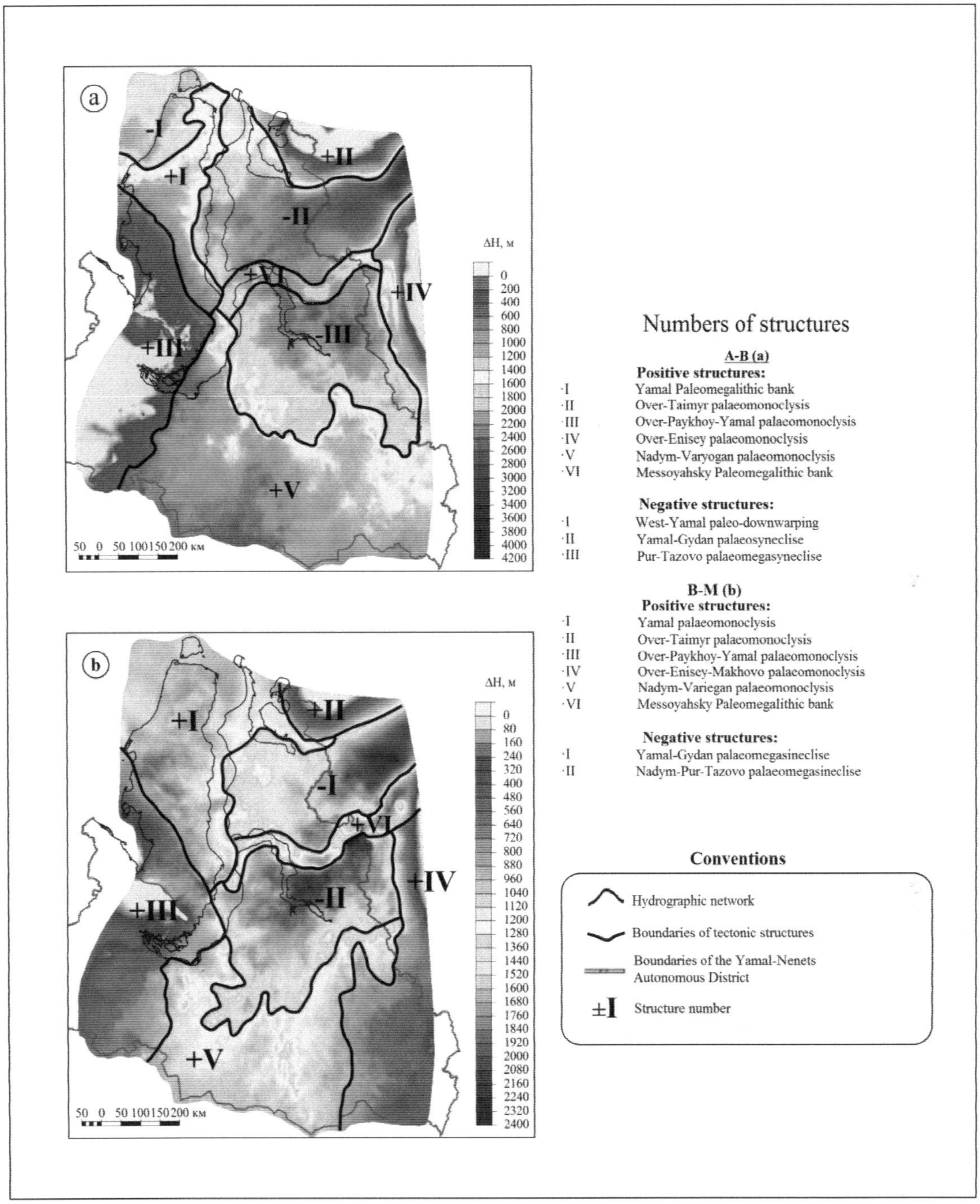

Figure 1. Paleotectonic maps of the north of Western Siberia in the Jurassic and Neocomian periods.

The first significant regional changes in the character of the tectonic development of the territory under consideration are recorded in the Early Cretaceous time (Figure 1, b). Its hypsometric expression was preserved by the over-Taymyr, over-Paykhoy-Yamal and over-Enisei paleomonoclays. At that time, the structures mentioned have been characterized by inherited relative uplift, which resulted in an increase in the size of the over-Paykhoy-Yamal and over-Enisei paleomonoclays.

At the considered time, the first significant changes in the direction of tectonic movements occurred that caused the dissolution and subsequent articulation of the western part of the Yamal-Gydan paleomegasineclise, the West Yamal paleobasin and the Yamal paleo-megalithic bank. As a result of the ascending tectonic movements to which the listed structures were subjected, a large inversion tectonic structure was formed, the Yamal paleomonoclysis within which sedimentation of the early Cretaceous age with the thickness from 600 to 1200 m took place. At the end of the Neocomian time, this structure conjugated the over-Taymyr and over-Paykho-Yamal paleomonoclays.

As a result of descending inversion tectonic movements, the expansion of the Pur-Tazovskaya paleomegasineclise to the south occurred. The thickness of the sediments accumulated within this structure during the Neocomian time reached 2,400 meters. The position of the main depocenter of this structure corresponds to the position of the Bolshekhetskaya cavity - the most contrasting negative structure of the northern part of the West Siberian plate.

Hypsometrically expressed in the Jurassic time, the Nadym-Variegan paleomonoclay located in the southern part of the study area significantly reduced its area due to the expansion of the over-Paykhoy-Yamal and over-Enisei paleomonoclay in the eastern and western directions, respectively. Also, this territory is characterized by a greater capacity of accumulated deposits, which indicates that the area experienced a more active dive in the Neocomian period, in comparison with the Jurassic period. The thickness of the Early Cretaceous sediments within the considered territory varies from 800 meters within a large part of the structure, up to 1500 m in individual depocenters.

Among the most significant moments of the Early Cretaceous stage of the tectonic development of the territory under consideration is the formation of Messoyakhsky paleomegalithic bank, which divided the Nadym-Pur-Taz and Yamal-Gydan paleomecasineclises. The activation of ascending tectonic movements is indicated by changes in the thicknesses of the Early Cretaceous (800-1300 m) and Jurassic (1600-2200 m) sediments within the structure under consideration.

Thus, the combination of ascending and descending inversion tectonic movements caused significant changes in the size and hypsometric severity of both positive and negative tectonic structures presented in the Paleorelief of the North West Siberian Plate during the Neocomian time. At the same time, when comparing maps of the Jurassic and Neocomian (Early Cretaceous) isopachites, transitions of positive forms of the Jurassic relief to negative in the neocomian were recorded, and vice versa - negative to positive ones.

The most significant transformations of the paleo-relief within the northern part of the West Siberian sedimentary basin occurred in the Aptian time (Figure 2, a). At this time, most of the study area was characterized by a predominance of ascending tectonic movements. One of the most significant transformations of the paleorelief, which occurred under the influence of activated positive vertical movements, was the articulation of the over-Taymyr and over-Enisey-Makhovskaya paleoanteclises, which led to the formation of massive over-Taymyr-over-Enisey paleomegaanteclise, occupying the entire eastern part of the study area. Within this structure, from 40 to 200 meters of Aptian sediments accumulates.

The second most important tectonic transformation that occurred in the Aptian period can be called the disbandment of the Messoyahsky Paleomegalithic bank. This structure, clearly distinguished in the Paleorelief of the Neocomian time, is not visible in the Paleorelief of the Aptian.

Significant changes occurred with Nadym-Pur-Tazovo and Yamal-Gydan paleomecasineclases. These negative structures, actively developing in the Jurassic and Early Cretaceous (Neocomian) time, as a result of the influence of inverse ascending tectonic movements, were partially connected between themselves and re-formed into intermediate forms of the paleorelief. So, in place of these structures, in the central part of the study area, the Nadym-Pur-Tazovo paleo-megaterrace was formed, where 200 to 500 meters of sediments accumulated.

IOP Conf. Series: Earth and Environmental Science **181** (2018) 012011 doi:10.1088/1755-1315/181/1/012011

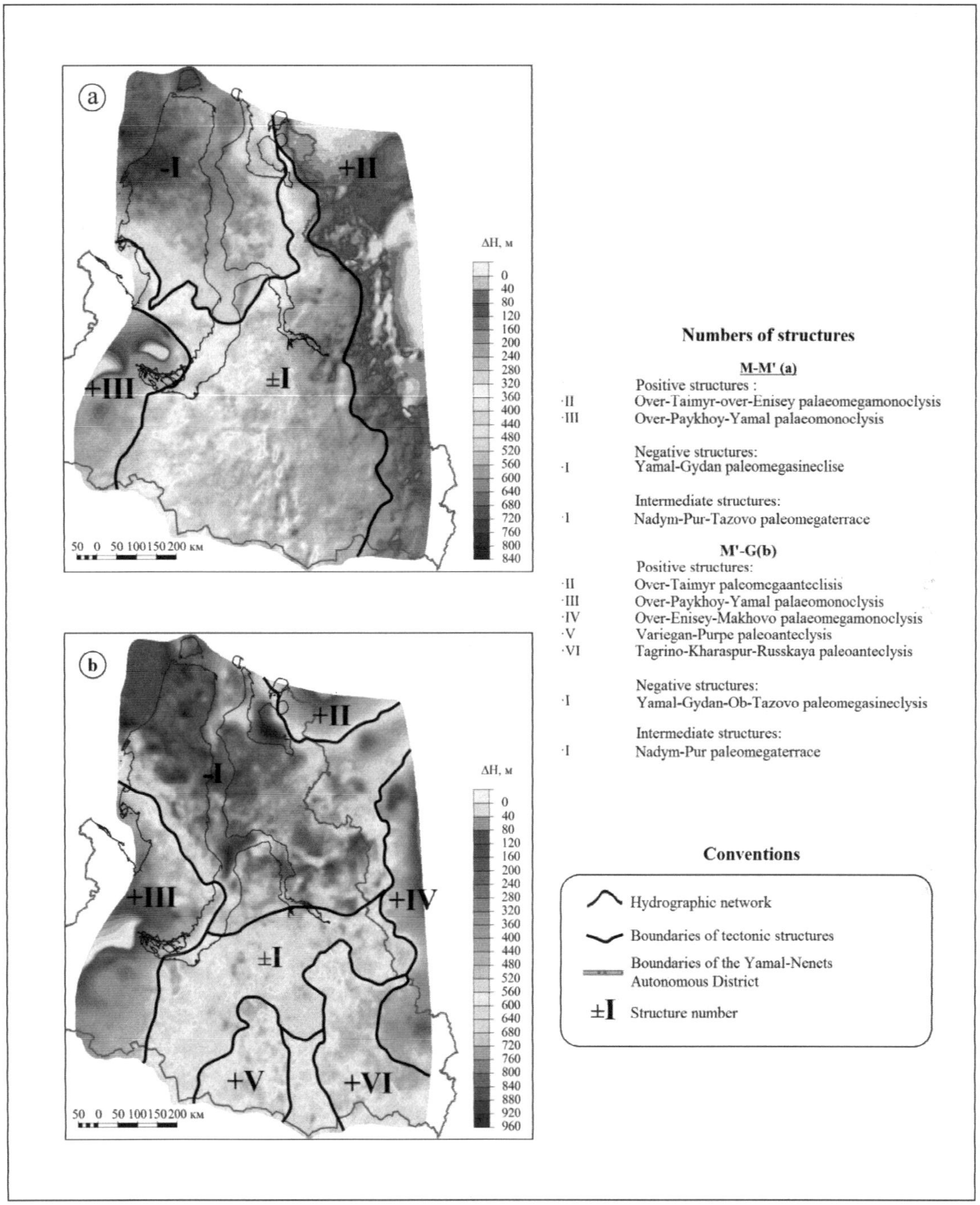

Figure 2. Paleotectonic maps of the north of Western Siberia in the Aptian and Alb-Cenomanian periods.

The complete conversion of the Neocomian paleorelief, caused by the change in the regime of ascending movements to the immersion in the Aptian, occurred in the northwestern part of the West Siberian Plate, within which the Yamal paleomonoclysis was transformed into the extensive Yamal-Gydan paleomegasineclise. The thickness of the Aptian age deposits in this part of the study area is up to 840 meters, which makes it the most contrasting negative structure of that time.

IOP Conf. Series: Earth and Environmental Science **181** (2018) 012011 doi:10.1088/1755-1315/181/1/012011

In the Alb-Cenomanian time (Figure 2, b), most of the territory of the north of Western Siberia experiences intense immersion. Thus, a clear tectonic inversion is observed - active uplift in the aphthus is replaced by immersion in the Cenomanian.

The north-western part of the investigated territory is most actively immersed. Yamalo-Gydan paleomegasineclise significantly increases its area, covering most of the territory of the Yamal and Gydan peninsulas, as well as the northern part of the Ob-Tazovo interfluve (the territory of the Semakovo, Yamburg and Severo-Urengoy fields). Such an area transformation allows us to name this structure as the Yamal-Gydan-Ob-Tazovo paleomegasineclise. Within this structure, from 800 to 1000 m of sediments are accumulated.

The central part of the Nadym-Pur interfluve also experiences an immersion. At this time, the authors distinguish here the structure of the intermediate type - the Nadym-Pur paleomegaterrace. This structure occupies an intermediate position between the actively submerged northern part of the sedimentary basin and its uplifting peripheral part. Here, 400-700 m of sediments accumulate.

Within the southern part of the territory of the Yamal-Nenets Autonomous District two major positive structures are distinguished: the Variogan-Purpei and Tagrinsky-Harampur-Russkaya paleoanteclises. Within them, the territory experiences relative growth. The thickness of sediments accumulating within these structures is estimated at 300-500 meters.

Ascending tectonic movements have already traditionally experienced the peripheral part of the territory under study. The over-Paykhoy-Yamal paleomegaanteclise, located in the west, is characterized by a reduced thickness of sedimentation, varying from 40 to 500 meters.

Over-Taimir -over-Enisey paleo-megamonoclysis at the end of the Cenomanian time is again divided into two separate positive structures - over-Taymyr paleomeganteclise and over-Enisey-Makhovskaya paleomegamonoclysis. This happens due to a significant immersion caused by tectonic inversion within the territory of the modern Ust-Yenisei-Agap belt of mega-downwarpings. Within the over-Taymyr and over-Enisey-Makhovskaya paleostructures, from 80 to 500 meters of sediments are accumulated.

4. Conclusion

Based on a comprehensive analysis of the results of seismic work, as well as exploration drilling data, the authors compiled regional paleotectonic maps at a scale of 1: 500,000, characterizing the main stages in the formation and development of the sedimentary basin in the Jurassic-Cretaceous period.

Based on the isopachite maps constructed, the authors identified large, regional tectonic structures that formed the face of the paleorelief of the northern part of the West Siberian Plate at the end of the Jurassic, Neocomian, Aptian and Alb-Cenomanian time.

The analysis of the constructed maps made it possible to identify and localize the regions of existence of inversion tectonic movements, and also to trace the transformation of tectonic structures in the course of geological time. In the course of the work it was established that during the Mesozoic period of the development of the area studied, the negative forms of the paleorelief changed to positive, and vice versa - positive forms passed into negative. Hence, such a transformation was not a local, but a regional phenomenon, covering the entire northern part of the West Siberian plate.

The most active effect of tectonic inversions during the Jurassic-Cretaceous period was experienced by the central and northern parts of the area under study. As is known, the main oil and gas content of the northern part of the West Siberian oil and gas province is just confined to the territory of the Nadym-Pur-Tazovo interfluve, as well as to the Yamal and Gydan peninsulas.

Thus, the territory of the Yamal Peninsula experienced two tectonic inversions - a partial immersion in the Jurassic was replaced by a rise in the Neocomian, then it was replaced by a dive that lasted during the Aptian-Albian-Cenomanian time.

The territory of the Gydan peninsula developed relatively successive, experiencing a predominant immersion in the Mesozoic. Hence, some parts of the peninsula periodically experience uplifting.

The territory of the Nadym-Pur-Tazovo interfluve was most actively affected by alternating tectonic movements. In the Mesozoic time, three tectonic inversions can be distinguished. In Jurassic

IOP Conf. Series: Earth and Environmental Science **181** (2018) 012011 doi:10.1088/1755-1315/181/1/012011

time, the territory under consideration experiences a predominant immersion, in the Neocomian, as a result of the change of signs of tectonic movements, the area of immersion of the territory increases significantly in a southerly direction.

Then, in the Aptian, there is a complete transformation of paleorelief - an active uplift of the territory takes place. In the subsequent Alb-Cenomanian geological era, the next major tectonic inversion occurs, entailing the immersion of the Nadym-Pur-Tazovo interfluve.

The inverse character of the development of the northern part of the West Siberian sedimentary basin, which caused a frequent change of ascending and descending tectonic movements, was accompanied by crushing and disturbing the continuity of sedimentary rocks, which are reflected in the formation of macro- and microcracking. The tectonically weakened zones were areas preferable for vertical jet migration of formation fluids. This becomes possible due to the reduction in the stress state of the rock massifs, which manifests itself in the form of a drop in both reservoir pressure and pore pressure, which makes these plots the most favorable for migration of fluids. Similarly, the existence of tectonically weakened zones can explain the movement of oil and gas in horizontal directions during the filling of traps..

The peripheral part of the investigated territory in the period under consideration experienced a predominant uplift and was deformed to a lesser degree than the central part of the basin, which makes them less promising for the localization of oil and gas fields.

Summarizing all the above mentioned, one can draw a conclusion about the influence of inversion tectonic movements on oil and gas potential at the regional scale. Inversion tectonic movements, according to the authors, predetermine the main directions of migration of hydrocarbon flows in deposits of the sedimentary cover.

The authors set a further task to develop a methodology that will allow a quantitative assessment of the effect of tectonic inversion movements on the sedimentary cover and on oil and gas potential. The methodology should allow the identification of new oil and gas-promising areas within the territory of the northern part of the West Siberian oil and gas province, which have not yet been tested by drilling.

References

[1] Surkov V S 1986 *Megacomplexes and the Deep Structure of the Earth's Crust of the West Siberian Plate* (Moscow: Nedra) p 149

[2] Rudkevich M Ya et al. 1970 *The main Stages of the History of Geological Development of the West Siberian Plate* (Tyumen: ZapSibNIGNI) 28 p 175

[3] Kontorovich V A 2001 Tectonic structure and history of development of the West Siberian geosyneclise in the Mesozoic and Cenozoic *Geology and Geophysics* **42** 11–12 pp 1832–45

[4] Markevich V P 1966 *History of Development and Oil and Gas Content of the West Siberian lowland* (Moscow: Nauka)

[5] Rudkevich M Ya 1971 *The Origin of the Structures of the West Siberian Plate* (Tyumen: ZapSibNIGNI) 45 p 219

[6] Aplonov S V 1989 Paleogeodynamics of the West Siberian plate *Soviet Geology* 7 pp 27-36

[7] Surkov V S 1970 Tectonic zoning of the Mesozoic-Cenozoic platform cover of the West Siberian plate *Soviet Geology* 4 pp 80-9

[8] Rudkevich M Ya 1974 *Paleotectonic Criteria of Oil and Gas Potential* (Moscow: Nedra) p 184

[9] Rudkevich M Ya 1972 *Structural-Formational and Paleotectonic Criteria of Oil and Gas Content* (Tyumen: ZapSibNIGNI) 61 p 143

[10] Neumann V B 1984 *Theory and Methodology of Paleotectonic Analysis* (Moscow: Nedra) p 80

[11] 1990 *Tectonic Map of the Mesozoic-Cenozoic Orthoplatform Cover of the West Siberian Geosyneclise. Scale 1:1000000* Ed I I Nesterov (Tyumen: ZapSibNIGNI)

IOP Conf. Series: Earth and Environmental Science **181** (2018) 012012 doi:10.1088/1755-1315/181/1/012012

Relationship of inversion tectonics and oil and gas potential within the Northern part of West Siberia

P A Gorbunov and S V Vorobyov

Industrial University of Tyumen, Tyumen

e-mail: razpas@mail.ru

Abstract. The paper presents the results of paleo-tectonic reconstructions of the formation features of Jurassic-Cretaceous deposits in the northern part of the West Siberian sedimentary basin, based on the method of constructing the isopach triangle. The paleo-tectonic reconstruction carried out served as the basis for creating a tectonic dislocation scheme for the Mesozoic-Cenozoic sedimentary cover of the northern part of West Siberia. The degree of tectonic dislocations is justified by the authors in this paper as one of the criteria for oil and gas content within the studied area. Thus, the designed scheme can be used to identify new and promising oil and gas potential areas and expand the hydrocarbon resource base in West Siberia.

1. Introduction

To date, to maintain a high level of oil and gas production, it is necessary to ensure a constant increase in the resource base of hydrocarbons. This is achievable only when carrying out a wide range of geological exploration. Taking into account the fact that most of the stock of un-drilled large anticlinal traps was exhausted to this day, the issue of identifying new oil and gas bearing criteria which can be used in the practice of geological exploration is acute.

In this paper, on the basis of paleo-tectonic reconstructions, the factor of the tectonic dislocation of the sedimentary cover is justified by the authors as one of the new criteria for oil and gas bearing within the northern part of West Siberia. The basis for identifying this criterion is a unique process of tectonic development of the northern part of the West Siberian oil and gas province, due to Triassic rifting. The existence of a massive system of rifts led to a frequent change of descending and ascending tectonic movements in the Jurassic-Cretaceous period, which contributed to the formation of "weakened" sections of the earth's crust favorable for migration and accumulation of reservoir fluids, including oil and gas.

2. Formulation of the problem

The main task set by the authors is the identification and subsequent mapping of areas with a high degree of tectonic dislocation, as well as the identification of their relationship with oil and gas content within the northern part of the West Siberian oil and gas province, which will allow us to justify the degree of tectonic dislocations as the most important criterion for the oil and gas potential of the studied area.

Content from this work may be used under the terms of the Creative Commons Attribution 3.0 licence. Any further distribution of this work must maintain attribution to the author(s) and the title of the work, journal citation and DOI.
Published under licence by IOP Publishing Ltd

IOP Conf. Series: Earth and Environmental Science **181** (2018) 012012 doi:10.1088/1755-1315/181/1/012012

3. Results and discussion

3.1. Analysis of the paleo-tectonic development of the north of West Siberia

During the Mesozoic and Cenozoic stages of tectonic development, the northern part of the West Siberian Plate was a region of stable submersion accompanied by the filling of a sedimentary basin with terrigenous rocks [1, 2]. The deposit and development of the rift system in the Triassic led to the fragmentation of the earth's crust and the emergence of a series of inter-rift blocks, to which the largest and most high-amplitude uplifts are assigned in the deposits of the sedimentary cover. Within their boundaries, the Jurassic incisions are reduced in relation to the adjacent rift depressions. In the Cretaceous, both the rifts and the inter-rift areas were intensively submerged. In this connection, in their limits more powerful sediments were accumulated than on stable blocks distant from the rifts [3, 4]. In this case, the deposit of tectonic structures controlling oil and gas fields occurred in the late Triassic – early Jurassic time. Subsequent tectonic and neo-tectonic movements of the Mesozoic and Cenozoic times caused the structures transformation, determining their morphology, sizes and amplitudes.

At various stages of the development of the sedimentary basin, the periods of intense deflection were repeatedly replaced by the stages of uplift, which is reflected in changes in the thickness of the sedimentary complexes and in the displacements of the boundaries of the basins and uplifts. At the same time, changes in vertical tectonic movements were caused by the block structure of the pre-Jurassic complex. Such a change in the direction of tectonic movements during geological time is called tectonic inversion, which in plain language is manifested in the change of the anticlinal forms of the paleo-relief to synclinal and vice versa.

According to M.Ya. Rudkevich and other researchers [5, 6, 7] it was the inversion of the tectonic movements that took place during the formation and development of the sedimentary cover which exerted a decisive influence on the structure of anticlinal structures controlling the deposits of multi-layer oil and gas fields and on the oil and gas content of the northern part of West-Siberian plate as a whole.

3.2. Construction of a tectonic dislocation scheme

To assess the influence of alternating tectonic movements on oil and gas potential in the northern part of West Siberia, the authors developed a scheme for the tectonic dislocation of the Jurassic-Cretaceous part of the sedimentary cover. The construction of this scheme was carried out on the basis of combining the results of seismic studies and data of prospecting drilling. The authors made regional structural maps of the scale 1:500000 on the sole of the sedimentary cover, as well as the roofs of the Middle, Upper Jurassic, Neo-Aqom, Aptian, Cenomanian and Turonian deposits. Based on the data obtained by the method of V.B. Neiman [8] a set of isopach maps was developed, which characterizes the accumulation of precipitation in the Jurassic, Neocomian, Aptian, Albian, Cenomanian and post-Cenomanian times. Then the isopach maps were renormalized in the range from -1 to +1. The first number corresponds to the most deeply submerged sections of the paleo-basins, and the second corresponds to the paleo-uplifts of the positive structures.

In the next step, all the transformed maps were sequentially summed, divided by the number of additions. As a result of the implemented transformations, a scheme was obtained which characterized the general orientation of tectonic movements in the Jurassic-Cretaceous stage of tectonic development of the northern part of the West Siberian plate. In this case, the maximum positive values of the integral parameter correspond to the areas where the uplifting processes of the territory prevailed, and the minimal ones correspond to the areas of immersion. The values of the parameter close to zero (from 0.1 to +0.1) correspond to the areas of inversion development characterized by the most frequent change of signs of tectonic movements in the course of development of the sedimentary basin. There are these areas which are characterized by the maximum degree of tectonic dislocation in the Mesozoic stage of the area development.

3.3. Relationship between oil and gas content and tectonic dislocations

Analysis of the distribution obtained (Figure 1) indicates that most hydrocarbon deposits discovered in the northern part of West Siberia tend to sites with the maximum degree of tectonic dislocations of deposits of the sedimentary cover. These include Kharasaveyskoe, Krusensternskoye, Medvezhye, Yamburgskoye, Urengoiskoye, Yuzhno-Russkoye and Kharampurskoye deposits. About 60% of the geological reserves of gas, gas condensate and oil found in the Yamal-Nenets Autonomous District are concentrated within these areas.

Figure 1. Map of the tectonic dislocation of the Jurassic-Cretaceous part of the sedimentary cover.

IOP Conf. Series: Earth and Environmental Science 181 (2018) 012012 doi:10.1088/1755-1315/181/1/012012

Localization sites of the Malyginskoye, Yuzhno-Tambeyskoye and Severo-Tambeyskoye giants deposits correspond to the values of the integral parameter, varying in the range of values from -0.2 to +0.2. These deposits are referred to areas characterized by an average degree of tectonic dislocation of sedimentary cover rocks. Under the same conditions, the Zapolyarny dome-shaped uplift developed, which controls the giant gas field of the same name.

The imposition of multidirectional vibrational motions, which conditioned the deposit, development and reformation of both positive and negative structures, makes it possible to assume the conjugacy of the plicative structures with disjunctive dislocations. The linear orientation of the areas with the maximum degree of tectonic dislocation of the Mesozoic part of the sedimentary cover, observed at the junction sections of the northern part of the West Siberian plate with the Siberian platform and folded formations of the Urals, and also within the Nadym-Pur-Taz interfluve, is most likely due to disjunctive dislocations. The northwestern and north-north-western stretch of disjunctive disturbances is a characteristic reflection of the fold-forming tectonic movements within the northern part of the West Siberian plate, which is reflected in the character of the area spread and the extent of the riftogenic areas.

Manifestations of disjunctive tectonics at various stages of formation of the sedimentary cover, accompanied by activation of more ancient tectonic disturbances of the pre-Jurassic complex, conditioned the limitations of large structural elements of the relief, vertical and horizontal displacements along the fault surfaces, thereby determining the features of sedimentation and thickness of sedimentary complexes, as well as their stretching and crushing in the folds. As a result, there was the emergence of linearly oriented anticlinal structures and the articulating cavities as well as intermediate structures.

The change of signs (directions) of tectonic movements also caused the renewal of the previously formed fault system. As a result, disjunctive dislocations spread upwards along the section, covering the basement rocks and deposits of the sedimentary cover, presumably up to the day surface.

Estimating the oil and gas control role of the factor of tectonic dislocations, it is necessary to note the following. The inverse nature of the development of the West Siberian sedimentary basin which caused frequent changes in ascending and descending tectonic movements, caused fragmentation and disruption of sedimentary rocks, resulting in the formation of areas of macro- and micro-cracking. The tectonically weakened zones were areas preferable for vertical jet migration of reservoir fluids, accompanied by the movement of oil and gas during the filling of the traps.

4. Conclusion

The nature of the area distribution of giant gas fields, as well as the majority of oil and gas fields, indicates that within the northern part of the West Siberian plate the parameter of the tectonic dislocation of the Jurassic-Cretaceous part of the sedimentary cover can be considered as a predictive criterion for localizing hydrocarbon deposits in the prospecting process for oil and gas. The tectonic dislocations scheme developed by the authors justifies the most promising areas for further geological exploration, which can include the central part of the Gydan Peninsula, the southwestern part of the Yamal Peninsula, and the peripheral part of the north of the West Siberian oil and gas province.

References

[1] Kontorovich A E et al 2014 Paleogeography of the West Siberian sedimentary basin in the Cretaceous period *Russian Geology and Geophysics* 5-6 **55** pp 745-76

[2] Kontorovich A E et al 2013 Paleogeography of the West Siberian sedimentary basin in the Jurassic period *Russian Geology and Geophysics* 8 **54** pp 972-1012

[3] Surkov V S et al 1986 *Megacomplexes and the Deep Structure of the Earth's Crust of the West Siberian Plate* (Moscow: Nedra) 149 p

[4] Fotiadi E E 1971 *Geological Structure of the Basement of the West Siberian Plate* ed Surkov V S (Leningrad: Nedra) 207 p

[5] Rudkevich M Ya, Glukhoedov Yu M and Maksimov E M 1976 *Tectonic Development and Oil and Gas Geological Zoning of the West Siberian Province* (Sverdlovsk: Sredne-Uralsk Publishing House)

[6] Rudkevich M Ya 1974 *Paleo-Tectonic Criteria of Oil and Gas Potential* (Moscow: Nedra) 184 p

[7] Rudkevich M Ya (ed) 1970 The main stages of the history of geological development of the West Siberian plate: an explanatory note to the atlas of structural, paleo-tectonic slice maps for the territory of the West Siberian lowland of scale 1:5000000 *Proc. of ZapSibNIGNI* 28 (Tyumen) 175 p

[8] Neiman V B 1984 *Theory and Methodology of Paleo-Tectonic Analysis* (Moscow: Nedra) 80 p

IOP Conf. Series: Earth and Environmental Science **181** (2018) 012013 doi:10.1088/1755-1315/181/1/012013

On goal setting in the sphere of Smart Grid

V I Kolesov, O F Danilov, G A Hmara and A L Portnyagin

Industrial University of Tyumen, 38 Volodarskogo st., Tyumen, 625000, Russia

E-mail: hmaraga@tyuiu.ru

Abstract: The article focuses on the synthesis of intelligent control systems for the development of power grid complexes. The system approach to designing is formulated, allowing justification of the target function of the project.

1. Introduction

For a long time in Russia, there was an industrial type of power supply for consumers of electrical energy. Power plants were built near large industries, and cities were connected to power supply systems on a residual basis. Simultaneous growth of energy needs from residents and non-residential municipal consumers and its reduction in large-scale industries compels to reconsider the approach to managing the efficiency of electrical systems and complexes (ESCs). In the world practice, modernization of power supply systems to the level of Smart Grid is performed to solve the problem of increasing the efficiency of the balance of generation, transmission and consumption of electrical energy. In this work, Smart Grid is considered as part of the Smart-City system.

2. Research

Management of the efficiency of electrical systems and complexes (ESCs) is aimed largely at improving the quality of life of the population and, therefore, directly related to the problem of sustainable development of cities where the bulk of the population is concentrated. This by default means that in modern terms the innovative development of Smart Grid ESCs falls directly within the scope of Smart-City.

The solution of this class of tasks for large and complex systems (such as Smart-City and Smart Grid) is, first, associated with correct goal setting. There are different conceptual models when developing large-scale urban projects. Their principal difference lies in the question of what is meant by the effectiveness of the system. Today, we can note at least four different approaches to goal setting:

- the goal is a sustainable development of the city (when the rate of development of negative processes is less than that of positive ones) [1-3];
- the goal is a sustainable innovation development (when the efficiency of the system has a positive derivative and grows due to innovations) [4];
- the goal is to maximize the vector performance evaluation in conditions of regulated individual key indicators [5-6];
- the goal is to minimize the risk of failure to meet assumed responsibilities [7-10].

The goal setting refers to the standard procedures in the tasks of managing technological and administrative processes, however, the logic of setting the goal is not always justified. This is usually

Content from this work may be used under the terms of the Creative Commons Attribution 3.0 licence. Any further distribution of this work must maintain attribution to the author(s) and the title of the work, journal citation and DOI.
Published under licence by IOP Publishing Ltd

associated with either the complexity of the process being serviced or simply with the reluctance to thoroughly understand its specifics, and urban management in this sense is no exception. The article attempts to eliminate the existing gap at least partially.

Solution. Managing complex processes is usually done in the form of a pyramid, the upper level of which declares the challenges caused by changes in the external environment. Challenges are treated as an invitation to action. And the first stage of the work is to assess their importance (i.e., prioritization). Unfortunately, the customer cannot always clearly identify them, so often the contractor must do this work.

The most common approach is the use of private STEP challenges [1] at the top level of the quadruple (tetrad), which characterize the social (S), technological (T), economic (E) and political (P) aspects of the problem, respectively (Table 1).

The social sphere is understood as a sphere aimed, firstly, for the realization of the constitutional rights of citizens and, secondly, for improving the quality of life of the social group of the population for which the system operates.

The technological sphere is focused on the success of exploitation of all types of resources. The modern specificity consists in attracting an unconventional measure - a unit of power, and as a result, the use of the coefficient of efficiency as a numerical indicator of the technological efficiency of the process or system [4].

The economic sphere is the most understandable and methodically worked out type to increase efficiency, the numerical characterization of which can be, for example, the profitability of the system.

As for the political sphere, according to the laws of classical management, it characterizes the success of the competition for markets. A broader interpretation presupposes the growth of the image of either the developer, the owner, or the user of the system.

This point of view is generally accepted, but it is more correct, in our view, to consider the five challenges (pentad) or STEEcP: with the addition of the ecological (resource) aspect (Ec). This fully agrees with the position of Hartmut Bossel who in his report to the Balaton group [2] identifies 6 subsystems in the anthroposphere (Figure 1).

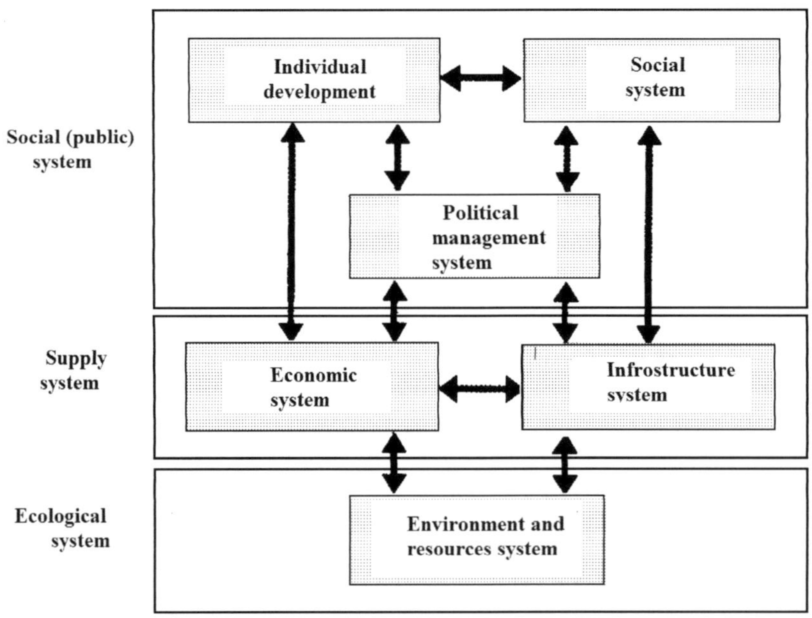

Figure 1. Six main subsystems of the anthroposphere [1].

IOP Conf. Series: Earth and Environmental Science **181** (2018) 012013 doi:10.1088/1755-1315/181/1/012013

The specificity of these subsystems is summarized in Table 1.

Table 1. Main subsystems of the anthroposphere.

No.	Name	Identifier	Content
1	Individual development	IR	Individual potential expresses the potential of a competent individual action.
2	Social	SP	Social potential means the ability to constructively manage social processes and their use for the benefit of the common system.
3	Political management	PP	Management potential (professional skills of the government, administration, business and management circles).
4	Infrastructure	IP	Infrastructure potential is the stock of constructed facilities, including cities, roads, water supply systems, schools and universities, etc.
5	Economic	EP	Production potential for the implementation of all economic activities.
6	Environment and resources	RP	Potential, representing reserves of renewable and non-renewable resources of materials, energy sources and biosystems, including the ability to absorb waste and regenerate.

Note: the economic subsystem of H.Bossel combines 2 blocks: technological and economic.

So, we identify 6 subsystems, the level of development of which predetermines the success of the urban development. These subsystems are interlinked with 5 key performance indicators (*KPIs*) as follows:

- KPI_1 - social performance (includes subsystems of individual development and social);
- KPI_2 – technological performance (technological block in the economic subsystem);
- KPI_3 – economic performance (economic block);
- KPI_4 – ecological performance (environment and resources subsystem);
- KPI_5 – political performance (a subsystem of political management).

Key indicators KPI_i are typically formed, as a result of which dimensionless quantities are formed ^

$$g_i = (KPI_i - KPI_{i\,min})/(KPI_{i\,max} - KPI_{i\,min}).$$

The standardized key indicators g_i predetermine the structure of the goal function (GF), which is a vector (generalized) estimate.

A generalized performance estimate is usually considered [13] either as a convolution or in the form of a weighted sum

$$\varphi = \sum_{i=1}^{n} \lambda_i \cdot g_i, \tag{1}$$

or as a weighted product

$$\varphi = \prod_{i=1}^{n} g^{\lambda_i}, \tag{2}$$

where λ_i - weight coefficients corresponding to the normalization condition $\sum_{i=1}^{n} \lambda_i = 1$;

IOP Conf. Series: Earth and Environmental Science **181** (2018) 012013 doi:10.1088/1755-1315/181/1/012013

g_i - normalized individual performance indicator.

The estimate (2), as we see, is more stringent than (1), since the zero equality of any g_i nullifies the estimate φ itself, therefore, the ratio (1) is often used.

There are two principal questions: what is the significance (weight λ_i) of each component, and which of them are the most priority? Such questions arise before any responsible party in the sphere of Smart-City. Of course, the customer of the executed project (program) must answer them. It is necessary to note with regret that the city administration, which by its status is competent to form "weights", has not yet indicated its position, so the decision will be entirely determined by the experience, professionalism and boldness of executors.

Formally, the "weights" and the priority of challenges can be estimated by the Analytic Hierarchy Process (AHP) proposed by T. Saati [10]. As a rule, the initial data for such analysis is formed by experts, and the final processing is performed by software, for example, MPRIORITY 1.0 [11], which implements the AHP. An example of such an analysis is given below (Figure 2). Of course, one can disagree with it, remembering, however, that, firstly, a plausible model is still better than no model at all, and, secondly, the result obtained depends on the specific nature of the problem being solved. Each problem has its own matrix of "weights".

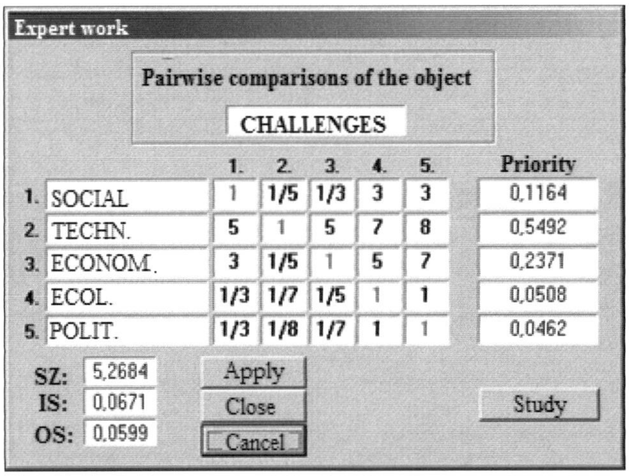

Figure 2. Estimation of weight coefficients.

Based on the data obtained, a Pareto chart is constructed and an ABC analysis is performed (Figure 3), whose goal is to identify priority areas that determine ~ 90% of success.

Priorities are those areas whose ranks fall into the region A + B. This means that not all individual indicators g_i, but only the most significant ones can be in the objective function. This explains the fact that researchers form the final list of individual performance indicators in different ways. So, in our case, these are studies aimed at increasing the technological and economic efficiency of the process.

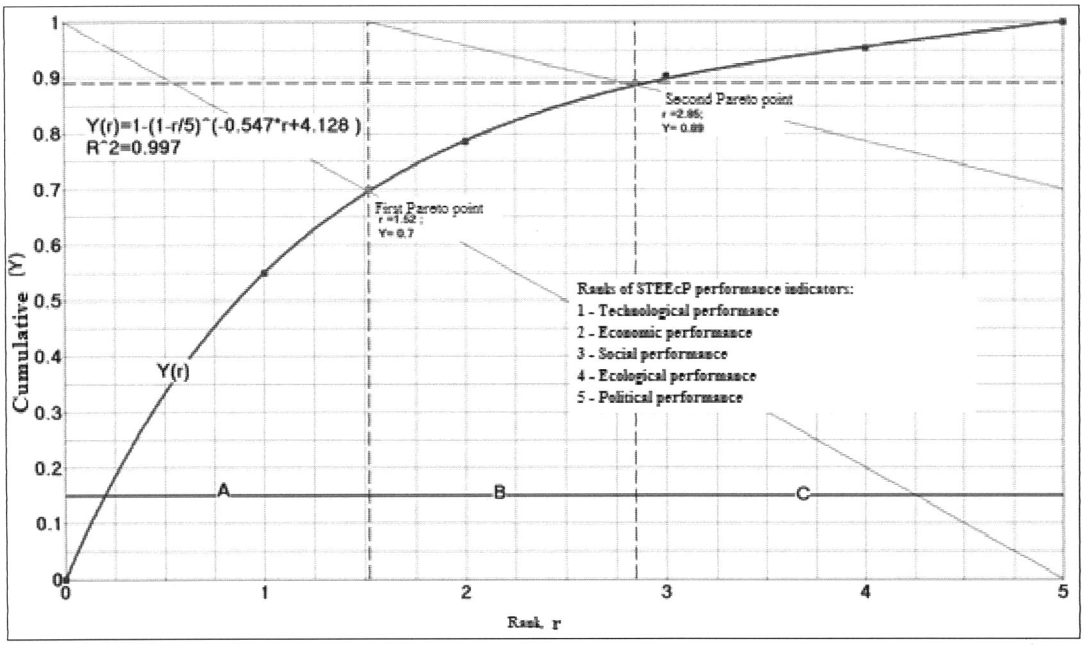

Figure 3. "Pareto" ABC analysis for STEEcP-indicators.

A logical result of the selection of priority g_i is the list of necessary indicators of the process since each indicator g_i is predetermined by the set of measured parameters (indicators x_{ij}) inherent only in it. Interconnection of g_i and x_{ij} is inherently a model of an individual, standardized indicator of effectiveness $g_i = F(x_{ij})$. Identification of these models is one of the main tasks of the project.

Creation of models $g_i = F(x_{ij})$ allows designing the goal function of the project (see relation (1)), aimed at achieving maximum performance ^

$$\varphi = \max \sum_{i=1}^{n} \lambda_i \cdot g_i = \max \sum_{i=1}^{n} \lambda_i \cdot F(x_{ij}) \cdot \quad (3)$$

If we consider Smart Grid (SG) as a subsystem of Smart-City, then it inherits all the specifics of the goal setting for the city in the methodological plan. The difference is only in the nomenclature of indicators x_{ij} and, as a consequence, in models $g_i = F(x_{ij})$.

Smart Grid as a management object. It is quite obvious that SG belongs to the class of large systems in which several subjects interact in their interests (Figure 4). Their main attributes determined by regional and functional features in the example of the Tyumen region are shown in Table 2.

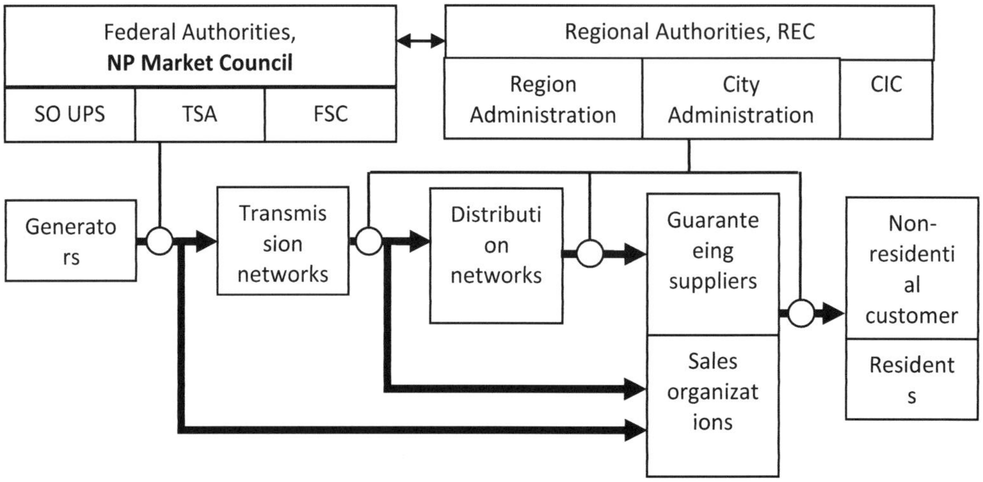

Figure 4. System structure.

The nature of the interaction of subjects. On the chart, thick lines show the flows of electricity and power, thin lines - information and financial flows. It is necessary to specify that the power here is a reserved quantity of electrical energy, and the electric power is the amount of electrical energy consumed.

Energy flows: Generators create electricity and power and transfer them to the wholesale market. Electricity and power are obtained by guaranteeing suppliers and energy sales organizations either directly or through network organizations. Suppliers provide electricity and power to consumers in the required quantities with the required technological parameters. Due to the peculiarity of electrical energy, it is consumed at the moment exactly as much as is produced.

Information flows: The wholesale market for electricity and power (WMEP) is managed by the Federal authorities through the Non-Profit Partnership Market Council (NP Market Council), and retail market (RMEP) - through regional authorities represented by the Regional Energy Commission (REC). The operational dispatching management is performed by the System Operator of the Unified Power System of JSC "SO UPS" or other entity of operational dispatch management of network organizations or guaranteeing suppliers.

Financial flows: The commercial operator of the WMEP is the Administrator of the wholesale electricity market trading system JSC "TSA"; calculations of requirements and obligations of wholesale market participants, JSC "SO UPS" and JSC "FSK EES" is carried out by the Financial calculation centre JSC "FSC". As for the RMEP, calculations are made with the network organization through the Calculation and information centre (CIC) or directly.

Table 2. Attributes.

No.	Subject	Names of organizations; Name of the department on the example of the Tyumen region	Functions	Key performance indicators (KPI)	Source
1.	Generator	OJSC Fortum Tyumen TPP-1	Generation of electricity. Sale of electricity.	*KPI* – profitability *Restrictions:* Reliability. Quality. Safety.	[14]
2.		OJSC Fortum Tyumen TPP-2			
3.		PJSC "Sibur" Tobolsk TPP			[15]
4.	Network	JSC "SO UPS" ODU of the Urals - Tyumen RDU	Operation-dispatching control	*KPI* - profitability *Restrictions:* Reliability. Quality.	[16]

5.		PJSC "FSK EES" - MES of Western Siberia - Southern PMES	Electric power transmission	Safety. Availability. Price-dependent consumption.	[17]
6.		PJSC "Rosseti" - Tyumenenergo			[18]
7.		LLC "Corporation STS" - PJSC "SUENCO"			[19]
8.	Sales and Distribution	JSC "EK Vostok" - SC "Tyumenenergosbyt"	A guaranteeing electricity supplier. Purchase and sale of electricity.	*KPI* - profitability *Restrictions:* Reliability. Quality. Availability.	[20]
9.		PJSC "Gazprom" - OJSC "Mezhregionenergosbyt"- JSC "TEK"			[21]
10.		OJSC "Siburenergomanagement"			[22]
11.	Regional administration	Regional Energy Commission of the Tyumen Region	Management of the electricity market. Development of investment programs. Coordination of the development of the power grid infrastructure, generating capacities. Tariff formation. Development of the power industry of the Tyumen region	*KPI* - economy *Restrictions:* Productivity. Reliability. Quality. Safety. Availability.	[23]
12.		Department of Housing and Communal Services of the Tyumen Region			[24]
13.		Department of Municipal Economy	Checking the readiness of power supply organizations for the heating season		[25]
14.	Consumer	Residents	Power consumption	*KPI* - economy *Restrictions:* Reliability. Quality. Safety.	
15.		Non-residential customers		*KPI* - profitability *Restrictions:* Reliability. Quality. Economy. Safety. Price-dependent consumption.	

3. Results and discussion

The analysis of Table 2 showed that in the field of performance management of the power system there is virtually no systemic paradigm and, as a result,

- missions of subjects are not formulated;
- there is no conceptual model of the process of providing a service to a consumer;
- goal functions of subjects are not formulated;
- there are no mathematical models of key performance indicators;
- there is no engineering algorithmic support for the process of network performance management;
- work of the organisation to maintain market equilibrium is non-transparent.

From the positions of the system approach, two aspects are principal today:

- conditions for achieving the agreed optimum between the interacting subjects [26];
- mechanisms for achieving the optimum and regulation of market equilibrium.

These tasks, in accordance with their status, should be carried out by the Regional Energy Commission. By variation of tariffs (prices), it must find a reasonable compromise between the five subjects (see Figure 4): generators, network, sales, consumers and city administration. Unfortunately, it was not possible to find a transparent solution to this problem in accessible sources.

4. Conclusion

The formalization of the problem, in our opinion, assumes an analytic assignment of the goal function of each subject $J_i(\boldsymbol{x})$ (here $\boldsymbol{x}$ is a vector of control parameters) followed by maximization of the weighted criterion $J = \sum_{i=1}^{5} \lambda_i \cdot J_i$ with respect to $\boldsymbol{x}$. Unknown weights λ_i are usually determined on the basis of the Lagrange principle [26].

Using the system approach will make it possible to introduce the necessary logic into the development of the conceptual model of the Smart Grid project, and, as a result, increase the chances of success.

References

[1] Tsibulsky V R et al 2004 *Tyumen: Indicators of Sustainable Development* ed V R Tsibulsky (Tyumen: IPOS SB RAS) p 306
[2] Bossel H 2001 *Indicators of Sustainable Development: Theory, Method, Practical Use. Report Submitted to the Balaton Group* (Tyumen: IPOS SB RAS) 123 p
[3] Tsibulsky V R 2001 Methodology of the United Nations. A systematic approach to selection of indicators for assessing the sustainability of development of territories *Problems of Interaction between People and the Nat. Envir.* (Tyumen: IPOS SB RAS) 2 pp 114-8
[4] Kuznetsov O L, Kuznetsov P G and Bolshakov B E 2001 *Sustainable Development: Scientific Foundations of Design in the System of Nature-Society-Man* (Dubna: The State Scientific Center of the Russian Federation VNIIgeosistem) 604 p
[5] New Models of Urban Development. The Point of View of McKinsey & Company [Electronic resource] Access mode: http://gtmarket.ru/laboratory/expertize/5974
[6] Kolesov V I 2014 Strategic management in fuzzy conditions *Innovations in the Management of Regional and Sectoral Development: Materials of the All-Rus. Sc. and Pr. Conf. with Intern. Participation* (Tyumen: Tyumen State Oil and Gas University) pp144-8
[7] Kaplinsky A I, Russman IB and Umyvakin V M 1991 *Modeling and Algorithmization of the Weakly Formalized Problems of Choosing the Best Variants of the System* (Voronezh: VSU Press) 168 p
[8] Bermant M A and Russman I B 1978 On the problem of quality assessment *Econ. and Math. Meth.* **14** 4 pp 697-9
[9] Russman I B 1978 Complex evaluation of the system and evaluation of subsystems *Proc. of the USSR Acad. of Sc. Ser. Techn. Cybernetics* 2 pp 201-4

IOP Conf. Series: Earth and Environmental Science **181** (2018) 012013 doi:10.1088/1755-1315/181/1/012013

[10] Kolesov V I 2012 Statement of the problem of quality management of passenger transportation in the city. *Problems of the Functioning of Transport Systems: Materials of the All-Rus. Sc. and Pr. Conf.* (Tyumen: Tyumen State Oil and Gas University) pp 206-10
[11] Saati, T L 1989 *Decision-Making. The Method of Analysing Hierarchies* (Moscow: Radio i sviaz) 316 p
[12] Abakarov A Sh and Sushkov Yu A 2005 Program support system "MPRIORITY 1.0" for making rational decisions *Electronic Sc. J. "Investigated in Russia"* pp 2130-46
[13] Venttsel E S 1964 *Introduction to the Study of Operations* (Moscow: Sovetskoe radio) 388 p
[14] OJSC Fortum [Electronic resource] Access mode: https://www.fortum.com/countries/ru/about/production/west-sib-region/pages/default.aspx
[15] PJSC Sibur [Electronic resource] Access mode: www.sibur.ru
[16] JSC SO UPS [Electronic resource] Access mode: http://so-ups.ru/
[17] JSC FSK EES [Electronic resource] Access mode: http://www.fsk-ees.ru/
[18] PJSC Rosseti [Electronic resource] Access mode: http://www.rosseti.ru/
[19] PJSC SUENCO [Electronic resource] Access mode: http://www.suenco.ru/
[20] JSC Tyumenenergosbyt [Electronic resource] Access mode: tyumen.vostok-electra.ru/
[21] JSC TEK [Electronic resource] Access mode: http://www.mrg-sbyt.ru/
[22] OJSC Siburenergomanagement [Electronic resource] Access mode: www.sibur.ru/siburenergomanagment/
[23] Regional Energy Commission of the Tyumen Region [Electronic resource] Access mode: http://www.rectmn.ru
[24] Administration of the Tyumen Region [Electronic resource] Access mode: http://admtyumen.ru
[25] Administration of the City of Tyumen [Electronic resource] Access mode: http://www.tyumen-city.ru/
[26] Volgin L N 1977 *Principle of the Coordinated Optimum* (Moscow: Sovetskoe radio) 144 p

IOP Conf. Series: Earth and Environmental Science **181** (2018) 012014 doi:10.1088/1755-1315/181/1/012014

Complex research of the application of the non-stationary waterflooding in interrelationship with other EOR methods

A V Klimov-Kayanidi, V N Gulyaev, S K Sokhoshko and N P Zakharova

The branch of OOO LUKOIL-Engineering - KogalymNIPIneft in Tyumen, Russia
Industrial University of Tyumen, Tyumen, Russia

E-mail: Klimov-KayanidiAV@tmn.lukoil.com

Abstract: The article is concerned with the application of non-stationary waterflooding (NW) on the example of «LUKOIL-Western Siberia» fields. The experience of selection of candidate wells for NW has been analyzed, and the criteria for selecting an area for the application of the method have been developed, taking into account the conduct of other oil recovery enhancement operations (EOR) in the near regions. Recommendations are given on the use of NW in conjunction with other methods of EOR.

1. Introduction

Currently, many oil fields are characterized by high water cut, due to both the recovery of reserves, and premature water cutting of reservoirs with injected water.

In nonuniform stratums, injected water breaks into producing wells through highly permeable interlayers, leaving oil unallocated in low-permeability zones [1].

An objective necessity to cover the less permeable part of the productive formation by flooding with progressive watering of the formation is the restriction of water filtration in washed interlayers. Additional coverage of flooded areas and areas of the formation that are not involved can increase the oil recovery of reservoirs during normal waterflooding, prolongation of the water-free period, and decrease in the volume of water produced. One of the effective ways to achieve this goal can serve a NW (like cyclic waterflooding), which allows to affect the formation, both in the lateral and in the vertical section.

In contrast to the traditional waterflooding, the NW allows to cover the oil layers and stagnant areas of the formation that are not involved in the development, which contributes to an increase in oil recovery and a decrease in the relative volume of produced water [2].

2. Research

The purpose of this work was to study the application of NW in conjunction with other methods of EOR in the LUKOIL-Western Siberia fields.

In fact all of the company's fields are being developed with the use of maintaining reservoir pressure by waterflooding. The waterflooding system used does not always take into account the features of the geological structure and does not provide the necessary coverage by lateral and vertical section, as a result of which the development of many fields is characterized by insufficiently high oil recovery factors, low recovery rates of oil production and large volumes of produced water.

An important condition for the effective application of NW is the correct choice of the method for the production target, and determining the conditions for its application. Criteria for the applicability

Content from this work may be used under the terms of the Creative Commons Attribution 3.0 licence. Any further distribution of this work must maintain attribution to the author(s) and the title of the work, journal citation and DOI.
Published under licence by IOP Publishing Ltd

of methods are determined on the basis of an analysis of the indicators of their implementation, the generalization of the experience of introducing technologies in various geological and physical conditions, the use of theoretical and laboratory research.

3. Results and discussion

3.1 Selection area for the waterflooding.
The choice of zones for NW is based on a multifactor analysis and is carried out using the methods developed in KogalymNIPIneft, which allows identifying areas that are likely to be suitable for the successful application of NW [3]. The analysis of information in the selection of zones includes the rationale for the duration of the half-cycle of waterflooding for the geologic and physical reservoir characteristics, the characteristics of the well stock, maps of the current state of development, accumulated samples map, isobars, permeabilities and immovable oil, and log data. The algorithm for selecting areas for NW is shown in Figure 1.

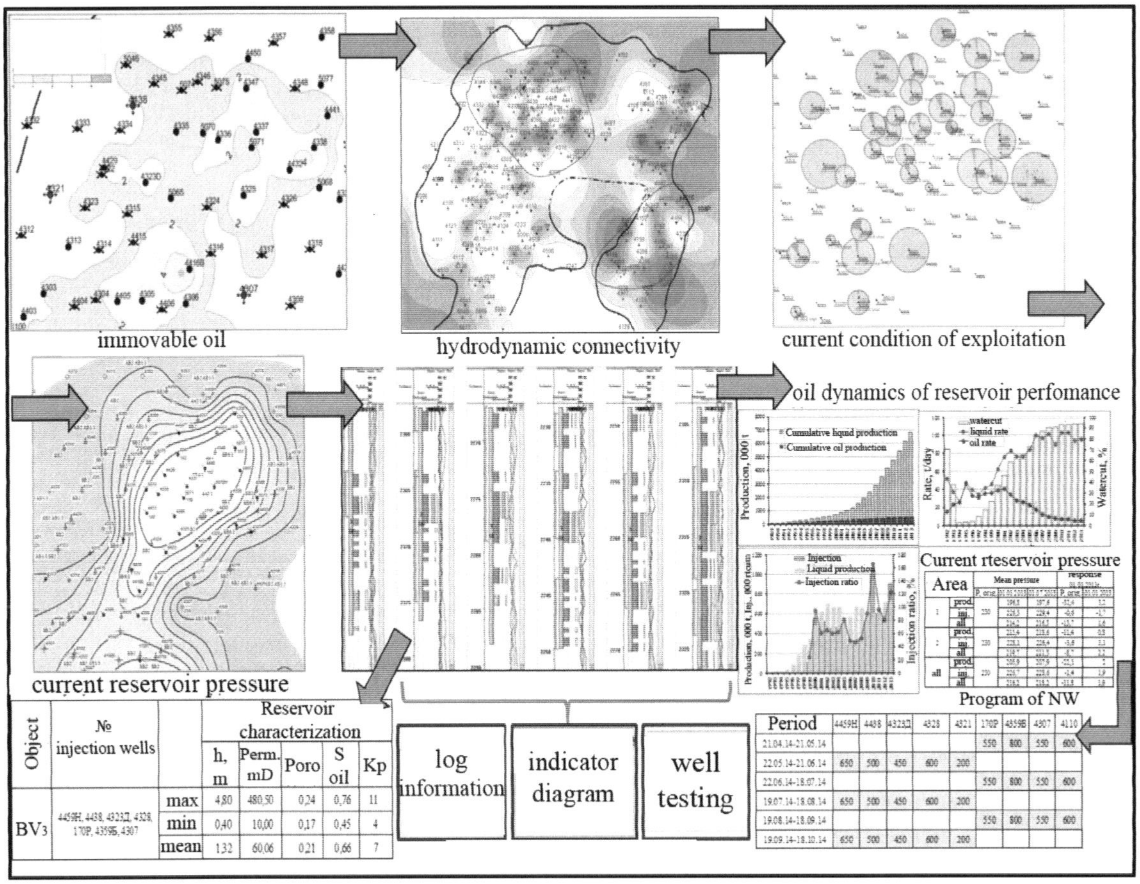

Figure 1. Algorithm for selecting areas of NW and program development.

At the first stage, the values of the complex parameter of the geological favorability (Fco) were calculated from all the injection and production wells of the selected zone, based on the available data. A map of the distribution of this parameter over the area of the object was constructed.

A geological model of the reservoir was developed for the selected areas, which takes into account the connected heterogeneous part of the formation, i.e. the part in which vertical overflows are possible due to uneven pressure redistribution.

For each layer, the following parameters are determined from the model: permeability k, effective thickness h, porosity m and oil saturation S (Table 1).

Table 1. Results of calculating half-cycle of waterflooding at the areas of the Klyuchevoe Field.

Category interlayer	Layer characteristic				half-cycle, days
	K, md	h, m	m	S	
Low-permeable coherent	28.3	0.45	0.21	0.65	30.8
Low-permeable insulated	33.6	6.69	0.21	0.59	
High-permeability insulated	219.6	3.59	0.23	0.64	
High-permeability connected	176.2	0.93	0.23	0.65	

Next, the optimal operating frequency of the cycle change is determined by the formula:

$$Wp = 2\chi / l2 \quad \text{or} \quad t = l2 / 2\chi, \tag{1}$$

Here, Wp is the operating frequency of the flow oscillations; t – duration of the half-cycle of waterfloding; $\chi = k/(\mu^*\beta_{pr})$ – the average piezo-conductivity of a low permeable coherent interlayer; β_{pr} – reduced coefficient of compressibility of rock and fluid; μ, m, l, k are the characteristic average viscosity, porosity, length and permeability of the formation, respectively [4].

Further, the current state of wells performance was analyzed, and the technical capabilities of the injection wells of the NW areas were considered.

The diagrams of injection wells stopping are selected according to the technology of NW implementation and the current state of the reservoir pressure maintenance system. At the company's fields in 2014 – 2015 during NW injection wells were stopped on a "one-by-one" basis or groups of wells operating in antiphase.

3.2 Approbation of the methodology

Approbation of this method of selection of areas for NW was carried out in 2013 - 2015.

The results of calculating the duration of the half-cycle of waterflooding and the effect of using NW on the company's fields using the methodology developed in the Branch are shown in Table 2.

Table 2. Results of calculating the half-cycle time and efficiency of NW in 2014.

Field	**Facilities**	**Number of zones NW**	**The accepted distance to the front of displacement , L, m**	**Average coefficient of piezo-conductivity of the reservoir, m^2/sec.**	**Duration of half-cycle, days**	**Viscosity of oil, mPa*sec.**	**Incremental ultimate recovery, thousand tons**
Klyuchevoe	BV	2	430/400	0.03	30	1.82	3.13
Nong-Eganskoe	BV	1	530	0.05	30.73	1.82	4.09
Pokachovskoe	AV	1	330	0.03	23.49	1.75	-0.33
Las-Yoganskoe	AV	1	800	0.11	32.3	2.27	3.60
Lokosovskoe	BV	1	100	0.07	80	2.78	-2.14
Urievskoye	AV	4	800	0.8	45	2.45	22.13
Vatjeganskoe	AV	3	550	0.03	45	2.55	1.76
Povkhovskoye	UV	6	500	0.05	28	0.71	-2.04
Yujno-Vyintoyskoe	BV	1	500	0.03	9	0.43	1.95
Tevlinsko-Russkinskoe	BS	2	600/775	0.05/0.07	38/53	1.43	4.00
Total for 2014		22					36.14

IOP Conf. Series: Earth and Environmental Science **181** (2018) 012014 doi:10.1088/1755-1315/181/1/012014

Along with NW, the measures for the formation of barriers in the pore space with washed reservoirs of a productive reservoir for water displacing oil technology (physical and chemical treatment) are relevant.

Tevlinsko-Russkinskoye field, where during the NW period, stops of high-water-producing production wells were conducted can be attributed to the experience of complex application of the methods.

Disabling high-water production wells contributes to a more uniform flooding of the formation. As a result of the application of such a scheme, more efficient displacement of oil from zones of the formation not covered by the exploitation is achieved. During the stopping of the injection wells of the second half-cycle, during the implementation of the NW production wells No. 7062 and No. 7064 were stopped with the aim of redistributing the injected water to the producing wells of the central row. The location of wells is shown in Figure 2 from the top. The regime of daily performance of these wells showed high efficiency: reduced water cut and increase in well performance (oil production rate) in surrounding wells. Figure 2 on the graphs shows the dynamics of the daily performance of these wells.

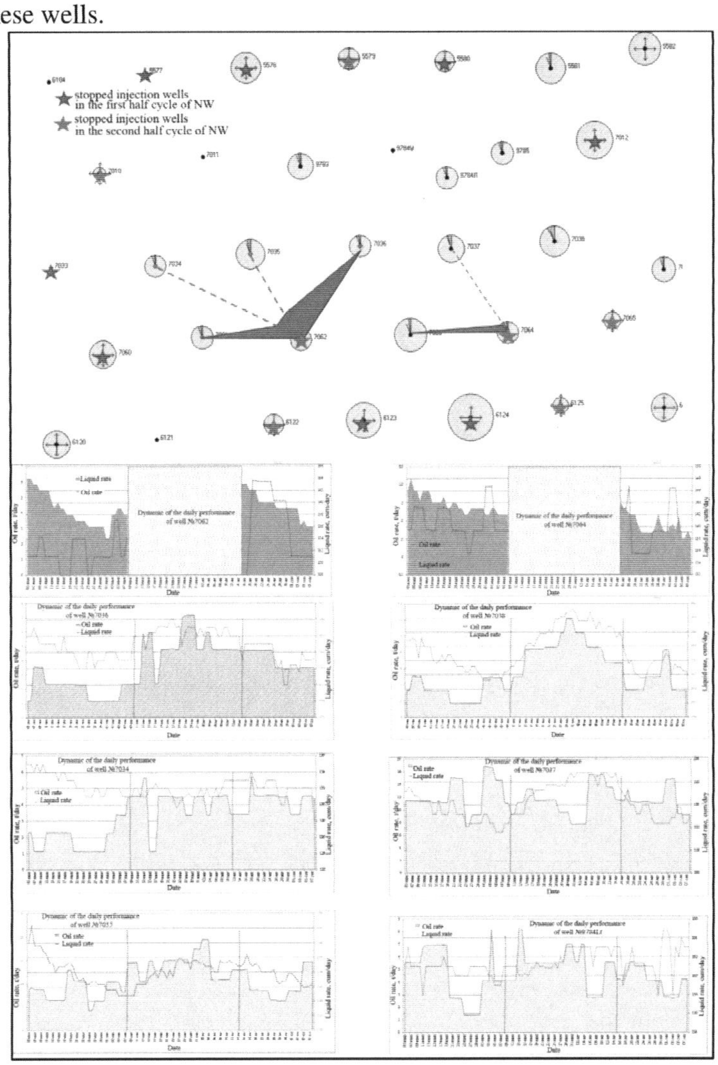

Figure 2. Fragment of the map of the zone with NW. Tevlinsko-Russkinskoye field. Change in the daily production of wells due the stop of production wells No. 7062 and No. 7064.

3.3 Integrated impact on formation

Consider the experience of the complex impact of NW and measures aimed at stimulation of production (forced fluid withdrawal, hydraulic fracturing, perforation of additional intervals). The implementation of such measures is carried out by increasing the pressure gradient in the zone of the formation affected by the well. The technologies of such integrated impact in the filtration process include the pressure readjustment between producing and injection wells with a change of direction of filtration flows in the formation.

The zone of Nivagalskoye field was used for pilot-industrial application of the method. The zone met all the criteria for efficient NW implementation and was characterized by good hydrodynamic cohesion of the reservoirs, their layer-by-layer and zonal heterogeneity, while the water cut in the area did not exceed 60%. The implementation of the NW, by stopping injection wells, began in June 2014. The use of (forced fluid withdrawal at the zone was carried out 1-3 months before NW, as shown in Figure 3.

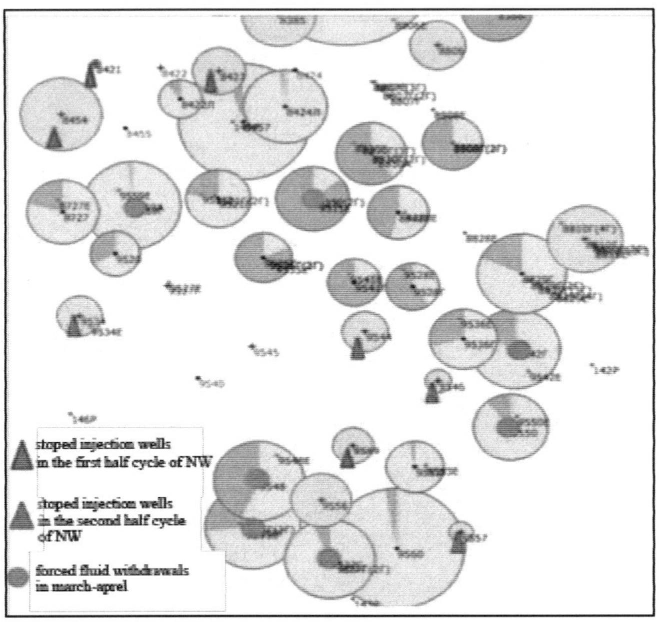

Figure 3. Scheme for the implementation of the forced fluid withdrawal and NW at the zone of the Nivagalskoe field.

Horizontal wells on the zone were put into operation in late 2012 and early 2013. Compensation for fluid withdrawals by injection at the time of wells forced was 54 %. During the implementation of the NW, water injection volumes were reduced, despite the fact that it was recommended to maintain the pumping volumes at a stationary level, but the fall is permissible and was 20%.

Due to the forced fluid withdrawal and the shutdown of injection wells, the direction of displacement of the fluids changed, which led to an increase in fluid withdrawals and an increase in water cut in horizontal wells.

In production wells, the combined use of forced fluid withdrawal and NW led to an increase in water cut in production and fluid production by increasing the inflow of water from the underlying aquifers. The dynamics of the change in the water cut and well flow rates are shown in Figure 4. In horizontal wells (HW), fluid flow rates decrease due to the limitation of the volumes of injected water into the formation at the site. Reducing the volume of injection leads to a drop in depression and, as a consequence, to a drop in fluid withdrawals. When carrying out NW measures on sections with horizontal wellbore, it is necessary to strictly adhere to the basic requirements for effective implementation of the NW for compliance with the levels of water pumped into the reservoir. The impact of NW on the operation of horizontal wells has not been adequately studied and requires further studies to determine the conditions for its successful application in areas that have such wells

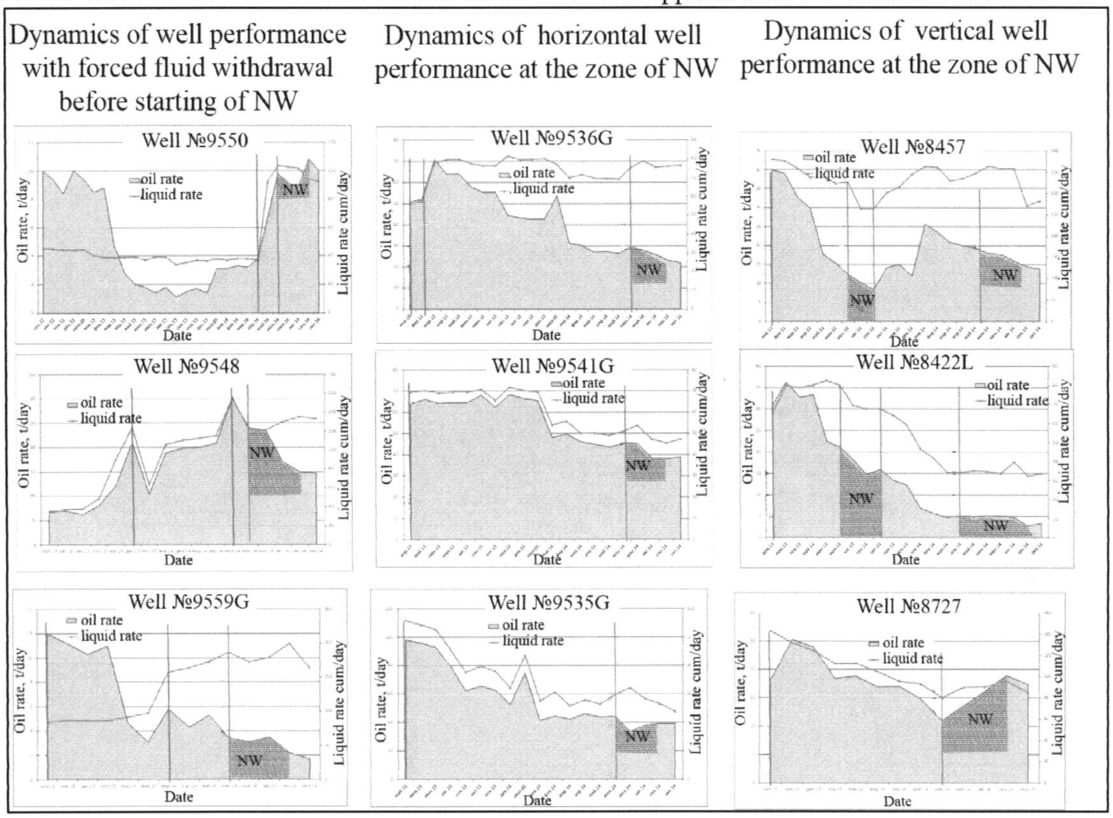

Figure 4. Characteristics of changes in water cut and oil production rates in the area of the forced fluid withdrawal and NW of the Nivagalskoe field.

In the majority of vertical wells, during the implementation of NW measures, the water cut of the products is periodically reduced. This indicates that during the periodic change in the water flood direction, oil is displaced to the faces of the operating oil wells from the stagnant and weakly drained zones.

As a result of the application of the NW method in combination with the forced fluid withdrawal in the section with horizontal wells in 2014, low efficiency was noted, additional oil production was 104 tons, the increase in water withdrawals at the area reached 18.8 thousand m^3.

NW at the company's fields has shown quite good efficiency. In 2014, the additional oil production amounted to 39.7 thousand tons. During the period of its carrying out, as a rule, there is a periodic reduction in water cut of production and reduction of water withdrawals.

4. Conclusion

Complex application of the methods turned out to be quite effective, since the processes of redistribution of pressure gradients and interstratial overflows during the NW implementation increase the efficiency of address treatments by flow deviation compositions. Therefore it is recommended to treat by physicochemical methods in injection wells 3-6 months before the start of the NW implementation.

Stops of highly watered production wells during the NW period also have high efficiency. When the production wells stop, the pressure gradient increases, and in the case of cyclic waterflooding, a condition is created in them to force the withdrawal of fluid from the formation with a change in the direction of the filtration flows due to the lithological heterogeneity of the formation as a zonal (lateral between the wells) and layered (between interlayers).

Measures to intensify oil production on the production well stock require enhanced reservoir energy in the drainage zone of the well.

Therefore, it is not recommended to carry out measures aimed at forced fluids withdrawal 2 months before the start of the NW implementation and during the period of its holding.

The proposed series of NW in conjunction with other workovers program makes it possible to increase oil extraction in complicated conditions for the exploitation reservoirs with hard-to-recover reserves, in highly-watered and low-productive reservoirs, related both to watering reservoirs with high reservoir properties and to the recovery of oil in low-permeability zones of facilities.

References

[1] Surgutchev M L 1985 *Secondary and Tertiary Methods of Increasing Oil Recovery* (Moscow: Nedra)

[2] Sokhoshko S K, Klimov A V 2015 Calculation of working parameters of injection wells for organization of cyclic waterflooding *Proc. of the All-Russian Conf. New Technologies for the Oil and Gas Region* (Tyumen: IUT) pp 368-71

[3] Gulyaev V N, Lanin N A and Yagafarov A.K 2009 Features of the choice of zones for the application of technology of non-stationary waterflooding *Proc. of the All-Russian Sc. and Techn. Conf. Oil and Gas of Western Siberia* (Tyumen: Tyumen State Oil and Gas University) pp 53-6

[4] Gulyaev V N, Kiprin I I and Zakharova N P 2015 Application of technology of non-stationary influence for increase of ORF of zones with hard-to-recover reserves *Geology, Geophysics of Oil and Gas Deposits* 10 pp 39-44

IOP Conf. Series: Earth and Environmental Science **181** (2018) 012015 doi:10.1088/1755-1315/181/1/012015

System aspects of price responsive energy consumption

V I Kolesov, G A Hmara and A L Portnyagin

Industrial University of Tyumen, 38 Volodarskogo st., Tyumen, 625000, Russia

E-mail: hmaraga@tyuiu.ru

Abstract: The article is focused on the synthesis of decision support systems in the field of Smart Grid. The structural and parametric identification of the model is made for assessing the effectiveness of the decision taken by the consumer to participate in demand management. The results of calculations are presented.

1. Introduction

The introduction of the mechanism of price responsive consumption (PRC), which increases the efficiency of electricity production in the regions and in Russia as a whole, is rapidly progressing. Its essence, as is known, is to maintain a balance of supply and demand in the electricity market both due to changes in the load of generation facilities, and by changing consumption. The abolition of the upper price limit for electricity in the wholesale market for electricity and power (WMEP) in the Russian Federation, scheduled for 2021, changes the principles of interaction between consumers and producers. Now, to manage the price in a deficit, a balanced decision should be taken on the part of consumers. Involvement of consumers in the process of demand management raises new problems for them, related to the adoption of reasonable decisions in the processing of applications. Due to the novelty of the topic, effective tools for consumers in the sphere of decision support (DS) have not been created yet. The task is set, at least partially, to eliminate the existing gap.

2. Research

The formalization of the task is reduced to determining the conditions for making a decision on price responsive consumption. The essence of the solution is to choose the amount of energy consumption reduction during the regional maximum hours by ΔV (MWh) by switching off a part of consumers N times per month m ($1 \le N \le 10$) [1]^

$$\Delta V_m = \sum_{N=1}^{10} P_N \cdot T_N , \quad (1)$$

where $P \ge 2$ MW – power of disconnected consumers; T – shutdown duration: 2, 4, 8 hours per day.

In this case, shutdowns are carried out at the direction of the System Operator (SO UES). In addition to reducing the cost of electricity, a consumer with a PRC receives an additional premium, in the form of a reduction in the cost of capacity, for the assistance of the SO UES in solving system tasks for ensuring a reliable and secure electricity supply to the region.

Today, the interaction between the consumer and the SO UES in the sphere of price responsive consumption is the planning and implementation of joint actions on the impact on power flows in the

Content from this work may be used under the terms of the Creative Commons Attribution 3.0 licence. Any further distribution of this work must maintain attribution to the author(s) and the title of the work, journal citation and DOI.
Published under licence by IOP Publishing Ltd

IOP Conf. Series: Earth and Environmental Science **181** (2018) 012015 doi:10.1088/1755-1315/181/1/012015

regional energy system. The interaction algorithm is the following: the consumer declares that his readiness for price responsive consumption during the regional maximum hours, indicating the volume, the System Operator accepts the application and schedules the power flows, and the Administrator of the trading system (ATS) of the wholesale electricity market calculates the cost of electricity.

Thus, the cost of electricity in the wholesale market is formed after the decision on price responsive consumption, and consumers have no opportunity to plan their economic activities taking into account the resulting effect from the PRC. Moreover, for enterprises with price responsive consumption, switching off part of the load with a capacity of more than 2 MW can lead to a decrease in productivity and product quality and, as a result, to financial losses not covered by the premium received. That is, there is the problem of deciding on price responsive consumption in the context of uncertainty of the benefits received.

3. Results and discussion

To solve this problem, the authors propose a method of decision support based on the interaction of the supply and demand functions.

The demand model is adopted in the form [1]:

$$\left(C_p \cdot V_p\right)^{m_p} = k_p \, , \qquad (1)$$

where C_p - price of service; V_p - volume of consumption; m_p - parameter; k_p - level of quality.

Under the level of quality, we will understand the efficiency of the production process of the consumer in question.

In accordance with (1), the demand function (Fig. 1) is equal to:

$$C_p = \frac{k_p^{1/m_p}}{V_p}, \qquad (2)$$

and the volume of consumption

$$V_p = \frac{k_p^{1/m_p}}{C_p}, \qquad (3)$$

The supply model is presented as follows:
the supply function is

$$Cs = k_s \cdot V_s^{m_s} \, , \qquad (4)$$

and the volume of supply

$$V_s = \left(\frac{C_s}{k_s}\right)^{1/m_s} . \qquad (5)$$

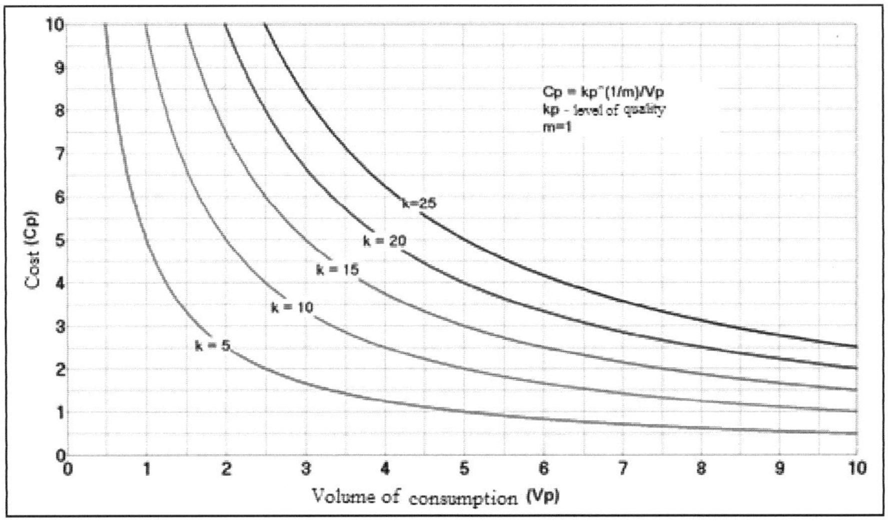

Figure 1. Demand function.

When the market is in equilibrium (Figure 2), the conditions are met $V_p = V_s = V_0$, $C_p = C_s = C_0$, i.e. $\frac{k_p^{1/m_p}}{C_0} = \left(\frac{C_0}{k_s}\right)^{1/m_s}$, from which follows

$$C_0 = \left(k_p^{1/m_p} \cdot k_s^{1/m_s}\right)^{\frac{m_s}{1+m_s}}. \quad (6)$$

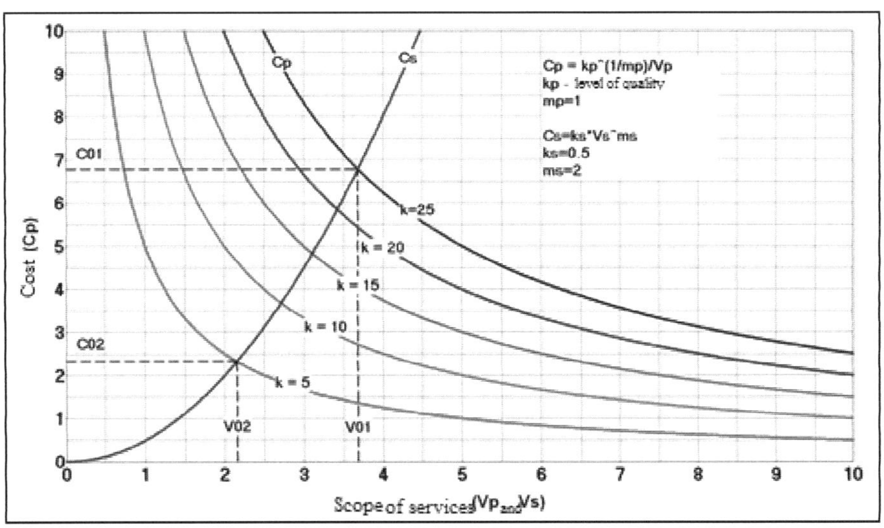

Figure 2. Market Equilibrium

The price responsive energy consumption changes the market equilibrium point from the coordinates (V_{01}, C_{01}) to the coordinates (V_{02}, C_{02}).

When $m_p = 1$; $m_s = 2$ and $k_s = 0.5$ we obtain $C_0 = 0.5^{1/3} \cdot k_p^{2/3}$. The graph $C_0(k_p)$ is shown in Figure 3.

IOP Conf. Series: Earth and Environmental Science **181** (2018) 012015 doi:10.1088/1755-1315/181/1/012015

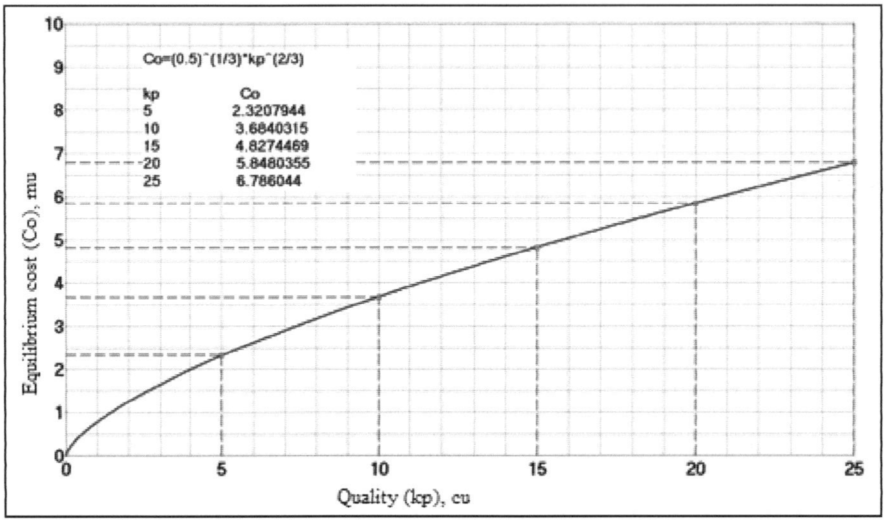

Figure 3. The price at the market equilibrium point.

The analysis of Figure 3 shows that the reduction in the equilibrium price is achieved by reducing the level of quality k_p. How to understand this decline and what really happens?

The decline in quality can be interpreted as reducing the attractiveness of the service to the consumer. In other words, a certain negative increases the difficulty of reaching the consumer's global goal. Indeed, on the one hand, the positive of the situation is to reduce the price for electricity, but on the other - there is an obvious negative, associated with the inability to fully use the energy resource during rush hours.

A way to assess the difficulty of achieving the goal and to parry its progress is offered by a methodology developed by the scientific school of I.B. Russman and his students [2]. It is widely used in the design of complex systems to solve problems of optimizing their effectiveness, taking into account the risk of not reaching their sub-goals with the components of the system. In general, the components of a system can be understood as key performance indicators (KPI_i), from which the objective function (*OF*) is constructed, which is a mathematical model of the accepted criterion. KPI_i for each specific consumer is determined in accordance with the characteristics of its activities and in this article are not considered in detail.

The key point of the methodology is the introduction of the concept of difficulty of achieving the goal d_i for each *i*-subsystem. It is postulated that each of them has a normative requirement for quality of ε_i, while in reality it is equal to μ_i, and then d_i is interpreted as the risk of not reaching the i-subsystem of its goal

$$d_i = \frac{\varepsilon_i \cdot (1-\mu_i)}{\mu_i \cdot (1-\varepsilon_i)}, \tag{7}$$

where $\mu_i = \dfrac{X_i - X_{imin}}{X_{imax} - X_{imin}}$; $\varepsilon_i = \dfrac{X_{in} - X_{imin}}{X_{imax} - X_{imin}}$;

X_{in}, X_{imin}; X_{imax} - the normative, minimum and maximum values of the level of quality.

With this approach, the objective function D (or its information equivalents) is expressed through the so-called qualitative function $F(d_1, d_2, .., d_n)$.

One of the typical functions is a quality function similar to the Cobb-Douglas production function, whose analytical representation is of the form [2]

IOP Conf. Series: Earth and Environmental Science **181** (2018) 012015 doi:10.1088/1755-1315/181/1/012015

$$J(D) = J(d_0)\cdot\prod_{i=1}^{n}[J(d_i)]^{\lambda_i}. \tag{8}$$

where $J(D) = ln\left[\dfrac{1}{1-D}\right]$; $J(d_i) = ln\left[\dfrac{1}{1-d_i}\right]$;

λ_i - weight coefficients corresponding to the normalization condition $\sum_{i=1}^{n}\lambda_i = 1$;

d_0 - risk determined by the quality of the resource consumed by the system itself.

4. Conclusion

The engineering formulation of the problem under consideration is as follows. A system is being created to which quality requirements E are imposed. Real quality levels of i-subsystems μ_i included in it are known. It is required to formulate normative quality levels ε_i for each i-subsystem. Before constructing the objective function, we consider the relation (2). Obviously, the lower the d_i is, the lower the risk that i-subsystems will not reach their partial targets is and, consequently, the higher system's efficiency is. Thus, the objective function should be focused on minimizing $\sum_{i=1}^{n} d_i$, which is equivalent to

$$\sum_{i=1}^{n} J(d_i) \rightarrow min. \tag{9}$$

with the restriction of the form $ln\,J(D) = ln\,J(d_0) + \sum_{i=1}^{n}\lambda_i\cdot ln\,J(d_i)$.

The problem is solved using the Lagrange principle, and the Lagrange function takes the form

$$\Phi = \sum_{i=1}^{n} J_i + \lambda\cdot\left(ln\,J(D) - ln\,J(d_0) - \sum_{i=1}^{n}\lambda_i\cdot ln\,J_i\right)$$

$$\begin{cases} \dfrac{\partial\Phi}{\partial J_i} = 1 - \dfrac{\lambda\cdot\lambda_i}{J_i} = 0; \\ \dfrac{\partial\Phi;}{\partial\lambda} = ln\,J(D) - ln\,J(d_0) - \sum_{i}^{n}\lambda_i\cdot ln\,J_i = 0 \end{cases}$$

From the first equation follows

$$J_i = \lambda\cdot\lambda_i, \tag{10}$$

then the second equation takes the form

$ln\,J(D) - ln\,J(d_0) - \sum_{i}^{n}\lambda_i\cdot[ln\lambda_i + ln(\lambda)] = 0$ or

$ln\,J(D) - ln\,J(d_0) - \sum_{i}^{n}\lambda_i\cdot ln\lambda_i - \sum_{i}^{n}\lambda_i\cdot ln(\lambda) = 0$, i.e.

$ln(\lambda)\cdot\sum_{i}^{n}\lambda_i = ln\,J(D) - ln\,J(d_0) - \sum_{i}^{n}\lambda_i\cdot ln\lambda_i$, but $\sum_{i}^{n}\lambda_i = 1$, consequently,

$$exp[ln(\lambda)] = exp\left[ln\,J(D) - ln\,J(d_0) - \sum_i^n \lambda_i \cdot ln\,\lambda_i\right].$$

It means that $$\lambda = exp\left[ln\,J(D) - ln\,J(d_0) - \sum_i^n \lambda_i \cdot ln\,\lambda_i\right]. \tag{11}$$

According to (10),

$$J_i = \lambda \cdot \lambda_i = \lambda_i \cdot exp\left[ln\,J(D) - ln\,J(d_0) - \sum_i^n \lambda_i \cdot ln\,\lambda_i\right], \tag{12}$$

but $J_i = J(d_i) = ln\left[\frac{1}{1-d_i}\right]$, from which it follows that

$$d_i = 1 - exp[-J(d_i)]. \tag{13}$$

On the other hand, $d_i = \frac{\varepsilon_i \cdot (1-\mu_i)}{\mu_i \cdot (1-\varepsilon_i)}$, which allows us to formulate a requirement for the normative quality of the *i*-subsystem ε_i:

$$\varepsilon_i = \frac{d_i \cdot \mu_i}{d_i \cdot \mu_i + 1 - \mu_i} = \frac{1}{1 + 1/(d_i \cdot \mu_i) - 1/d_i}. \tag{14}$$

Substitution of (13) into (14) makes it possible to estimate the initial quality requirements for each *i*-component of a complex system.

Moreover, the relation (14) can be used to estimate the required quality margin $k_i = \mu_i / \varepsilon_i$ for each *i*-subsystem

$$k_i = \mu_i / \varepsilon_i = \mu_i \cdot (1 - 1/d_i) + 1/d_i. \tag{15}$$

The received working algorithms have passed the software testing.

References

[1] Kolesov V I 2006 Structural identification of economic models of the city bus route. *Operation of Vehicles and Special Oil and Gas Equipment: Interuniversity Collection of Scientific Papers* (Tyumen: Publishing House "Vector Buk") p 71-80

[2] Kaplinsky A I, Russman I B and Umyvakin V M 1991 *Modeling and Algorithmization of Weakly Formalized Problems of Choosing the Best Variants of the System* (Voronezh: VSU) 168 p

[3] Appendix 19.9. Regulation of participation in the wholesale market of buyers with price responsive consumption [Electronic resource] Access mode: http://so-ups.ru/index.php?id=dr_doc

IOP Conf. Series: Earth and Environmental Science **181** (2018) 012016 doi:10.1088/1755-1315/181/1/012016

On modeling of non-stationary two-phase filtration

V A Korotenko, S I Grachev, N P Kushakova, S A Leontiev, M I Zaboeva and M A Aleksandrov

Industrial University of Tyumen, 38 Volodarskogo st., Tyumen, 625000, Russia

E-mail: nel56-33@mail.ru

Abstract: Based on the basic mechanics equations of continuous medium, the author's model of non-stationary two-phase oil filtration and water pumped into the reservoir was developed, based on the dependence of the water saturation coefficient on the pressure of water entering the reservoir. The substantiating for applying the proposed dependence was given. To solve the system of non-linear differential equations, the integral relations method was used, which allowed obtaining an approximate analytical solution of the water pressure disturbance distribution in the two-phase filtration area. The radius of the perturbation area (front of the displacement of oil by water) was defined, and the boundaries of the perturbation region were determined for the given boundary conditions at the bottom of the injection well. Examples allowing the quantify assessment the application of the model proposed by the authors were considered.

1. Introduction

The justification of the methods for extracting residual mobile oil is an urgent problem of the development of oil deposits. Realization of the paper results of N.N. Lisovsky, R.Kh. Muslimov and other researchers devoted to the problems of the application of hydrodynamic methods of increasing oil recovery, allowed to extend the terms of field development. N.N. Mikhailov studied the problem of deteriorating the properties of the previously considered recoverable part of the reserves in the process of long-term development [1]. For this, it is necessary to know the nature of the distribution of mobile oil reserves in the pore volume of the deposit and the degree of its mobility. The implementation of the ongoing development system has a significant impact on the ratio of non-recovered oil reserves and the amount of water pumped into the reservoir. In order to bring project and actual production data and to extract the unreached by approved development system of residual oil reserves, enhanced recovery methods are used. The choice is influenced by geological, technological and economic factors.

2. Research

By constructing the computational models based on known physical fluid properties and reservoir porosity and permeability, a search for the areas of existence of unrecovered oil is made [1, 2]. When the pressure and temperature of the oil change, an associated hydrocarbon gas can release and one-, two-and three-phase filtration areas arise in the reservoir. For the construction of physical models of two and three phase filtration, the basic mechanics equations of continuous medium are used - equations of continuity, motion and rheological equations. After substituting the motion equations into the continuity equations, we have a system of differential equations describing the processes of multiphase filtration, including the displacement of oil by water. Equations describing non-stationary

Content from this work may be used under the terms of the Creative Commons Attribution 3.0 licence. Any further distribution of this work must maintain attribution to the author(s) and the title of the work, journal citation and DOI.
Published under licence by IOP Publishing Ltd

three-phase filtration in a porous and fractured-porous medium were considered in papers [3-12]. Differential equations include unknown pressures as well as water and oil saturation coefficients, respectively. It is believed that oil and water do not mix in reservoir conditions.

3. Results and discussion

Let us consider a two-phase filtration water-oil at a pressure above the saturation pressure [5-12]: For isothermal two-phase filtration, the equations of continuity and motion have the form [5-12]:

$$\frac{\partial}{\partial t}(m\rho_i s_i)+div(\rho_i\bar{v}_i)=0,\quad \vec{v}_i=-\frac{k_0k_i^*}{\mu_i}gradp_i,\ i=1\text{-}b,\ 2\text{-}n\ \text{(water, oil)},\tag{1}$$

where k_i^* – relative phase permeability, k_0 – coefficient of absolute permeability, μ_i – dynamic viscosity; v_i – filtration rates, m – coefficient of porosity, ρ_i – phase densities, p_i – current phase pressures.

To solve the equations system (1), rheological equations are needed, which construction is based on the experimental study of reservoir fluids and media. In underground hydromechanics, parameters determined from rheological equations, as a rule, are functions of pressure and temperature. If in the downhole reservoir area the pressure value becomes lower than the saturation pressure value, the rheological equations of liquids change. As the pressure decreases from the initial pressure to the saturation pressure, the density and viscosity of the oil first decrease, and when the pressure drops below the saturation pressure, these parameters begin to increase. The reverse pattern is observed when the volumetric oil ratio is changed. The amount of gas dissolved in oil is accounted for by the gas content factor therefore the composition of oil and the content of light hydrocarbons significantly affect the shape and boundaries of the application of rheological equations. When pumping water, the pressure, the density and viscosity increase. Dependences of the rheological parameters of porous media saturated with liquids on temperature are mainly needed when designing the application of thermal methods for enhancing the oil recovery. For isothermal processes, rheological equations which parameters obey an exponential law have the form:

$$\begin{gathered}\rho_i=\rho_{i0}\,exp(\alpha_{i\rho}\Delta p_i),\ m=m_0\,exp(\alpha_m\Delta p_i),\quad \mu_i=\mu_{i0}\,exp(\alpha_{\mu i}\Delta p_i),\\ k_i(p_i)=k_{i0}\,exp(\alpha_{ki}\Delta p_i),\end{gathered}\tag{2}$$

where $\alpha_{i\rho}$ – coefficients of fluid compressibility, $\Delta p_i=p_i\text{-}p_0$ – pressure change (depression or repression). It is usually assumed that $\alpha_m=\beta_c$ is the compressibility factor of the reservoir, the skeleton. $\alpha_{i\mu}$ – are the coefficients of viscosity depending on pressure, α_{ki} – are the coefficients of permeability change depending on pressure, $i=1$ *or b, 2 or n* are water and oil.

A necessary condition for solving the system (1) is the correspondence of the number of equations to the number of unknown parameters. To solve the direct problem of two-phase filtration required unknowns are the pressures p_i and the saturation coefficients of phases s_i. By substituting the motion equations into the system continuity equations (1), we obtain two partial differential equations for the determination of four unknowns. Therefore, for a correct solution, two more closing equations are needed, connecting the unknowns p_i and s_i. The volumetric content of oil and water in the reservoir is characterized by the coefficients of oil saturation s_n and water s_b, which are interrelated by the ratio

$$s_n=1-s_b\,.\tag{3}$$

The residual mobile oil saturation can be estimated from the change in the water saturation coefficient in various parts of the deposit. Changes in the saturation coefficients are possible as a result of a decrease in the volume of fluids in the pore space and due to a change in the volume of the pore space itself. And, if it is talked about mobile oil reserves, then in the two-phase filtration area, the saturation coefficients vary within

$$s_0 < s_b < s^* < 1 \text{ and } 1 - s^* < s_n < 1 - s_0, \tag{4}$$

where s_0 is the residual water saturation; s^* is the limiting value of the water saturation coefficient at which oil filtration ceases.

Another closing equation is the relationship between the phase pressures taken into account by the capillary pressure p_k, which is determined experimentally.

$$p_1 = p_2 \pm p_k(s_1). \tag{5}$$

It should be noted that (5) is valid for the stationary state equilibrium of the medium, but in the absence of a better one, it is also used to describe dynamic processes (an analog of Coulomb's law for dry friction in theoretical mechanics.)

Let us consider some models of two-phase filtration of oil displacement by water.

The Buckley-Leverett model is one of the most common physical models describing non-piston oil displacement by water [6, 7, 11].

It is believed that the density of water, oil and porosity are constant. The rheological equations (2) are simplified (in fact they are equations of state), capillary pressure is neglected, the closing equation (5) takes the form $p_1=p_2$. It follows from the continuity equations that the total phase velocity is only a function of time. After substituting the motion equations and the Buckley-Leverett function, which characterizes the fraction of water in the total fluid flow, in the continuity equations one partial differential equation, connecting the derivative of the Buckley-Leverett function, coordinates and time is obtained. For constant injective capacity (flow rate of injected water) Q after integration and transformations for flat-radial filtration, the following relation is satisfied

$$t = \frac{\pi m h r^2}{Q f'(s_f)}, \tag{6}$$

where s_f is the water saturation at the displacement front, determined graphically for the known values s_0 and s^*, $f'(s)$ is the derivative of the Buckley-Leverett function, Q is the water flow rate (injection capacity), m is the porosity coefficient, and h is the effective thickness of the reservoir. The water cut coefficient and the current oil recovery index are determined by the Buckley-Leverett function and its derivative. On the displacement front, the derivative of the Buckley-Leverett function has a jump. From our point of view, jump occurs depend on the used hypothesis. Therefore, despite the elegance and mathematical beauty of the solution, this model has physical drawbacks due to the chosen rheological equations.

In the Muskat-Mires model [5, 7, 8], when describing the rheological properties of a porous medium, the following hypotheses are accepted: the incompressible collector, the porosity coefficients m and the absolute permeability k are constant, the change in density ρ_i of the liquid phases is expressed in terms of the volume coefficients b_i. Rheological equations in this case have the form:

$$b_i = \frac{V_i}{V_{i0}}, \quad i=b,n \text{ (water, oil)}, \quad \rho_i = \frac{\rho_{i0}}{b_i},$$

where V_i is the volume of fluids with dissolved gas under the reservoir conditions, V_{i0} is the fluid volume under normal conditions (degassed liquids), ρi0 is the density of fluids under normal conditions. The volume coefficients are determined experimentally. If it is assumed that the pressures in the phases are equal (capillary pressure is not taken into account), then the closing equation (5) takes the form $p=p_i$, $p_k=0$.

Substituting the motion equations into the continuity equations, we obtain two differential equations for the two-phase oil-water filtration without gas in the free state

$$\frac{m}{k_0}\frac{\partial}{\partial t}\left(\frac{s_i}{b_i}\right)+div\left(\frac{k_i^*}{\mu_i b_i}gradp\right)=0 \quad . \tag{7}$$

Two equations (7) and relation (1) form a closed system of equations for determining unknowns p, s_n, and s_b. The solution of the system of differential equations system (7) is carried out by numerical methods. Here, as in the previous Buckley-Leverett model, the basic position of continuum mechanics is used, that in an infinitesimal volume of a porous formation, there are simultaneously movements (filtration) of two fluids due to a single pressure gradient. The specified rheological equations do not take into account the deformation of the porous medium, which can lead to a mismatch between calculated and actual development indices, for example, in high-compressible bituminous reservoirs of the Bazhenovskaya suite.

More general is the model of two-phase filtration of compressible and deformable media, considered in the papers of V.N. Nikolayevsky, K.S. Basnieva, A.T. Gorbunov and other researchers [9-10]. As a result of substitution of the equations of motion and rheological relations into the continuity equations, the following system of differential equations is obtained, taking into account the deformation of the solid phase, compressibility of fluids, porous media and capillary pressure.

$$\frac{\partial}{\partial t}\left[s_i\varphi_i(s_i)\exp\left(\beta_i(p_i-p_0)\right)\right]=D_i^2 div\left[F_i(s)grad\left(\exp(\alpha_i(p_i-p_0))\right)\right], \tag{8}$$

$$\varphi_i(s_i)=exp\left[\alpha_m s_i p_k(s_b)\right],\ \beta_i=\alpha_m+\alpha_{i\rho},\ D_i^2=\frac{k_0}{\mu_{i0}m_0\alpha_i},\ p_1=p_2\pm p_k(s),$$
$$F_i(s)=k_i^*(s)exp\left[\alpha_i s_i p_k(s)\right],\quad \alpha_i=\alpha_k+\alpha_{i\rho}-\alpha_{\mu i}. \tag{9}$$

In the derivation of equations (8), it was assumed (hypotheses) that the rock pressure is constant, and the changes in permeability and porosity are functions of the effective pressure $P=p_1s_1+p_2s_2$. From the condition (hypothesis) of the conservation of the mass of one of the phases in an elementary volume, if the flow of this phase is zero, the closing equation will take the form

$$(1-s)\varphi_2(s)\exp\left[\beta_2(p_2-p_0)\right]=A(x),\quad s=s_1, \tag{10}$$

where $A(x)$ is the function determined by the initial distribution of the second phase in the porous medium, that is, the distribution of the saturation coefficient as a function of the coordinate. The closing equation (10) relates the pressure to the water saturation coefficient and is based on the assumption that the flow rate of one phase in the elementary volume is zero, that is, the filtration is not carried out. The solution of the equations system (8) is performed by numerical methods. It was found in the paper [9] that, the solution of system (8) reduces to solving of a non-linear differential equation. In contrast to the Muscat-Mires equations (7), the system of three equations (8), (10) takes into account the deformations of the collector, changes in permeability, porosity and capillary pressure. Hence, it is a more accurate model for describing the two-phase filtration. Thus, the differential equations system (8) in the presence of the closing equations (9), (10) makes it possible to determine the distribution of phase pressures and saturation coefficients in the two-phase filtration area. It is believed that the filtering-capacitive parameters are functions of effective pressure, which, in turn, is a function of unknown pressures and phase saturations. The two-phase filtration model described by the system (8) is more informative, in comparison with the Muscat-Mires model.

The model proposed by the authors of the paper, as well as the previous one, is constructed on the basic equations of continuum mechanics. An element of the operational object is considered in which two-phase filtration occurs as a result of the injection well operation. Changes in phase pressures and saturation coefficients in the field of non-piston oil displacement by water are carried out under the following assumptions: the increase in water saturation occurs due to water entering the reservoir, hence, it is a function of water pressure; the phase pressure is greater than the saturation pressure; the two-phase filtration region is determined by the radius of the displacement front.

As a closing equation, we take the dependence of the water saturation coefficient on repression at the downhole of the injection well Δp_c and the current repression in the reservoir $\Delta p_b = p(r,t) - p_0$ in the area of oil displacement by water [13].

$$s_b(r,t) = s_0 + (s^* - s_0)\frac{\Delta p_b}{\Delta p_c} = s_0 + s_w(r,t),$$
$$s_w(r,t) = (s^* - s_0)\frac{\Delta p_b}{\Delta p_c},\ 0 \le s_w(r,t) \le s^* - s_0. \quad (11)$$

The coefficient $s_w(r,t)$ is the increment of the water saturation coefficient $s_b(r,t)$ as a result of water pumped into the reservoir. Equation (11) is obtained under the following boundary conditions: at the downhole of the injection well, the water saturation coefficient $s_b(r,t)$ takes the value s^*, and at the front of oil displacement by the injected water it is equal to the residual water saturation s_0. Equation (11) is valid for both plane-radial and linear filtration.

After substituting the motion equations and the closing equation (11) into the continuity equations of the system (1), we obtain two partial differential equations

$$\frac{\partial \Phi_{1i}}{\partial t} = \frac{a_{0i}}{\beta_{2i}} div\, \Phi_{2i},$$
$$\Phi_{1i} = s_i\, exp(\beta_{1i}(p_i - p_0)),\ \Phi_{2i} = k_i^*(s_i) grad\, exp(\beta_{2i}(p_i - p_0)), \quad (12)$$
$$a_{0i} = \frac{k_0}{m_0 \mu_{0i}}, \beta_{1i} = \alpha_m + \alpha_{\rho i},\quad \beta_{2i} = \alpha_{\rho i} + \alpha_k - \alpha_{\mu i}.$$

where a_{0i} is a coefficient which takes into account the initial physical properties of the reservoir and fluid. Omitting the index $i=1=b$ of the water phase, for flat-radial filtration, the first equation of system (12) takes the form:

$$\frac{\partial \Phi_1}{\partial t} = \frac{a_0}{\beta_2}\frac{1}{r}\frac{\partial (r\Phi_2)}{\partial r},\quad r\frac{\partial \Phi_1}{\partial t} = \frac{a_0}{\beta_2}\frac{\partial (r\Phi_2)}{\partial r}. \quad (13)$$

To solve the equation (13), we use the G.I. Barenblatt's method of integral relations [10]. The functions entering into the differential equation (13) depend on the water pressure p. Moreover, the function Φ_2 is the product of two functions of scalar $k^*(s(p))$ and vector $grad\, \exp(\beta_2(p - p_0))$. After substituting the functions Φ1 and Φ2 in (13) and transformations, we obtain a non-linear partial differential equation related to p. For the function p, we set the following boundary conditions: on the displacement front, the pressure gradient is zero, the pressure is constant

$$\left.\frac{\partial p}{\partial r}\right|_{r=\rho(t)} = 0,\quad p(\rho(t)) = p_0 \quad (14)$$

Restricting ourselves to the three terms of the series, we seek the solution in the form

$$p = c_0 + c \ln\frac{r}{\rho(t)} + c_k \frac{r^k}{\rho^k(t)}.$$

Here k is a number, not necessarily an integer, determined from the comparison of fishing data with the calculated indicators; c_0 and c_K, are the parameters determined from the boundary conditions. For the boundary conditions (14), we obtain

$$p(r,t) = p_0 - \frac{c}{k}\left(k\,ln\frac{\rho(t)}{r} + \frac{r^k}{\rho(t)^k} - 1\right) = p_0 - \frac{c}{k} f_k(r,\rho(t))$$
$$f_k(r,\rho) = \left(k\,ln\frac{\rho(t)}{r} + \frac{r^k}{\rho(t)^k} - 1\right) \tag{15}$$

We integrate the equation (13) with respect to r from r_c to $\rho(t)$, taking into account the boundary conditions at the displacement front and the bottom of the injector set for the pressure p. We obtain a differential equation for determining the radius of the displacement front $\rho(t)$

$$A_k \frac{d}{dt}\left[\rho^2(t)\left(\frac{\Delta s}{f_k(r_c,\rho(t)} + \beta_1 s_0 \frac{c}{k} + \beta_1 \frac{\Delta s}{f_k(r_c,\rho(t)} \frac{c}{k}\frac{D_k}{A_k}\right)\right] = -a_0 k_m^* r_c \frac{\partial p}{\partial r}\bigg|_{r=r_c}, \tag{16}$$

$$A_k = \frac{k^2}{4(k+2)},\ D_л = \frac{k^5 + 7k^4 + 20k^3 + 24k^2 + 16k + 8}{4(k+1)(k+2)^2}.$$

To determine the parameter c, a third boundary condition is necessary, depending on the choice of the operating mode of the well. As a rule, two boundary conditions at the downhole of the injection well are considered: with constant pressure at the bottom or with constant injection capacity.

Let us consider the case when the discharge pressure at the bottom is constant, $p(r_c)=p_c$, where $\Delta p_c=p-p_0$ is the repression at the bottom of the injection well. And p_c should not exceed the ultimate strength of the rock, otherwise the rock will break down, remove mechanical impurities, increase the radius of the well, etc. The parameter c is equal

$$c = -k\frac{p_c - p_0}{f_k(r_c,\rho)} = -k\frac{\Delta p_c}{f_k(r_c,\rho)}. \tag{17}$$

The right-hand side of equation (16) in this case, is equal $a_0 k_m^* \dfrac{k\Delta p_c}{f_k(r_c,\rho)}$, k_m^* – is the maximum value of the coefficient of relative phase permeability of the reservoir on water, determined experimentally.

The radius of the displacement front $\rho(t)$ is determined by integrating the differential equation (16)

$$A_{k1}\rho^2\left(1 - \frac{1}{\rho^2}J_1(\rho) - \frac{2D_{k1}}{A_{k1}\rho^2}J_2(\rho) + \frac{2D_{k1}}{A_{k1}\rho^2}J_3(\rho)\right) = k_m^* a_0 k \Delta p_c t,$$
$$A_{k1} = A_k(\Delta s - \beta_1 s_0 \Delta p_c),\quad D_{k1} = D_k \Delta s \beta_1 \Delta p_c, \tag{18}$$
$$J_1(\rho) = \int_{r_c}^{\rho} u^2 \frac{\partial f_k(u)}{\partial u}\frac{1}{f_k(u)}du,\ J_2(\rho) = \int_{r_c}^{\rho} u \frac{1}{f_k(u)}du, J_3(\rho) = \int_{r_c}^{\rho} u^2 \frac{\partial f_k(u)}{\partial u}\frac{1}{f^2{}_k(u)}du.$$

It follows from (18) that the radius of the displacement front depends on the values of the water saturation coefficients s_0 and s^*, which determine the area of two-phase filtration, and also from the given boundary condition Δp_c. For single-phase filtration, when determining the radius of influence of a well, the effect of the change in pressure at the bottom of the disturbing well, established by V.N. Shchelkachev, was considered in detail in the papers of A.P. Telkova and S.I. Grachev [12, 14].

The pressure of the injected water in the region of two-phase filtration is determined by the relation

$$p(r,t) = p_0 + \frac{\Delta p_c}{f_k(r_c,\rho)} f_k(r,\rho(t)),\ f_k(r,\rho) = \left(k \ln \frac{\rho(t)}{r} + \frac{r^k}{\rho(t)^k} - 1 \right). \tag{19}$$

The distribution of the water saturation coefficient in the injection region is determined from the closing equation (11), which, taking into account the obtained results, is written in the form

$$s_b(r\ ,t) = s_0 + (s^* - s_0)\frac{f_k(r,\rho(t))}{f_k(r_c,\rho(t))} \quad . \tag{20}$$

It follows from (20) that the dependence of the water saturation coefficient on the boundary conditions is realized implicitly in terms of the displacement radius $\rho(t)$, determined from the solution of equation (18).

Example. Let us consider how the radius of the displacement front depends on time for different values of the parameter *k*.

Assumed:

p_0=22MPa, Δp_c=4MPa, k_0=250mD, k_m*=0.6, MPa, α_m=10^{-4}1/MPa, α_μ=10^{-4} 1/MPa, α_ρ=2*10^{-4} 1/MPa, m_0= 0,2, Q_0=300m^3/day.

Table 1. Changing in the radius of the displacement front depending on time.

	t, days			
$\rho(t)$, m	k=1	k=1.5	k=2	k=3
5	0.003	0,004	0.005	0.007
10	0.013	0.018	0.02	0.03
50	0.4	0.5	0.5	0.7
100	1.5	1.9	2.2	2.9
400	24	31	36	46
1000	150	194	226	289
2000	604	779	909	1157
3000	1362	1757	2051	2603
4000	2426	3130	3653	4628
5000	3797	4897	5716	7231

From Table 1 it follows that at the initial period of operation of the injection well, the time for reaching the radius $\rho(t)$=100 meters does not exceed 3 days. And for $\rho(t)$=400 meters, which corresponds to the distance between the injection and production wells, the displacement time varies from 24 to 46 days. Therefore, the choice of the parameter *k*, determined by comparing the actual and calculated parameters, significantly affects the rate of propagation of the water flooding area. It should be noted that with increasing *k*, the propagation velocity of the flooding area slows down.

Let us compare the obtained results with the piston displacement of oil by water. Before the displacement front, the water saturation coefficient is s_0, and behind the displacement front it is s=s*. The radius of the displacement region is determined in this case from equation

$$\rho^2(t)\left(1 - \frac{J_1}{\rho^2(t)}\right) = \frac{k_m^* a_0 k}{\beta_1 s * A_k} t \tag{21}$$

The results of calculations using formula (21) are shown in Table 2. From a comparison of the results of the calculations of Tables 1 and 2, it is clear that in this case the propagation time of the flooding area is much less than for non-piston displacement. As the parameter *k* grows, the time increases. If we neglect the integral in (21), then the displacement time remains practically unchanged.

Table 2. Change in the radius of the displacement front depending on time for piston displacement.

	t, days			
t,m	k=1	k=1.5	k=2	k=3
5	0.0000	0.0000	0.0000	0.0000
10	0.0000	0.0000	0.0000	0.0000
50	0.0004	0.0005	0.0006	0.0007
100	0.002	0.002	0.002	0.003
400	0.027	0.034	0.040	0.048
1000	0.168	0.217	0.254	0.305
2000	0.677	0.873	1.019	1.225
3000	1.53	1.97	2.30	2.76
4000	2.72	3.51	4.10	4.92
5000	4.26	5.49	6.41	7.70

Unlike the model of Muskat-Mires and the deformation model, the author's version allows to obtain an analytical solution of the required problem.

The fraction of water in the total liquid flow at identical phase-pressure gradients corresponds to the Buckley-Leverett function

$$f\ (s)=\frac{k^*(s)}{k^*(s)+\mu_0 k_n^*(s)} \tag{22}$$

It follows from (22) that the watering of the production well located at a distance $r=r_f$ from the injection well begins at $s>s_0$, that is, the water saturation coefficient depends on time and coordinates. According to the author's model, from the closing equation (20) it follows that the growth in the water saturation coefficient at the bottom of the production well increases gradually over time Therefore, the water cut of the production well also increases gradually and depends on the specified boundary conditions and the numerical parameter *k*.

The fraction of water *f* in the total flow of the filtering liquid, taking into account the capillary pressure, is determined by the formula given in [15]

$$f=\frac{k^*(s)}{k^*(s)+\mu_0 k_n^*\left(1+\frac{\partial p_k}{\partial r}/\frac{\partial p}{\partial r}\right)}, \tag{23}$$

here μ_0 – is the ratio of water viscosity to oil viscosity. Capillary pressure depends on the water saturation through the Leverett function $J(s)$, the wetting angle Θ and the surface tension [16]

$$p_к(s)=J(s)\frac{a}{\cos\Theta}\sqrt{\frac{m}{k_0}}.$$

The derivative of the capillary pressure along the coordinate is

$$\frac{\partial p_k}{\partial r}=\frac{\partial p_k}{\partial s}\frac{\partial s}{\partial r}=\frac{\partial s}{\partial r}\frac{\partial J(s)}{\partial s}\frac{a}{\cos\Theta}\sqrt{\frac{m}{k_0}}.$$

It follows from (21) and (25) that

$$\frac{\partial p}{\partial r}=-\frac{c}{k}\frac{\partial f_k(r,\rho(t))}{\partial r}=c[1-\left(\frac{r}{\rho(t)}\right)^k] \qquad \frac{\partial s}{\partial r}=\frac{\Delta s}{f_k(r_c,\rho)}\frac{\partial f_k(r,\rho)}{\partial r}.$$

From (23) it follows:

$$\nu = f = \frac{k^*(s)}{k^*(s) + \mu_0 k_n^*(s)\left[1 \mp \frac{\partial p_k(s)}{\partial s}\frac{k\Delta s}{cf_k(r_c, \rho(t))}\right]}. \quad (24)$$

Consequently, the closing equation (20) plays an essential role in determining f and the water-cut coefficient. To determine f from (24), it is necessary to have experimental dependences $p_k(s)$ or $J(s)$. The parameter c corresponds to a given boundary condition (17). Since the relative phase permeability k_n^* and k^* are functions of the water saturation s, the value of the water saturation coefficient depends on the coordinates, the choice of the boundary conditions, $\rho(t)$ and the time to reach a given point of the reservoir, that is, from the selected method of operating the injection well. For an infinite reservoir, $\rho(t)\to\infty$, $k^* \to k^*_{m}$, $k^*_n(s) \to 0$, the expression in square bracket (24) tends to 1 and coincides with the classical Buckley-Leverett formula (22) and the product is watered. Thus, according to the author's model of non-stationary two-phase filtration, the water cut at the downhole of the producing well increases continuously, and not abruptly as follows from the Buckley-Leverett model.

For anomalous reservoirs containing high-viscosity oil or having low filtration properties, the initial pressure gradient in the equations of system motion (1) and expression (24) should be taken into account.

4. Conclusion

1) The known two-phase filtration models are based on the methods of continuous medium mechanics and reduce to a numerical solution of a system of nonlinear differential equations. The process of non-piston oil displacement by water depends on the selected rheological equations and the given boundary conditions. For a correct solution of a system of nonlinear differential equations, a closing equation is needed, which construction is based either on experimental studies or on accepted hypotheses which do not contradict physical sense. Each of the considered models has its advantages and disadvantages, which reliability can be confirmed only by comparing the calculated indicators with the actual ones.
2) The proposed author's model is based on the fact that when the reservoir is flooded, the change in saturation caused by the influx of injected water depends on the change in the pressure of the water pumped into the reservoir.
3) Unlike other models solved by approximate numerical methods, the method of integral relations of G.I. Barenblatt was used to solve the problem, by means of which an approximate analytical solution was obtained.
4) The closing equation (20) makes it possible to establish a relationship between the water fraction in the liquid flow and the water saturation coefficient.

References

[1] Mikhailov N N 1992 *Residual Oil Saturation of the Developed Reservoirs* (Moscow: Nedra) p 270
[2] Gazizov A A 2002 *Increase in Oil Output Heterogeneous Layers at the Last Development Stage* (Moscow: Nedra-Bisnesscentr) p 639
[3] Leibenzon L S 1947 *Movement of Natural Liquids and Gases in a Porous Medium* (Moscow-Leningrad: OGIZ Gos.izd.tekh.-theor. Literature) p 244
[4] Shchelkachev V N 1959 *Development of Oil Reservoirs under the Elastic Regime* (Moscow: Gostoptekhizdat) p 467
[5] Masket M 2004 *Physical Basis of Oil Production Technology* (Moscow-Izhevsk: IKI) p 606
[6] Zheltov Ju P 1998 *The Development of Oil Fields* (Moscow, «Nedra» JSC) p 365
[7] Basniev K S, Vlasov A M, Kochina I N and Maksimov V M 1986 *Underground Hydraulics* (Moscow: Nedra) p 303

[8] Kanevskaya R D 2002 *Mathematical Modeling of Hydrodynamic Proc. of Development of Hydrocarbon Deposits* (Moscow-Izhevsk: IC) p 140
[9] Nikolaevsky V N, Basniev K S, Gorbunov A T, Zotov GA 1970 *Mechanics of Saturated Pore Media* (Moscow: Nedra) p 336
[10] Barenblatt G I 1984 *Movement of Liquids and Gases in Natural Seams* (Moscow: Publishing House Nedra) p 211
[11] Telkov A P and Grachev S I 2009 *Hydrodynamics of the Reservoir in Relation to Application to Develop Oil and Gas Fields* (Tyumen: TSOGU) 1 p 240
[12] Telkov A P and Grachev S I 2009 *Hydrodynamics of the Reservoir in Relation to Application to Develop Oil and Gas Fields* (Tyumen: TSOGU) 2 p 380
[13] Korotenko V A, Kushakova N P, Leontyev S A, Zaboeva M I and Alexandrov M A 2016 Depenence of coefficients of saturation on time and coordinates *Bulletin of Higher Educational Institutions. Oil and Gas* 6 pp 74-81
[14] Telkov A P, Yagafarov A K, Sharapov A U and Kleshchenko I I 1993 *Interpretation Models of the Oil Deposit at the Development Stage* (Moscow: VNIIOENG) p 72
[15] Grachev S I, Korotenko V A and Zonova O P 2016 On the question of two-phase filtration in porous medium *Drilling and Oil* 5 pp 50-4
[16] Charniy I A 1963 *Underground Hydro Gas Dynamics* (Moscow: Gostoptekhizdat) p 396

Information about authors

1. Korotenko Valentin Alekseevich, Candidate of Technical Sciences, associate professor of the Department of Development and operation of oil and gas fields, Industrial University of Tyumen (IUT).
2. Grachev Sergey Ivanovich, Professor, Doctor of Technical Sciences, Head of the Department of Development and operation of oil and gas fields, Industrial University of Tyumen (IUT).
3. Nelly Petrovna Kushakova, Associate Professor, Candidate of Technical Sciences, Department of Automobile transport and road and construction machines of Tyumen Industrial University (TIU).
4. Leontyev Sergey Alexandrovich, Doctor of Technical Sciences, Professor, Department of Development and operation of oil and gas fields, Industrial University of Tyumen (IUT).
5. Zaboeva Marina Ivanovna, Candidate of Technical Sciences, associate professor of the Department of Development and operation of oil and gas fields, Industrial University of Tyumen (IUT).
6. Alexandrov Mikhail Alekseevich, Candidate of Technical Sciences, Associate Professor, Industrial University of Tyumen (IUT).

IOP Conf. Series: Earth and Environmental Science **181** (2018) 012017 doi:10.1088/1755-1315/181/1/012017

Subminiature eddy-current transducers for thickness studies

S F Dmitriev, A V Ishkov, A O Katasonov, V N Malikov and A M Sagalakov

Altai State University, Barnaul

E-mail: osys11@gmail.com

Abstract. On the basis of an eddy-current transducer of a transformer type, a measuring system is made, which makes it possible to estimate the possibility of using the eddy current method for measuring the thickness of conductive and dielectric coatings placed on a conductive base. The design of the measuring system and measurement techniques is described. The article presents data demonstrating the dependence of the amplitude part of the signal on objects of different thicknesses; limiting object dimensions at which these measurements are expedient are experimentally established.

1. Introduction

To control the thickness of electrically conductive sheets, films, plates, coatings on them, walls of pipes, cylindrical and spherical vessels, and determine the air gap between plates of the same material, thickness gauges based on the eddy-current method are used. Conductive and dielectric coatings are widely used in such industries as aerospace, shipbuilding, oil and gas. Such coatings are often used as a protection against thermal and corrosive effects, for electrical insulation [1]. However, in the process of application and operation of such coatings, problems arise with different thicknesses of the applied coating. Another potentially dangerous aspect is corrosion under the coatings themselves in case of their poor-quality application. Therefore, non-destructive testing of coatings applied to metals used in extreme conditions and in equipment operating under extreme conditions is especially urgent.

At present, thermal, microwave, ultrasonic and eddy-current testing methods are used for non-destructive testing of coatings [2, 3]. Eddy-current control has significant advantages over other methods: it is resistant to extreme operating conditions, the measurement is noncontact, and its cost is relatively low [4]. For example, to estimate the conductive coating thickness, Takashi et al [5] established the possibility of estimating the thickness of a coating based on nickel sputtered on stainless austenitic steels (type 304). The limit of the measured coating thickness was 22 μm.

Barbosa [6] derived a formula that allows obtaining the value of the thickness of galvanized steel, depending on the impedance of the coil of an eddy-current transducer. Yang [8] and Tai [3, 7] used pulse control methods to determine conductivity and coating thickness on a metal base plate under magnetization conditions.

In the eddy-current thickness studies of an object, which is a dielectric or conductive top layer (coating) located on a conducting base (base plate), control based on phase φ measurements of the signal introduced into the transducer is usually used [9]. The thickness of the top layer d is determined by the phase changes during scanning. An attachable transducer, as a rule, is a transducer of a transformer type with three windings. The exciting winding creates an excitation field, and the measuring and compensating windings are intended for measurements. The phase is measured relative to the sinusoidal excitation current of the transducer; in [10], in order to increase the accuracy of measurements, the EMF

Content from this work may be used under the terms of the Creative Commons Attribution 3.0 licence. Any further distribution of this work must maintain attribution to the author(s) and the title of the work, journal citation and DOI.
Published under licence by IOP Publishing Ltd

induced in the compensating winding of the transducer was used as the reference signal. The phase of the EMF introduced into the measuring winding depends on the geometric parameters of the transducer, the excitation current frequency *f*, the gap between the transducer and the object of control (OC), the electrical conductivities of the coating σ_1 and the base σ_2, and also the relative magnetic permeability of the base.

In practice, variations in the value of σ_1 are usually observed when moving from point to point along the surface of the OC, in the transition from one OC to another. This leads to variations in the value of φ with an unchanged value of the coating thickness *d*, which causes an error in determining *d*. At the same time, the amplitude, phase, amplitude-phase detuning from the influence of the variations of σ_1 on the results of measurements of *d* is practically impossible [11]. The influence of variations of σ_1 on the eddy-current thickness gauge can be reduced by lowering the frequency *f* of the excitation current of the transducer, but in this case, the interfering factors can be variations in the thickness of the base plate and the magnetic permeability of its material. As a result, with a small thickness of the base plate, it is necessary to choose a relatively high operating frequency *f* of the thickness gauge and apply the method of detuning from the influence of the variations of σ_1 on the measurement results of *d*. In this way, the value of σ_1 is determined on the basis of measuring the amplitude *A* or (and) phase φ of introduced into the attachable EMF transducer at such a relatively high frequency of the excitation field *f* at which the electromagnetic parameters of the base plate do not influence the measurement results. After this, the measurements of φ or *A* of the introduced EMF are made at a lower frequency *f*, when the measured parameters contain information on the thickness of the coating *d*. At the same time, when analysing the measurement results of φ or *A* in order to determine the value of *d*, the found value σ_1 is taken into account. An example of the implementation of this method of control is given in [12].

The purpose of this work was to evaluate the possibility of using only an amplitude eddy-current method for determining the thickness of a conductive or dielectric coating on a conductive base and estimating the error of such measurements. The possibility of determining the thickness of conductive objects which consist of several alternating conductive and nonconductive layers and monolithic conductive objects was also investigated.

2. Description of the installation

The subminiature eddy-current transducer [13-17] was developed for local studies of the thickness of various coatings in experiments, as well as to determine the effect of various coatings on the magnitude of the output signal.

The developed subminiature eddy-current transducer is a core with the following windings applied to it: exciting, measuring and compensating. The core and winding structure is impregnated with the compound and is enclosed in a corundum washer, so it is resistant to mechanical influences.

For testing on model objects from various conductive materials, a subminiature eddy-current transducer was used, which is connected to a personal computer via a sound card. The sound card is used as a generator and converter of a signal sent to the exciting winding.

The software allows one to control the value of the signal applied to the exciting winding, and also allows reading the voltage values from the measuring winding, which, taking into account the calibration, are converted into conductivity values. The developed software allows measuring the thickness of conductive and dielectric non-ferromagnetic coatings and conductive materials

To wind the turns, copper wire 5 microns thick is used. The core is made of ferrite 2000 NMZ with an initial magnetic permeability value of 500 and has a pyramidal shape. The characteristics of the developed transducer make it possible to achieve high localization of control, namely, to localize the field within 2500 μm2. This device provides a significant depth of penetration of the field into the object under investigation up to the values of ~ 5 mm (at frequencies of about 500 Hz).

Special software written in the C++ language for the Windows operating system allows controlling the signal on the exciting coil and receiving the signal from the measuring winding. By means of a special mixer, voltage is applied to the exciting winding of the transducer and the level and frequency of the sinusoidal signal of the generator are set.

The use of a sound card allows broadening the frequency limits of the signals fed to the exciter winding significantly, which in turn allows significantly expanding the measurement range.

3. Results and discussion

In order to test the developed measuring system, the aluminium coating having a different thickness and placed on a copper base 3 mm thick was scanned. Figure 1 shows the dependence of the signal value on the thickness of the aluminium coating on a non-ferromagnetic basis. As the thickness of the non-ferromagnetic coating is increased to 1200 μm, the signal value decreases from 28 to 22 cu. In the range from 750 to 1500 μm, the signal is much smaller than the signal from the monolith, which can be explained by the insufficient thickness of the coating. However, for coating thicknesses from 1500 to 2500 μm, the magnitude of the signal is constant and corresponds to the magnitude of the field from the monolith.

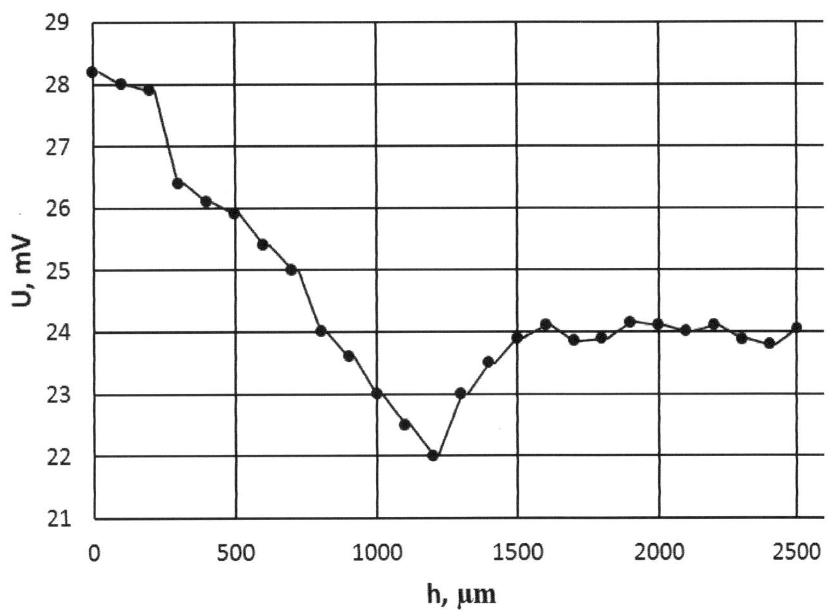

Figure 1. Dependence of the signal value on the thickness of the aluminium coating on a copper base.

The following experiment shows the dependences illustrating the change in the signal from the thickness of the layered coating with the alternation of layers of foil and polyethylene placed on a copper base. The object of the study was the alternation of layers of aluminium foil 20 μm and polyethylene 20 μm.

Figure 2 presents the dependence of the signal intensity on the thickness of the layered coating with the alternation of layers of foil and polyethylene placed on a copper base. At thicknesses from 0 to 100 μm, the magnitude of the signal from the base varies from 29 to 24 cu, and at thicknesses from 100 to 250 μm the character of the signal change is gentler. In the range from 250 μm to 400 μm the signal varies from 26 to 23 cu, which is due to the contribution of the signal from the layered coating and to the reduction of the contribution of the signal from the copper base plate.

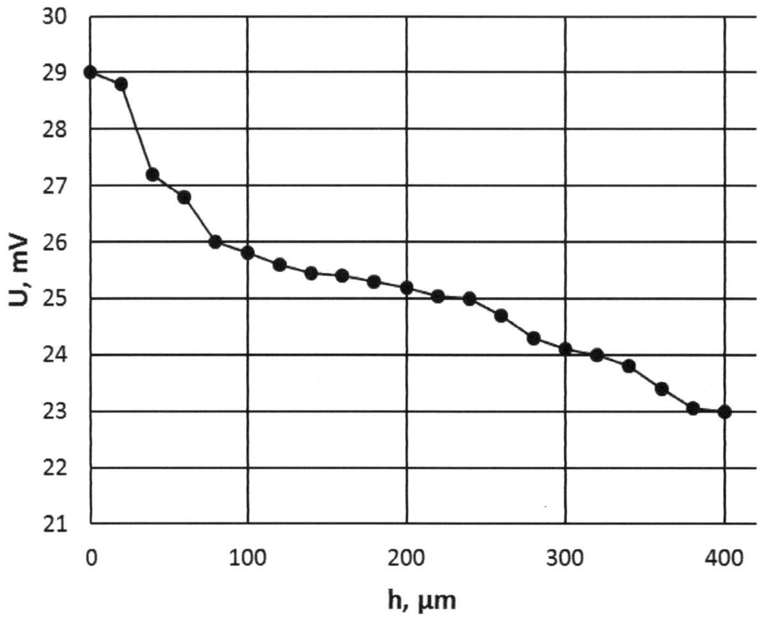

Figure 2. Dependence of the signal intensity during the scanning of a layered structure.

In the third test experiment, a sample was taken representing a single object of aluminium of different thicknesses. Figure 3 shows the dependence of the value of the signal on the thickness of the sample of aluminium. With the increasing sample thickness, the contribution of the deeper layers of the sample to the signal amplitude increases. When the thickness varies from 100 to 1200 µm, the signal value increases from 7 to 25 cu. When the thickness varies from 1200 to 2200 microns, the magnitude of the signal is constant and corresponds to the value of the amplitude from the monolith (25.5 cu).

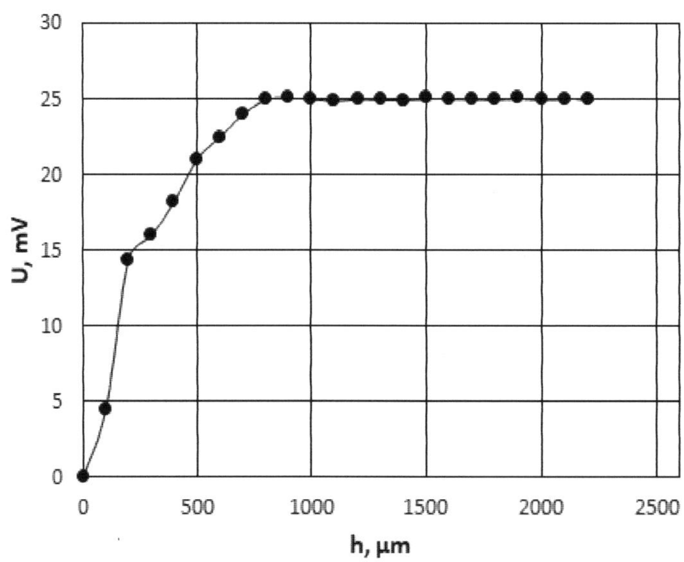

Figure 3. Dependence of the signal on the thickness of the aluminium sample.

IOP Conf. Series: Earth and Environmental Science **181** (2018) 012017 doi:10.1088/1755-1315/181/1/012017

In the final study, the control object was a copper plate and a dielectric coating. As a dielectric coating, a layer of varnish was used. The copper sample was made of a solid piece of copper. A layer of dielectric coating was applied to the controlled object.

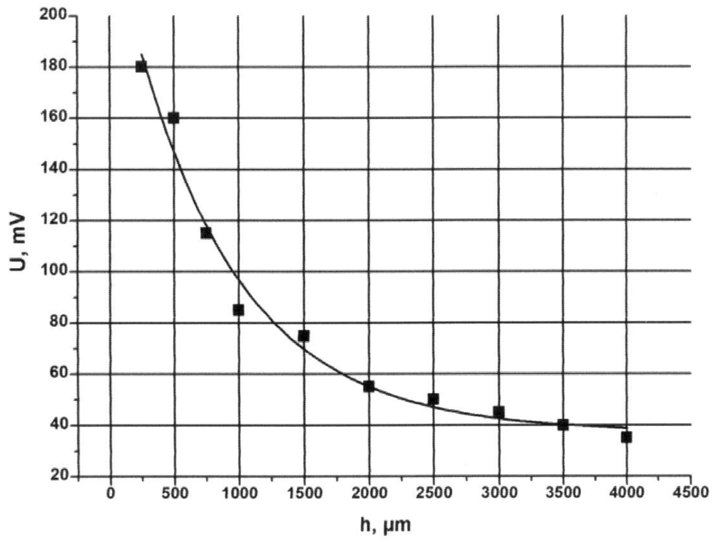

Figure 4. Dependence of the eddy-current transducer response on the thickness of the dielectric coating.

A series of experiments was also conducted with copper samples and a paint and varnish coating. On the pre-cut samples of copper, a layer of paint and varnish coating was applied. As can be seen from the dependence of the signal amplitude on the thickness of the dielectric coating (Figure 4), the magnitude of the output signal decreases rapidly with increasing coating thickness. This dependence can be approximated by an exponential function:

$$y=A_1e^{-\frac{x}{t}}+y_0.$$

The graph of this dependence and the parameters of the approximating function are shown in Figure 4. As can be seen from the Figure, the signal decreases exponentially with increasing thickness of the dielectric coating.

4. Conclusion

The developed measuring system was used to study objects representing conductive and non-conductive coatings placed on a conductive base, and also to measure the thickness of the monolithic conductive objects. It is established that the thickness of the coating affects the signal of the eddy-current transducer. This allows using a perspective amplitude control method of such objects for accurate measurements of the thickness of conductive and non-conductive coatings, as well as other objects.

In conclusion, we note that these test experiments were conducted to determine the effect of various objects on the output signal of an eddy-current transducer. The distortions of the eddy-current transducer signals during scanning of coatings are determined and it is established that these distortions depend on the type of coatings and the plate base.

References

[1] Ptchelintsev A and Halleux B 1998 Thickness and conductivity determination of thin nonmagnetic coatings on ferromagnetic conductive substrates using surface coils *Rev. Sci. Instrum.* (AIP Publishing) 69(3) pp 1488-95

[2] Ghasr M T, Horst M J, Lechuga M and Rapoza R 2015 Accurate one-sided microwave thickness evaluation of lined-fiberglass composites *IEEE Trans. Instrum. Meas.* (IEEE Xplore) 64(10) pp 2802-12
[3] Tai C C, Rose J H, Lechuga M and Moulder J C 1996 Thickness and conductivity of metallic layers from pulsed eddy-current measurements *Rev. Sci. Instrum.* (AIP Publishing) 67(11) pp 3965-73
[4] Halleux B, Ptchelintsev A and Limburg S B 1997 Thickness and conductivity determination of thin coatings on ferromagnetic substrates in the case of cylindrical symmetry *Rev. Sci. Instrum.* (AIP Publishing) 68(9) pp3533-40
[5] Takahashi Y 2012 Thickness evaluation of thermal spraying on boiler tubes by eddy current testing *Int. J. Appl Electromagn. Mech.* (IOS Press) 39(1) pp.419-25
[6] Barbosa C F 2014 A simple formula for calculating the coating thickness of galvanized steel wires *IEEE Trans. Magn.* (IEEE Xplore) 50(7) pp 1-5
[7] Tai C C 2000 Characterization of coatings on magnetic metal using the swept-frequency eddy current method *Rev. Sci. Instrum.* (AIP Publishing) 71(8) pp 3161-8
[8] Yang H C and Tai C C 2002 Pulsed eddy-current measurement of a conducting coating on a magnetic metal plate *Meas. Sci. Technol.* (IOP Publishing) 13(8) pp 1259-66
[9] Neumeier P 2008 Eddy current phase method for measuring the thickness of galvanic coatings. *V Mire Nerazrushayushchego Kontrolya* 2(40) pp 29–30
[10] Sjas'ko V A, Bulatov A S, Koroteev M Ju and Solomenchuk P V 2011 *Eddy Current Meter* RF Patent 2384839
[11] Bakunov A S and Kaloshin V A 2016 Evolution of thickness coating measurement *Kontrol. Diagnostika* 1(211) pp 27–31
[12] Bogdanov N G, Prihod'ko V A and Suzdal'cev A I 2002 *Eddy Current Method of Dual Frequency Control of Products* Patent RF 2184931
[13] Dmitriev S, Ishkov A, Katasonov A, Malikov V and Sagalakov A 2014 Subminiature eddy current transducers for studying metal-dielectric junctions *Instr. and Exper. Techniques* (Springer) 57(6) pp751-4
[14] Dmitriev S, Ishkov A, Katasonov A, Malikov V and Sagalakov A 2017 Non-destructive testing of Al-Mg alloys by using the eddy-current method *MATEC Web of Conf.* (EDP Sciences) 106 pp 1-7
[15] Dmitriev S F, Ishkov A V, Katasonov A O, Sagalakov A M and Shevtsova L I 2017 Measurement system for studying flaws in alloy slabs by means of subminiature eddy-current transducers *Meas. Techniques* (Springer)60(4) pp 372–5
[16] Dmitriev S F, Katasonov A O, Malikov V N and Sagalakov A M 2016 Flaw detection of alloys using the eddy-current method *Rus. J. of Nondestr. Testing* (Springer) 52(1) pp 32–7
[17] Dmitriev S F, Katasonov A O, Malikov V N and Sagalakov A M 2016 Eddy-current measuring system for analysis of alloy defects and weld seams *Rus. Eng. Res.* (Springer) 36(8) pp 626-9

Evaluation of well interference by correlation analysis

I I Nasyrov, E I Mamchistova and A I Nasyrova

Industrial University of Tyumen, Tyumen , Russia

E-mail: nasyrov87@mail.ru

Abstract: The analysis of well interference is understood as an assessment of the presence or absence of pressure communication, the calculation of the response time of production wells to the change in pumping of injection wells, as well as the evaluation of the contribution of injection wells to production wells flow rate. The problem of determining the interference of wells at the moment is solved by several methods, which can be divided into 2 groups:
1. Traditional methods. Methods of data analysis that require any technological operations on the investigated wells (observation well testing, tracer studies).
2. Alternative methods. Various methods for analyzing field data using mathematical statistics or by numerical modeling of the behavior of pressure or flow in specialized software.
The optimal option is a comprehensive approach based on the principles of the correlation method of analysis of field data, the main advantage of which is the speed of analysis and decisions, as well as the lack of large capital investments needed.

1. Introduction

Since the middle of the last century, when developing oil fields in our country, the method of artificial flooding was widespread. Its high efficiency is well known. However, the full extent of formation coverage and oil recovery decrease with an increase in the degree of formation heterogeneity. In heavily inhomogeneous reservoirs, injected water breaks into producing wells through highly permeable interlayers, leaving oil unallocated in low-permeability layers. In other cases, the production well area may not be sufficiently covered by the injection water front. To make effective decisions on adjusting the operation of the wells of the operational fund and determining additional measures for pumping water, it is necessary to have an actual, clear understanding on the wells interference. This need is related to the development of a mathematical approach to the analysis of field data that would correspond to high quantitative accuracy and efficiency in an active, dynamic system with the possibility of a complex description of the interaction of a whole group of wells, which is impossible without determining the contribution from the participation of injection wells.

2. Research

- Determination of the presence or absence of a hydrodynamic connection between injection wells and adjacent production wells.
- Determination of the contribution of injection wells to the production of adjacent production wells.
- Determination of the reaction time of the extraction fund for the operation of the formation pressure maintenance wells.

Content from this work may be used under the terms of the Creative Commons Attribution 3.0 licence. Any further distribution of this work must maintain attribution to the author(s) and the title of the work, journal citation and DOI.
Published under licence by IOP Publishing Ltd

IOP Conf. Series: Earth and Environmental Science **181** (2018) 012018 doi:10.1088/1755-1315/181/1/012018

3. Results and discussion

Correlation analysis is related to the search for dependencies between the behavior of two objects. The result of the analysis is the value of the calculated coefficient, indicating the presence or absence of some dependence. In the oil and gas business, the practice of applying this approach is connected with the use of the Spearman's rank coefficient [1, 2]. To evaluate its applicability and confirm the effectiveness, the Kendall correlation coefficient was also studied in addition to the general form of the correlation ratio [3].

The degree of dependence of the two parameters (parameters) of the operation of arbitrary wells No. 1 and No. 2 can be described on the basis of an analysis of the obtained value of the Spearman's ratio, which in the general case varies in the range from -1 (inverse dependence) to 1 (direct dependence). The zero value of the coefficient indicates that the parameters are independent. The Spearman's ratio is calculated by the formula:

$$R_s = \frac{\frac{1}{6}\cdot n(n^2-1) - \sum d - T(X) - T(Y)}{\sqrt{\left(\frac{1}{6}\cdot n(n^2-1) - 2T(X)\right)\cdot\left(\frac{1}{6}\cdot n(n^2-1) - 2T(Y)\right)}},$$

$$T(X) = \frac{1}{12}\sum_{1}^{M_X}(N_X{}^3 - N_X), \tag{1}$$

$$T(Y) = \frac{1}{12}\sum_{1}^{M_Y}(N_Y{}^3 - N_Y),$$

where M_X, M_Y - the number of groups of the same values for the parameters of the X and Y objects, N_X, N_Y - the number of identical elements in the corresponding group.

To estimate the correlation dependence, it is also possible to use other coefficients: the Kendall's rank coefficient (2), as well as the correlation ratio (3).

$$R_k = \frac{\sum P - \sum I}{\sqrt{(n(n-1)/2 - U_X)\cdot(n(n-1)/2 - U_Y)}},$$

$$U_X = \frac{1}{2}\sum_{1}^{M_X}(N_X{}^2 - N_X), \tag{2}$$

$$U_Y = \frac{1}{2}\sum_{1}^{M_Y}(N_Y{}^2 - N_Y),$$

where P - number of proversions; I - number of inversions.

$$R_{XY}^2 = \frac{S_{\bar{y}(x)}^2}{S_y^2},$$

$$S_{\bar{y}(x)}^2 = \frac{1}{n}\sum_{i=1}^{k} m_i\left(\bar{y}_i - \bar{y}\right)^2, \tag{3}$$

$$\bar{y} = \frac{1}{n}\sum_{i=1}^{k} m_i \bar{y}_i,\quad \bar{y}_i = \frac{1}{m_i}\sum_{j=1}^{m_i} y_{ij},$$

$$S_y^2 = \frac{1}{n}\sum_{i=1}^{k}\sum_{j=1}^{m_i}\left(y_{ij} - \overline{y}\right)^2,$$

where k - number of grouping intervals; m_i - the number of sampling points that fall in the i-th grouping interval, ($i=1, 2, ..., k$).

We note that the Spearman's and Kendall's rank coefficients well describe linear dependences. The correlation relation of the form (3) is not burdened by any additional assumptions regarding the general form of the regression dependence [3], i.e. it allows taking into account, among other things, nonlinear dependencies. It takes values from 0 (no connection) to 1 (strong interconnection).

Thus, to solve the problem of determining the hydrodynamic connection between the injection and production wells, the correlation coefficients are applicable which was noted a long time ago. However, to determine the contribution of pumping of injection wells into oil well production, calculations of such coefficients were not previously applied. To determine the applicability of the correlation analysis to this task, an experiment was performed.

The idea of the experiment is to simulate the production rate of a fictitious well with different injection ratios of real injection wells (Table 1), in calculating the correlation coefficient for each case and comparing them. It is assumed that there is a dependence between the correlation coefficients and the weight ratios of injectability.

As the investigated objects, real injection wells were taken of one of the fields of Western Siberia № 757, № 857. These wells participated in the cyclic flooding program, which only contributes to better characterization of responses (Figure 1).

Table 1. Modeling options.

Options	Well № 757	Well № 857
1	0.6	0.4
2	0.38	0.62
3	0.22	0.78

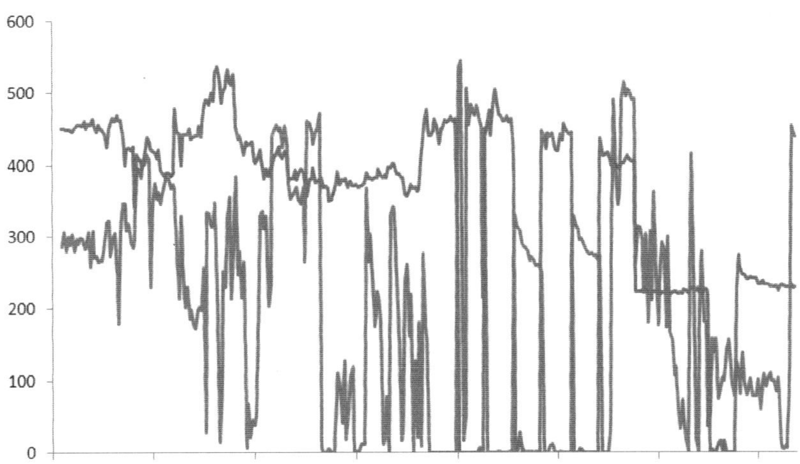

well № 757 injection capacity, m^3/day; well № 857 injection capacity, m^3/day.

Figure 1. History of injection of wells № 757, № 857.

Modeling was carried out in a hydrodynamic simulator. To minimize the influence of various factors that may affect the results, the model was created under the following conditions:

1. An infinite homogeneous formation was considered.

2. The fictitious production well was equidistant from the injection wells and was located between them.
3. The bottomhole pressure of the producing well was assumed to be unchanged.

The results for the simulated conditions are presented in Figures 2, 3, 4.

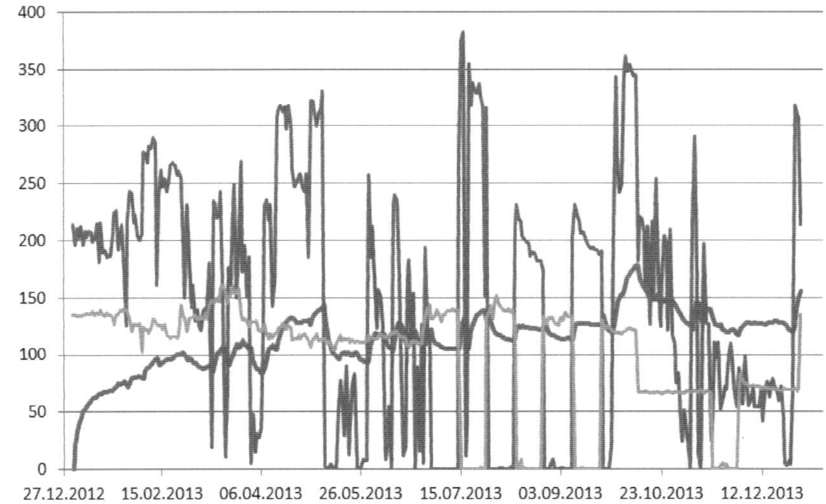

simulated production rate of a fictitious well, m^3/day;
well № 757 injection capacity (the share of self-injection - 0.7), m^3/day;
well № 857 injection capacity (the share of self-injection – 0.3), m^3/day.

Figure 2. The simulated production curve at the injection ratio of 0.6/0.4 of wells № 757 and № 857, respectively.

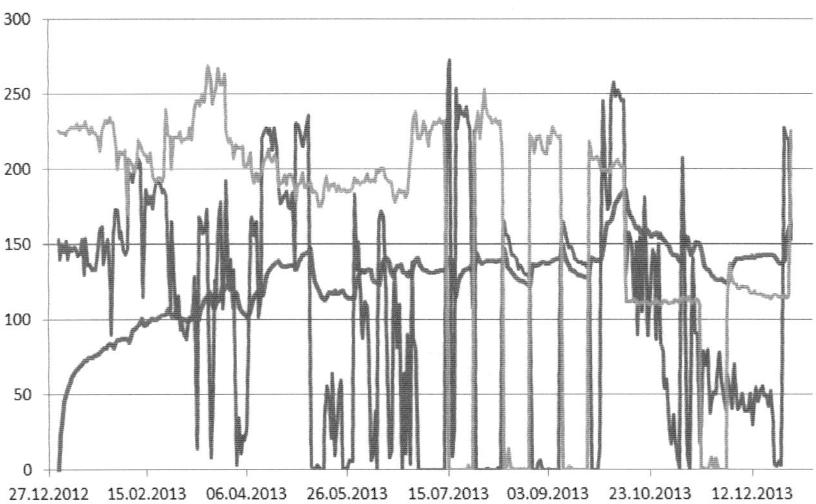

simulated production rate of a fictitious well, m^3/day;
well № 757 injection capacity (the share of self-injection – 0.5), m^3/day;
well № 857 injection capacity (the share of self-injection – 0.5), m^3/day.

Figure 3. The simulated production curve at the injection ratio of 0.38/0.62 of well № 757 and № 857, respectively.

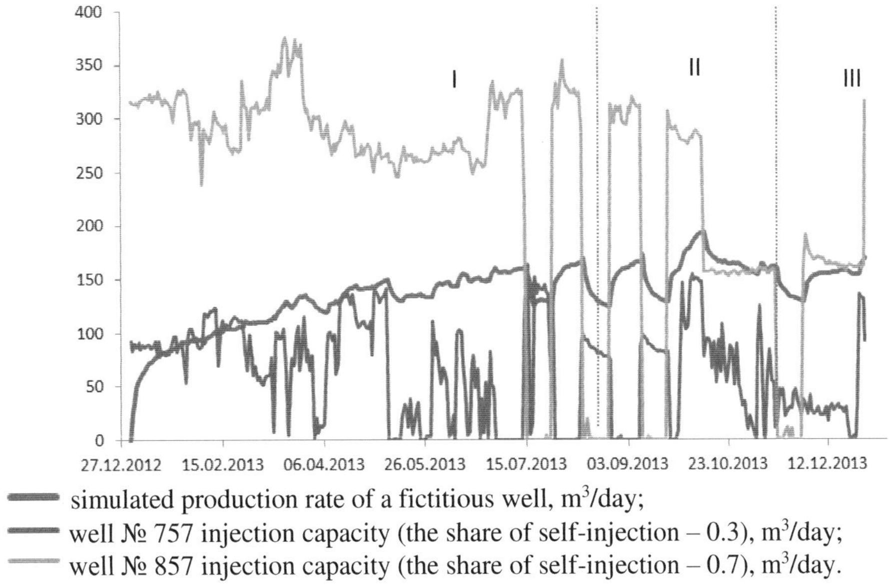

Figure 4. The simulated production curve at the injection ratio of 0.22/0.78 of wells № 757 and № 857, respectively.

On the presented diagrams it is possible to distinguish 3 characteristic sites:

1. The first half of the data (before cyclic flooding) - exponential growth of production rate is associated with simultaneous launch of all wells (Figure 4 - I).
2. Period of cyclic operation of injection wells (Figure 4- II).
3. Period after the cyclic flooding (Figure 4- III).

It is obvious that the values of the calculated correlation coefficients may differ for different parts of the data. To identify the most representative period, the calculations were carried out for different periods of time:

1. The entire period of the data under consideration.
2. Period from the beginning of the cyclic flooding to the end of the data.
3. Period with cyclic injection only.

The calculation results are presented in Tables 2, 3, 4.

Table 2. Correlation coefficients calculated at the ratio of injections of 0.6/0.4 of wells № 757 and № 857, respectively.

	Well	Spearman	Kendall	Cor. Ratio (k-means)
The entire period	757	0.18	0.13	0.47
	857	-0.59	-0.39	0.596
From the beginning of the cycle to the end of the entire period	757	0.66	0.49	0.67
	857	-0.16	-0.08	0.26
From the beginning to the end of the cycle	757	0.75	0.56	0.8
	857	-0.49	-0.34	0.21

Table 3. Correlation coefficients calculated at the ratio of injections of 0.38/0.62 of wells № 757 and № 857, respectively.

	Well	Spearman	Kendall	Cor. Ratio (k-means)
The entire period	757	-0.069	-0.04	0.47

	857	-0.39	-0.26	0.49
From the beginning of the cycle to the end of the entire period	757	0.22	0.185	0.4
	857	0.406	0.23	0.61
From the beginning to the end of the cycle	757	-0.109	-0.06	0.68
	857	0.572	0.374	0.64

Table 4. Correlation coefficients calculated at the ratio of injections of 0.22/0.78 of wells № 757 and № 857, respectively.

	Well	Spearman	Kendall	Cor. Ratio (k-means)
The entire period	757	-0.27	-0.19	0.48
	857	-0.21	-0.15	0.37
From the beginning of the cycle to the end of the entire period	757	0.014	0.006	0.28
	857	0.59	0.387	0.8
From the beginning to the end of the cycle	757	-0.33	-0.193	0.74
	857	0.702	0.465	0.85

The obtained coefficients were compared with the given injection ratios. The graph below (Figure 5) presents the average errors of different calculation methods. It can be seen from the graph that the most accurate tool is the calculation of the correlation ratio.

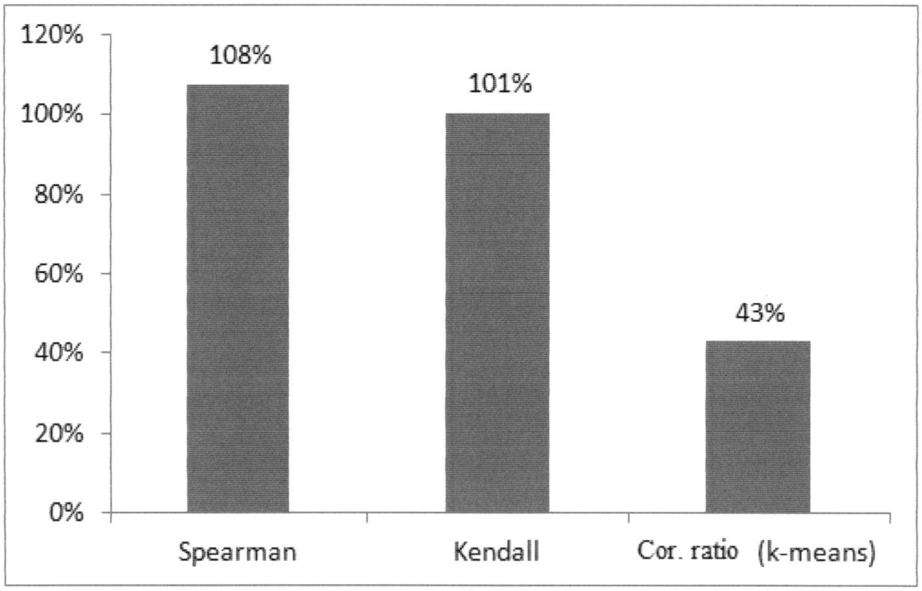

Figure 5. Average errors in the coefficients calculated in different ways.

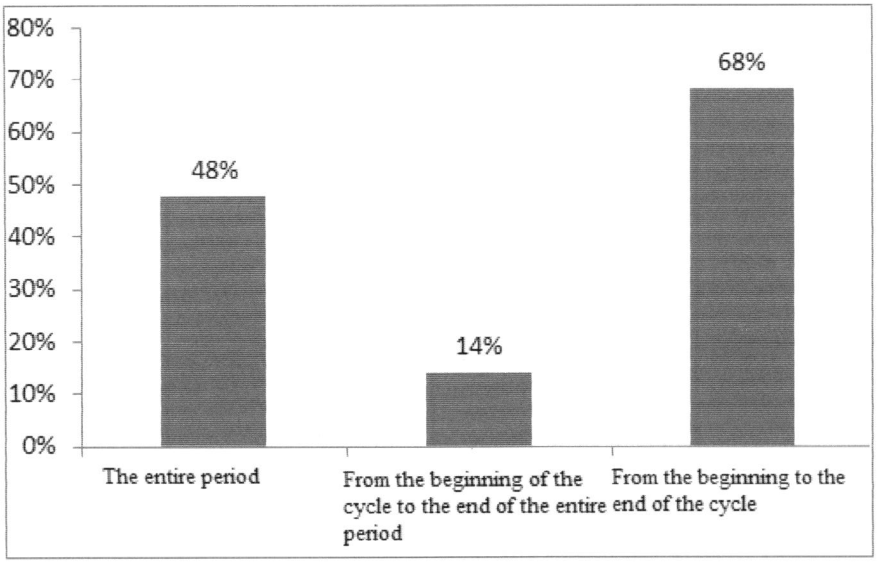

Figure 6. The average error of the correlation ratio, depending on the period of time under consideration.

Figure 6 shows the distribution of errors in the correlation ratio, depending on the data section. It can be seen that the best results were obtained in the data period from the beginning of the amplitude change in the operation mode of the injection wells to the end of the period under consideration. The average error in determining the contribution of injection wells to the productivity of the producing well by calculating the correlation ratio was about 14%.

Despite the positive nature, it should be noted that a small error and the results, in general, were obtained under fairly "ideal" conditions. When analyzing real conditions, it should be considered that, according to the basics of hydrodynamics, increase in the distance between the wells and decrease in permeability leads directly to a decrease in the connection and the quantitative estimate of the contribution.

Calculation of the correlation ratio can be considered not only an indicator of the interaction of two wells, but also a parameter with a quantitative characteristic by their interference, but without an estimate of the time of their response, the conditions will be incomplete.

One of the solutions to this problem is the grapho-analytical correlation of the operating conditions of each well, but without a well-developed mathematical base, the results are largely subjective.

To understand the mechanism for determining the response time, let us turn to the classical basics of hydrodynamic interpretation by the observation well testing. The actual result for the observation well testing is the Response time (Δt) of pressure and the Response amplitude (ΔP) of pressure (Figure 7) [4].

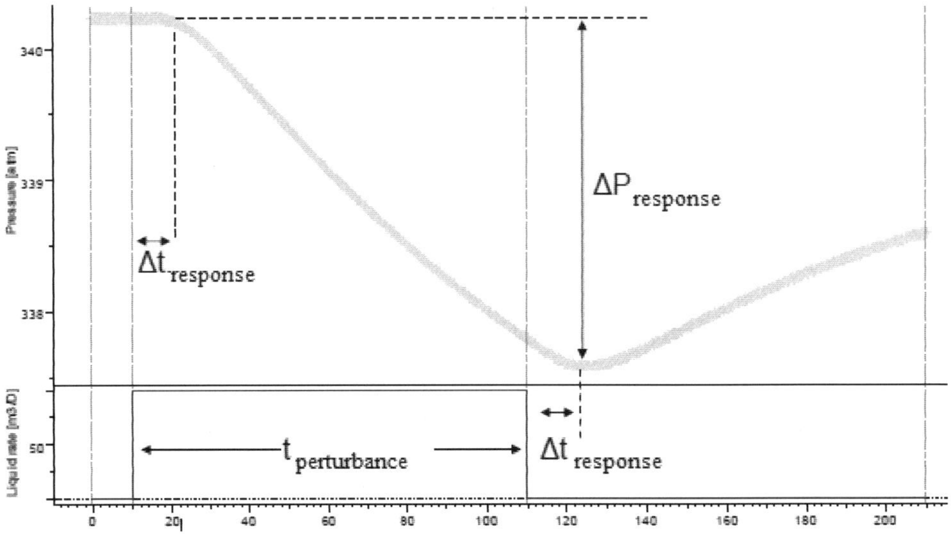

Figure 7. Response Time and Response Amplitude during the observation well testing.

It is known that the rate of pressure distribution in the reservoir is characterized by the coefficient of piezoelectric conductivity and depends on the filtration properties of the formation and fluid. In other words, the time at which the pressure perturbation front passes through the distance *r* is directly proportional to the piezoconductivity of the formation. In this case, the analytical calculation of the response time (5) also depends on the production rate and the chosen value of the pressure change *dP*, which can be fixed by measuring instruments.

$$\Delta t = \frac{1{,}781}{0{,}00144} \frac{\varphi \mu c_t r^2}{k} e^{\frac{dP \cdot kh}{9.205 q B \mu}}, \tag{5}$$

where *k* - phase formation permeability, mD; *t* - response time, h; φ - formation porosity; μ - dynamic viscosity, cPs; *ct* - total compressibility of the system, atm-1;
h - thickness h of the formation, m; *q* - flow rate, m3/day; *B* - volume ratio.

Thus, under the condition that the fixed change in pressure is very small (*dP~0*), the response time will be characterized only by the rate of pressure distribution in the formation, i.e. piezoconductivity. In this case, the estimate of the reaction time only by the piezoconductivity will be minimal.

It follows from the earlier discussion that low values of the correlation ratio can indicate both a weak pressure communication and, respectively, a small contribution, and a small contribution with a sufficiently good pressure communication.

In connection with this, it is theoretically possible to determine the reaction time of the producing well to change the injection well mode using the correlation ratio. Following is the algorithm for finding the most probable estimate of the reaction time:

1. A time interval is selected on which the accuracy of calculating the correlation ratio is maximal, i.e. field data have less noise.
2. An iterative calculation of the correlation ratios of a pair of wells (production-injection) is performed, in which, at each iteration step, the production data of the examined wells are shifted relative to each other by a certain time step. The number of iterations depends on the preliminary evaluation of the response time.
3. The values of the calculated coefficients at each step are compared with each other and the maximum value is calculated.
4. The value of the time lag at which the maximum correlation ratio was determined is considered to be the response time.

5. The results are compared with an estimate of the response time on the basis of the $\Delta t \sim \varphi \mu c_t r^2/k$ dependence using the results of the well test data.

As a confirmation of the efficiency of the developed approach, a typical analysis of a section of one of the fields in Western Siberia with the injection wells № 757, № 857 is performed. Figure 8 shows a map of the corresponding section of the formation drained by the wells №№ 269Б, 282, 283, 1048, 758, 61П, 270.

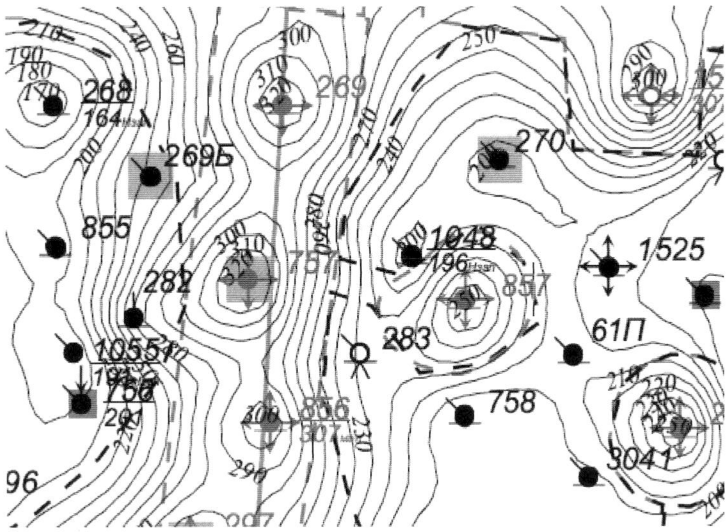

Figure 8. The site under study.

To carry out the study, the data on the surface flow rate of well № 282 (gusher) and the injectivity of wells №№ 757, 857 for 2013 (Figure 9) were taken as the initial data.

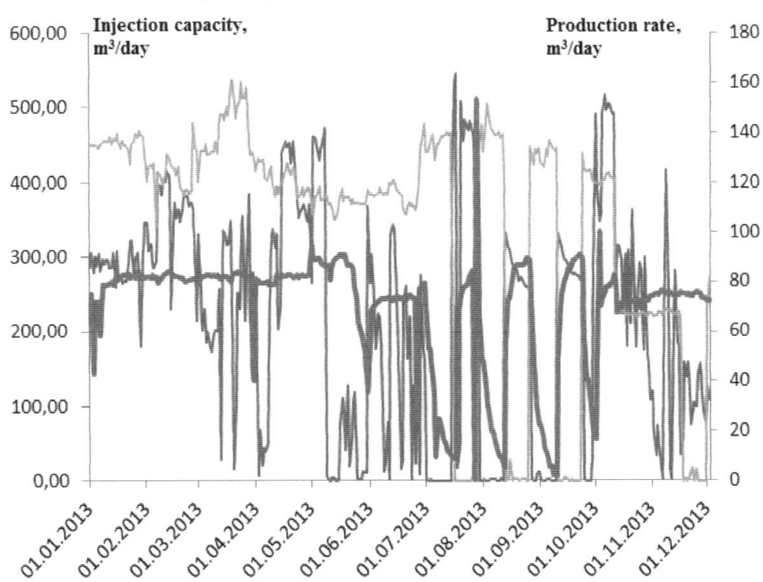

well № 757 injection capacity, m^3/day;
well № 857 injection capacity, m^3/day;
well № 282 production rate, m^3/day.

Figure 9. Dynamics of production of well № 282 and injection of wells №№ 757 and 857.

IOP Conf. Series: Earth and Environmental Science **181** (2018) 012018 doi:10.1088/1755-1315/181/1/012018

On the production schedule, there are jumps that are most likely related to the work of injection wells. This assumption confirms the coincidence of the periods of fluctuations in production rates and injectivity. To determine which of the two injection wells has the strongest effect, a correlation analysis was performed to determine the time delays

The preliminary evaluation of the maximum response time for wells № 857 and № 757 through the piezoelectric conductivity of the formation was 324 and 45 days, respectively.

To determine the actual response time, the corresponding calculations were performed based on the previously developed algorithm. Below is a graph of the dependence of the calculated correlation coefficients on the time delay (Figure 10).

According to the calculations, the following results were obtained:

Table 5. Calculation results.

Injector	Correlation ratio	Response time, d.
№ 757	0.721	2
№ 857	0.461	16

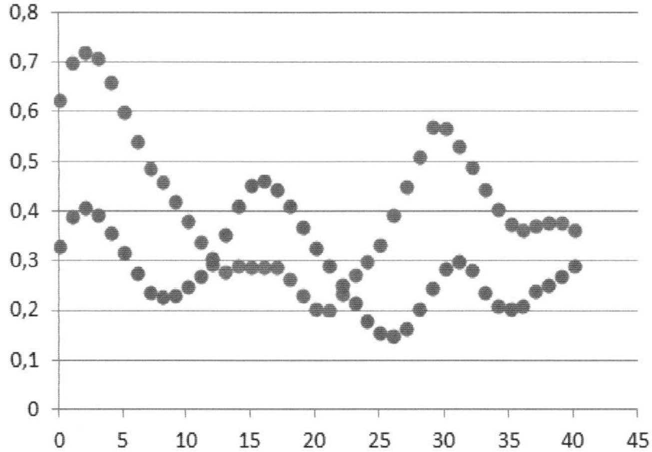

● correlation ratio between production rate of well № 282 and injection capacity of well № 757, m^3/day;

● correlation ratio between production rate of well №282 and injection capacity of well № 857, m^3/day;

Figure 10. Dependence of the correlation ratio on the response time.

The correlation ratio between production rate of well № 282 and injection capacity of well № 757 was 0.721, which indicates a good pressure communication between these wells and the predominant contribution of well № 757 to the production of well № 282. The low value of the correlation coefficient (0.461) between wells № 282 and № 857 may indicate a weaker effect of the injection well, while the pressure communication may be good enough. As a percentage, the contribution of well № 757 is about 61%, well № 857 - 39%, which is consistent with their location relative to each other.

The results of calculations for the remaining wells in the section under consideration are shown in Table 6.

Table 6. The calculation results for a group of wells.

Injector / observation well	№ 757		№ 857	
	Correlation ratio	Response time, d.	Correlation ratio	Response time, d.

283	0.41	17	0.6	32
282	0.62	2	0.5	47
269Б	0.4	18	0.54	102
1048	0.86	24	0.91	4
758	0.57	60	0.73	18
61П	0.47	62	0.57	32
270	0.4	70	0.43	33

The obtained results, generally, are consistent with the operation modes and their mutual position. Figure 11 summarizes and graphically depicts the interference of wells in the area under consideration.

In the graph, the arrows indicate the direction of connection between the wells. The difference in color indicates the gradation of the calculated correlation ratios. It should be noted that it is most useful to compare the values in the case when the calculated correlation coefficients refer to one production well.

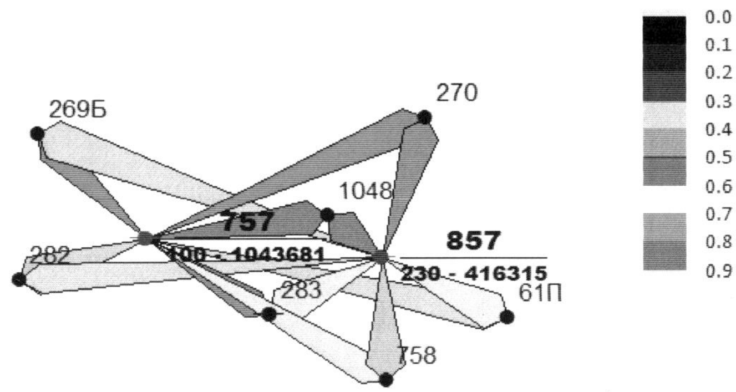

Figure 11. Wells interference.

4. Conclusion

In solving the problem of analyzing well interference, a complex mathematical analysis approach was developed to analyze and process field data, based on the principles of mathematical statistics.

A) A numerical experiment was performed to substantiate the approach, based mainly on the mathematical principles of correlation methods of data processing.

B) An algorithm was developed for estimating the reaction time, estimating the pressure communication and the contribution of injection wells to the production of adjacent oil wells (correlation ratio).

C) Response time and correlation ratio for wells of the selected section of the Western Siberia deposit were estimated.

References

[1] Heffer Kes J, Fox Richard J, McGill Colin A and Koutsabeloulis Nick C 1997 *Novel Techniques Show Links between Reservoir Flow Directionality, Earth Stress, Fault Structure and Geomechanical Changes in Mature Waterfloods* SPE № 30711-PA.

[2] Refunjol Belkis T and Lake Larry W 1997 Reservoir characterization based on tracer response and rank analysis of production and injection rates *Forth International Reservoir Characterization Technical Conference* (Houston, Texas)

[3] Aivazyan S A, Enyukov I S and Meshalkin L D 1985 *Practical Statistics. Dependencies Study* (Moscow. Finance and Statistics) 487 p

[4] Houze Olivier et al 2012 *Dynamic Data Analysis - v.4.12.03.* (Kappa)

A method of modelling phase permeability in the region of a three-phase filtration

M G Lozhkin[1], S F Mulyavin[2] and E A Mulyavina[2]

[1]TyumenNIIgiprogaz, Tyumen, Russia
[2]Industrial University of Tyumen, Tyumen, Russia

Abstract: The article analyses the models of relative phase permeabilities in the region of a three-phase core saturation. Existing models of relative phase permeabilities in the region of a three-phase core saturation yield a result much rougher than the model described in the works of other authors. The proposed model in this study gives a deviation from the experimental data up to 5%, and the Stone model - up to 70%.

1. Introduction

The verification of the adequacy of the mathematical model of relative phase permeabilities in the three-phase core saturation region implies an experimental determination of permeability values at points with a three-phase core saturation.

The experimental determination of the relative phase permeabilities with a three-phase core saturation still remains an insufficiently studied field of petrophysics. Publications about such studies are few, and their results are ambiguous due to the effect of core drainage [1]. In this paper, we propose a method for obtaining relative core permeabilities in the three-phase saturation region, taking into account the permeability hysteresis as a function of the direction of saturation variation, and an analysis is made of the applicability of various mathematical models for these permeabilities.

2. Research

At present, the main method used to determine the relative phase permeabilities in the region of three-phase core saturation is the method of combined stationary filtration [2]. Despite the fact that the concept of phase permeabilities, laid in the foundations of this method, was formulated for the first time in 1943 and has since been repeatedly confirmed, the physical foundations of this phenomenon are still not fully understood. As a consequence, studies of relative phase permeability in the three-phase core saturation region do not reflect the actual processes occurring in the formation. Below are the arguments in favour of the previous statement.

M. Muskat and others [3] hypothesized that in the simultaneous flow of immiscible liquids in a porous space, two essentially different filtration modes can exist.

In one of these modes, a series of channels of the porous medium remain occupied by the rock wetting fluid, while others are occupied by the fluid immiscible with the first one. Different permeability for the first and second fluids corresponds to different saturations of these fluids, while the fluids in the pores form a continuous flow for each of the filtering phases. It follows from this theoretical assumption that for each of the fluid saturations there is only one permeability for a given fluid. It does not matter which fluid or fluids are filtered together with the studied one, since the interaction between them is assumed only through the fraction of pore of the porous medium occupied

Content from this work may be used under the terms of the Creative Commons Attribution 3.0 licence. Any further distribution of this work must maintain attribution to the author(s) and the title of the work, journal citation and DOI.
Published under licence by IOP Publishing Ltd

IOP Conf. Series: Earth and Environmental Science **181** (2018) 012019 doi:10.1088/1755-1315/181/1/012019

by each of them. They flow, as if each is in its own system of tubes, without interacting with each other.

The concept of the second, slug filtration mode, in which intermittent portions of immiscible fluids move in a porous medium, presupposes a radically different flow pattern, which implies the emulsification of fluids into each other and the filtration of the resulting emulsion. The degree of dispersion of this emulsion depends on the structure and properties of the porous medium and the pressure gradient arising during filtration. So, the higher the hydrodynamic pressure gradient, the more likely the formation of fluid slugs with high capillary pressure gradients, which depends on the size of the slug and the capillary pressures arising on its front and end. When slugs with gradients of capillary pressures above the gradient of hydrodynamic pressure appear, the pore with such a slug ceases largely to be permeable.

Thus, experiments in which processes that actually occur on the objects under study cannot be taken into account and cannot be used to describe them. So, when determining the relative phase permeabilities in the region of three-phase saturation by the method of stationary filtration [2], a three-phase flow of fluids is fed into the core sample and their stationary filtration is expected. However, the fact that in natural conditions this is impossible is ignored. If for two-phase filtration there is no probability that the saturation of the porous medium does not correspond to the displacement process under investigation, for a three-phase flow this is not so. Since in reality, for example, an oil field is developed first in the regime of dissolved gas and then using water flooding, it is necessary to carry out measurements in which oil is displaced first by gas and then by water. The case of joint flow will give very high pressure gradients, which in reality are impossible.

For the same core sample, the relative phase permeabilities for the same saturations will differ during draining and impregnation. If the samples were initially saturated with a wetting fluid, and it is displaced by a nonwetting fluid, then this process is called drainage (for example, the same process occurred during the formation of oil or gas deposits). When draining, filtration of the wetting fluid in small capillaries lags behind the nonwetting fluid in large capillaries, since it is established [6] that, during draining, the nonwetting fluid moves first along the large pores. Therefore, initially, during draining, predominantly separate phase flows are formed which do not substantially affect each other. The relative phase permeabilities for the described process are located more to the right then during impregnation, and the permeability values for the nonwetting fluid are generally higher, at the same saturation values, the reverse situation is observed for the wetting fluid (Figure 1).

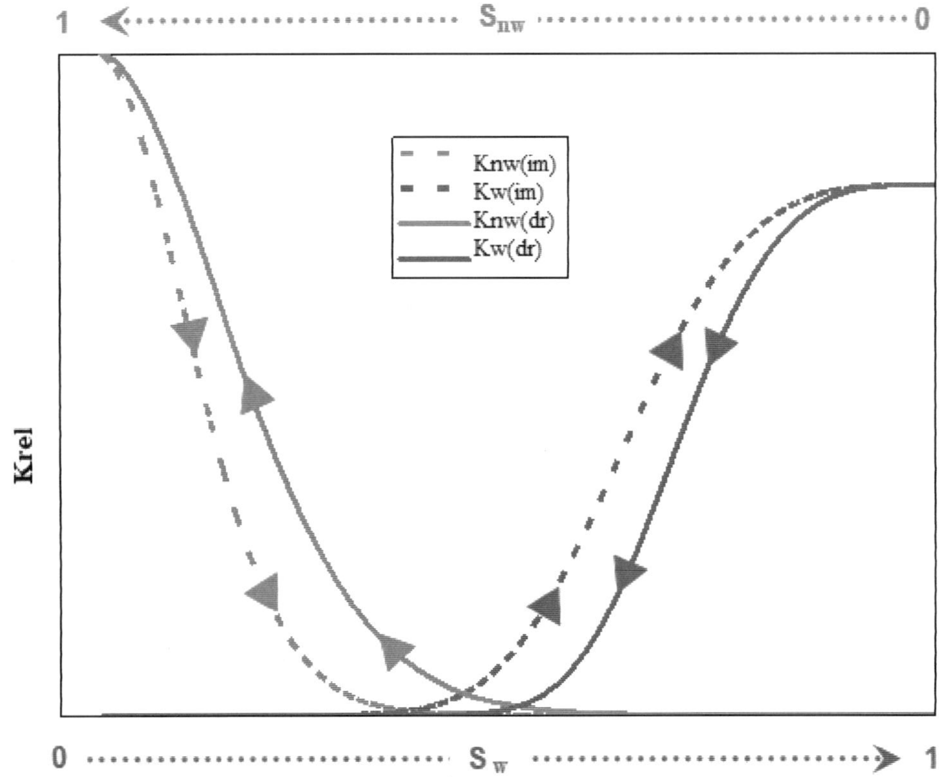

Figure 1. Hysteresis of relative phase permeabilities during impregnation and drainage. Sw - saturation with wetting phase, Knw - saturation with nonwetting phase, Krel - relative phase permeability (RPP), Knw (impr) - RPP for nonwetting phase during impregnation, Kw (impr) - RPP for wetting phase during impregnation, Knw (dr) - RPC for nonwetting phase during drainage, Kw (dr) - RPP for wetting phase during drainage [6].

The experiments to determine the relative phase permeabilities in the gas-condensate-water system are as follows. At the first stage, the RPP is determined by the method of stationary filtration, with different proportions of water and condensate. Further, on the same core column, oil is displaced with water, pumping 5-10 pore volumes of the displacing agent [5-12].

1. The oil model was a mixture of 64% of the volume fraction of degassed oil in the Urengoy field deposits and 36% of the volume fraction of normal hexane. The proportion in the mixture was selected according to the density of the reservoir oil. As a gas model, chemically pure nitrogen moistened in a gas container with distilled water was used. For the experiments, the oil model was saturated with nitrogen, and after loading into the separator, it was filtered through core samples (selected for preparation of fluids) together with nitrogen. The filtration was carried out under given thermobaric conditions simulating the formation. As a result, gas and oil in the core came to a state of thermodynamic equilibrium, that is, the mass transfer between the fluids and the core became negligible. Then, the viscosity of the fluids was determined by pumping them through the capillaries. Thus, preparation of the fluids for the experiment was completed.
2. Preparation of core samples, creation of residual water saturation on them. Samples were prepared as follows. Samples were cut from the core with a crown 30 mm in diameter. The samples were then extracted in Soxhlet apparatus with an alcohol-benzene mixture prior to extracting all hydrocarbons. The samples were then dried at a temperature of 105 °C. After

that, they were washed from the salts in distilled water and dried again. The samples were then saturated with a model of formation water, mineralized with sodium chloride at a concentration of 12 g/cm3. After saturation of the samples, their electrical resistance and permeability at 100% water saturation were determined to obtain the dependence of resistance on water saturation (first point). To create a residual water saturation, a Jouan KR-25i centrifuge was used. Residual water saturation was determined at a capillary pressure of 0.3 MPa, which corresponds to a rotation speed of 6310 rpm. The sample, previously weighed in a dry state, was completely saturated with a model of formation water and weighed in a saturated state. After that, it was centrifuged for 1 hour at the above-mentioned rotation frequency. After the centrifugation, the sample was weighed again. Then the sample was saturated with kerosene. Subsequently, after all kerosene was replaced by oil, the electrical resistance of the core was determined with the residual water saturation (second point). Two values of the resistance at a certain saturation are sufficient for constructing the dependence of resistance on saturation.

3. Loading the core into the installation.
4. Measurement of the permeability of the core for gas at zero saturation with gas.
5. Injection of 0.2 pore volume of condensate into the core and its distribution by volume of the core sample, pumping gas until the pressure drop stabilizes.
6. Phase permeabilities for gas are obtained by the method of pseudostationary filtration.
7. Gas displacement by condensate, determination of permeability for condensate, with residual gas saturation. Saturation by condensate is determined by the method of material balance.
8. Preparation of core samples with the creation of water saturation, exceeding the previous value by 5%, using the semipermeable membrane in an individual capillaryimeter.
9. Control the immobility of water saturation by pumping gas and determine the electrical resistance of core samples. The electrical resistance of samples during the oil filtration process should not increase.
10. Repeat steps 3-8 until mobile water appears. The mobile water is determined by the growth of electrical resistance.
11. The relative phase permeability for the gas-water system is determined on the same core sample. Based on the results of the determination, a three-phase model of relative phase permeabilities is constructed [4].

Carrying out experiments in this order will allow obtaining the relative phase permeabilities in the gas-condensate-water system and a mathematical model of the relative phase permeabilities in the same system for their mutual comparison.

3. Results and discussion

Based on this, it is required to carry out experiments in such a way that they exactly repeat the processes occurring in the deposits under natural conditions. Thus, for the displacement of oil with gas, water saturation at the same distance from the oil-water contact remains constant. And when approaching the oil-water contact, water saturation increases. Therefore, in order to predict the extraction of gas and oil from wells in the oil and gas condensate field (for example, the Urengoy OGCF), it is necessary to determine the relative phase permeabilities for oil and gas at various stationary saturations by water, that is, using the stationary phase method, where the stationary phase is water.

Thus, the experiments on determining relative phase permeabilities in the gas-oil-water system, when developing an oil and gas condensate field, are as follows.

On the core, the relative phase permeabilities for the oil-water system are measured with residual water saturation. On the same core, the relative phase permeabilities for the oil-gas system are then determined with residual water saturation. The third experiment (on the same core) consisted in determining the relative phase permeabilities for the oil-gas system, with various stationary water saturations (0.167, 0.207, 0.233) created by capillarimetry.

Phase permeabilities in the gas-condensate-water system for three-phase filtration are constructed from the results of the first two experiments in the form of functional (exponential) dependences :

$$k_{ro} = e^{E_{ow}\left(\frac{S_w - S_{wi}}{1-S_{wi}}\right)^{P_{ow}} + E_{og}\left(\frac{S_g}{1-S_{wi}}\right)^{P_{og}}}$$

where: k_{ro}– relative phase permeability for oil, in the oil-gas-water system, fr.unit; S_w – water saturation, fr.unit; S_g – gas saturation, fr.unit; S_{wi} – residual water saturation, fr.unit; E_{ow} – exponential coefficient of dependence for the oil-water system; P_{ow} – power-law coefficient of dependence for the oil-water system, E_{og} – exponential coefficient of dependence for the oil-gas system, P_{og} – power-law coefficient of dependence for the oil-gas system.

According to the method presented above, relative phase permeabilities were determined on the core of the Urengoi OGCF. Experiments were conducted to determine the relative phase permeabilities in the oil-water and oil-gas system for the Urengoy OGCF (Table 1). Based on the measurement results, a functional three-phase model of relative phase permeabilities is constructed, which has the form

$$k_{ro} = e^{-22{,}21\left(\frac{S_w - 0{,}13}{1-0{,}13}\right)^{2{,}126} - 20{,}33\left(\frac{S_g}{1-0{,}13}\right)^{1{,}7}}$$

$$k_{rg} = e^{-88{,}34\left(\frac{S_w - 0{,}13}{1-0{,}13}\right)^{4{,}228} - 17{,}06\left(\frac{S_o}{1-0{,}13}\right)^{4{,}135}}$$

$$k_{rw} = 0{,}741 e^{-5{,}945\left(\frac{S_o}{1-0{,}13}\right)^{0{,}836}} e^{-26{,}31\left(\frac{S_g}{1-0{,}13}\right)^{2{,}662}}$$

Table 2 provides a comparison of the results corresponding to the phase permeabilities constructed using the proposed method and the Stone method with the results of experimental determination of the relative phase permeabilities in the oil-gas system for various stationary water saturation.

Table 1. Calculation of the relative permeability model for oil in the three-phase saturation region in the oil-gas-water system.

Determination of relative phase permeabilities and coefficients P_{ow} and E_{ow} in the oil-water system							
S_w, fr.unit	S_o, fr.unit	k_{ro}, fr.unit	k_{rw}, fr.unit	S'_w	K'_o	S'_o	K'_w
0.130	0.870	1.000	0.000				
0.268	0.732	0.650	0.000	-1.844	-0.842		
0.311	0.689	0.442	0.004	-1.570	-0.202	-0.233	1.631
0.377	0.623	0.219	0.008	-1.260	0.419	-0.334	1.507
0.440	0.560	0.085	0.013	-1.033	0.902	-0.440	1.395
0.506	0.494	0.024	0.020	-0.839	1.317	-0.566	1.280
0.558	0.442	0.007	0.028	-0.710	1.591	-0.677	1.191
0.609	0.391	0.002	0.036	-0.596	1.831	-0.801	1.104
0.683	0.317	0.000	0.052			-1.010	0.979
1.000	0.000		0.741				
				P_{ow}	E_{ow}	P_{wo}	E_{wo}
Model Parameters				2.126	-22.21	0.8356	-5.947
Determination of relative phase permeabilities and coefficients P_{og} and E_{og} in the oil-gas system							
S_g, fr.unit	S_o, fr.unit	k_{ro}, fr.unit	k_{rg}, fr.unit	S'_g	K'_o	S'_o	K'_g
0.000	0.870	1.000	0.000				
0.027	0.843	0.947	0.000	-3.457	-2.918		
0.068	0.802	0.738	0.000	-2.550	-1.192	-0.081	2.472
0.108	0.762	0.552	0.000	-2.082	-0.519	-0.133	2.318
0.149	0.721	0.388	0.000	-1.765	-0.054	-0.188	2.137
0.189	0.681	0.247	0.001	-1.525	0.336	-0.246	1.914
0.230	0.640	0.129	0.006	-1.331	0.719	-0.307	1.628
0.258	0.612	0.059	0.020	-1.214	1.039	-0.352	1.363

0.287	0.583	0.000	0.066			-0.400	1.001
				P_{og}	E_{og}	P_{go}	E_{go}
Model Parameters				1.700	-20.33	4.498	-18.72

Table 2. Comparison of the relative permeability model for oil in the three-phase saturation region in the oil-gas-water system with experimental data.

S_w, fr.unit	S_o, fr.unit	S_g, fr.unit	k_{ro}, fr.unit	$k_{ro}(L)$, fr.unit	$k_{ro}(S1)$, fr.unit	$\Delta k_{ro}(L)$, %	$\Delta k_{ro}(S1)$, %
0.167	0.833	0.000	0.958	0.973	0.982	1.6	2.5
0.167	0.728	0.105	0.572	0.557	0.688	2.7	20.2
0.167	0.701	0.132	0.417	0.425	0.559	2.0	34.1
0.167	0.687	0.146	0.361	0.365	0.496	1.1	37.4
0.167	0.660	0.173	0.267	0.265	0.385	0.9	44.1
0.167	0.644	0.189	0.204	0.213	0.324	4.3	58.6
0.167	0.593	0.240	0.101	0.100	0.177	0.4	75.9
0.167	0.566	0.267	0.000	0.000	0.000	-	-
0.207	0.793	0.000	0.858	0.880	0.862	2.5	0.5
0.207	0.723	0.070	0.650	0.667	0.763	2.5	17.2
0.207	0.692	0.101	0.536	0.524	0.640	2.3	19.5
0.207	0.678	0.115	0.457	0.461	0.581	0.9	27.2
0.207	0.650	0.143	0.344	0.343	0.463	0.2	34.5
0.207	0.631	0.162	0.278	0.274	0.388	1.5	39.4
0.207	0.577	0.216	0.132	0.132	0.216	0.3	63.4
0.207	0.559	0.234	0.000	0.000	0.000	-	-
0.233	0.767	0.000	0.789	0.788	0.797	0.1	1.1
0.233	0.720	0.047	0.702	0.682	0.746	2.7	6.4
0.233	0.688	0.079	0.578	0.559	0.652	3.2	12.9
0.233	0.674	0.093	0.491	0.500	0.601	1.9	22.6
0.233	0.644	0.123	0.377	0.380	0.489	0.8	29.6
0.233	0.624	0.143	0.306	0.308	0.415	0.6	35.6
0.233	0.569	0.198	0.144	0.152	0.237	5.0	63.9
0.233	0.557	0.210	0.000	0.000	0.000	-	-

Where k_{ro} – experimental value of permeability; $k_{ro}(L)$ – value of permeability obtained during modelling using a method described in [4]; $k_{ro}(S1)$ – value of permeability obtained during modelling using a Stone method; $\Delta k_{ro}(L)$, $\Delta k_{ro}(S1)$ – deviations of the corresponding values of the model from the experimental data .

4. Conclusion

Thus, the existing models of relative phase permeabilities in the region of three-phase core saturation yield a result much rougher than the model described in [4]. The proposed model gives a deviation from the experimental data up to 5%, and the Stone model - up to 70%.

References

[1] Lozhkin M G 2016 Core draining in determining its relative phase permeability by the method of combined stationary filtration *Exposition Oil Gas* p 38

[2] OST 39-235-89 Oil. Method for determination of phase permeabilities in laboratory conditions with simultaneous stationary filtration

[3] Muskat M, Chatenever A and Hacherman N 1953 The displacement of oil from porous media. *Proc. A.P.I., The Oil and Gas J.* VI 33(28) p 238

[4] Lozhkin M G 2015 Model of relative phase permeabilities for gas displacement by condensate and water and displacement of oil by water and gas *Exposition oil gas* 1 39-41

[5] Lozhkin M G 2015 Method for determining the relative phase permeabilities in laboratory conditions under sequential pseudostationary filtration *Exposition oil gas* 7 51-53

[6] Tjolsen C B, Scheie A and Damsleth E A study of the correlation between relative permeability, air permeability and depositional environment on the core-plug scale *SCA Symposium on*

Core Analysis (London, UK) 1 (9110Euro) 17 p
[7] Dobrynin V M, Kovalev L G and Kuznetsov A M 1988 *Phase Permeabilities of Oil and Gas Reservoirs* (Moscow: VNIIOENG) 55 p
[8] Collins R 1964 *Flow of Liquids through Porous Materials* (Moscow: Mir) 350 p
[9] Surgutchev M L, Zheltov Yu V and Simkin E M 1984 *Physicochemical Microprocesses in Oil and Gas Bearing Strata* (Moscow: Nedra) 215 p
[10] Barenblatt G I, Entov V M and Ryzhik V M 1984 *Movement of Liquids and Gases in Natural Formations* (Moscow: Nedra) 207 p
[11] Efros D A 1956 *Determination of Relative Phase Permeabilities and Distribution Functions in the Displacement of Oil by Water* (DAN SSSR) 110(5)

IOP Conf. Series: Earth and Environmental Science **181** (2018) 012020 doi:10.1088/1755-1315/181/1/012020

Sound vibration modulator to control turbodrill operation

Kh N Muzipov, Yu Ovchinnikova and N V Zykova

Industrial University of Tyumen, 38 Volodarskogo Str., Tyumen, 625000, Russia

E-mail: muzipovhn@tyuiu.ru

Abstract: The paper describes a sound vibration modulator to control turbodrill revolutions during turbine motor drilling. The principle of the modulator operation is based on the quarter-wave acoustic resonator property to suppress noises at the resonant frequency. A resonant chamber, which is a quarter-wave acoustic resonator, is embedded into the sound vibration modulator to control the number of turbodrill revolutions. The resonant chamber is a hollow cylinder with two cavities: an internal closed annular cavity and a central cylindrical cavity. During each revolution of the turbodrill shaft absorption of sound energy occurs at a frequency equal to the resonant chamber frequency, i.e. it creates a wave break. The number of these breaks will be equal to the number of turbodrill revolutions.

1. Introduction

Drilling efficiency depends on a group of factors: axial load on the bit, bit rotation frequency, drilling mud circulation rate and parameters of the drilling mud quality, a bit type, geological conditions, mechanical properties of rocks.

They distinguish the drilling practices parameters which can be changed from the driller's panel during the bit operation in the bottom hole, and the factors established at the well construction design stage, some of which cannot be quickly changed. The first ones are called managed ones. Their certain combination, when mechanical well drilling is carried out, is called drilling practices.

During well drilling there is a necessity to monitor the drilling site day-and-night; and, if required, immediately bring in necessary adjustments to the process of well drilling.

The information obtained from the bottomhole telemetering means of control which are under specific, quite hard conditions, takes an important place in the control parameters solution during turbine motor drilling.

For many years, the main obstacle for practical use of measurements in the drilling process was the communication channel. It is the main and decisive factor, as it determines the MVD (measure while drilling) design, arrangement, information content, reliability, convenience of operation, as well as the conditions of passing signals.

The main requirements to the downhole information include increasing the reliability of the measuring signal, which is transmitted to the surface through the drill string and the flow of drilling fluid under conditions of high vibration, pulsation and noises generated by the drilling rig injection pump line.

Therefore, the problem of improving measurement and transmission of operating parameters when drilling wells has always been relevant.

The search for solutions to this problem aims at a development of new technical means for measuring operating parameters and development of a noise-proof communication channel, which is

Content from this work may be used under the terms of the Creative Commons Attribution 3.0 licence. Any further distribution of this work must maintain attribution to the author(s) and the title of the work, journal citation and DOI.
Published under licence by IOP Publishing Ltd

IOP Conf. Series: Earth and Environmental Science **181** (2018) 012020 doi:10.1088/1755-1315/181/1/012020

to provide telemetering of downhole parameters with the required accuracy and to be reliable for a long operating time to control drilling parameters.

In the course of turbine motor drilling, two sources of sound occur in the bottomhole: one - from the turbodrill and the other - from the bit. The turbodrill generates the sound into the inner cavity of the pipe; and the bit generates it not only into the drill string, but also into the drilling fluid of the annular space.

2. Research

The currently available range of the communication channels is very wide and represented by hydraulic, electromagnetic, acoustic, electrically conductive and many other types of communication channels.

As a result of long-term research work and practical use in real drilling conditions the following three communication channels are widely used:

- electrically conductive (ECCC);
- hydraulic (HCC);
- electromagnetic (EMCC).

Each of these communication channels has its advantages and disadvantages. The diversity of drilling conditions and economic feasibility designate its field of application for each communication channel.

By today various methods of information transfer have been developed, both in our country and abroad [1], [2], [3], [4]. Some of them are based on the analysis of vibration displacement of the drill string upper part and do not require any use of downhole automation devices.

The analysis of the existing channels of communication with the wellhead has shown that the most promising in terms of reliability and efficiency is the hydraulic communication channel (HCC), with relative simplicity of transmitting and receiving devices, and absence of any additional communication line built in the drilling tool.

MWD with HCC differs from others having a device that creates pressure pulses in the drilling fluid flow which represent information about the drilling practices. Several devices of different type are used to generate pressure pulses in the drilling fluid. The signal generated by them is divided into three types: positive pulse, negative pulse or continuous wave.

Positive pulses are generated by creating a short-term partial overlap of the drilling fluid downflow. Negative pulses are generated by short-term liquid bypasses into the annular space through the side valve. Hydraulic signals, close to harmonic ones, are generated by an electric motor that rotates the pulsator valve. Hydraulic pulses come to the surface through the drilling liquid column at a speed of about 1250 m/s, where the information, encoded in various ways, is decoded and displayed in a form acceptable for the operator.

The preference in use of MWD with HCC is based on both a relative simplicity of communication compared to other communication channels and the fact that this channel does not break down (compared to ECCC) technological operations during drilling and does not depend on the geological cross-section (compared to EMCC). The disadvantages of this communication channel are: the low information content due to the relatively low transmission rate, low noise immunity, consistency in information transmission, requirement of an electric power source (battery, turbo-generator), hydraulic energy outfeed for transmitter and turbo-generator operation and impossibility to work with air blowing and aerated liquids.

In this paper the authors suggest to use the hydraulic communication channel to transmit the acoustic signals that carry information about drilling parameters rather than pressure waves.

The authors of the paper [1] suggest to form a bit vibration noise frequency absorption band by means of an acoustic resonator at the bottomhole in the frequency spectrum of annular space elastic vibrations, and the sound vibration of the drilling unit body should be absorbed by a local vibration absorber. In this case the frequency absorption band corresponds to the downhole information transmission band.

Theoretical and experimental studies of the acoustic communication channel carried out by Dozorova T. A. [5], [6], Mikhailovsky V. N. [7], Shumilovsky N. M. [8] and Yamaliev V. U. [9] allowed to establish a frequency range of sound waves (0 – 10000 Hz) in the well and to estimate the longitudinal waves absorption coefficient in the drill string filled with clay mud (with density of 1.15 g/cm3), which ranges from 4.2 to 5.2 dB/1000 m with an mean value of 4.7 dB/1000 m. The sound level of operating roller cone bits is in the range of 75 – 95 dB and the channel coverage is 5 – 7 km at frequencies of 150 – 200 Hz.

The occurrence of sound waves generated by the drilling tools can be used to transmit information about the operating conditions of the turbine motor drilling.

But the information about drilling parameters is overlapped by interference in the form of noises generated by the bit and drilling pumps.

To suppress the noise generated by the bit a quarter-wave acoustic resonator, the resonant frequency of which is equal to the bit noise frequency, is suggested to be used.

Acoustic resonators can be used to amplify or attenuate a certain range of acoustic frequencies. The most vivid example of the acoustic "amplifier" is the acoustic system phase inverter which is the same Helmholtz resonator excited "from inside". If the Helmholtz resonator is excited from outside, it becomes an elimination (suppressive) filter that absorbs the energy of external vibrations.

3. Results and discussion

One of the new trends in the application of acoustics in the oil industry is the use of technological noises as useful signals and energy sources. As it is known, operation of any equipment is accompanied by noises that are not useful.

Every mechanical elastic system has its own vibration frequency. If any force brings this system out of balance and then ceases to function, the system will oscillate for some time near its equilibrium position. The frequency of these oscillations is called natural oscillation frequency of the system. The rate of its attenuation depends on the elastic properties, mass and friction forces and does not depend on the force that caused the oscillations.

Full resonance occurs at exact coincidence of the force oscillations frequency and the natural oscillations frequency of a structure and equal positive and negative amplitudes. Partial resonance occurs at incomplete coincidence of the frequencies and unequal amplitudes [1].

The analysis of the existing channels of communication with the wellhead has shown that the most promising in terms of reliability and efficiency is the hydraulic communication channel which transmits acoustic signals and has relative simplicity of transmitting and receiving devices, absence of any additional communication line built in the drilling tool.

4 ways of obtaining information have been developed as sources of information about the turbodrill shaft revolution frequency:

- a formation of a zone of noise suppression with shock pulses and radiation of elastic vibrations between the shock pulses [10];
- a formation of a wave break with acoustic elimination filter (resonator) [11];
- a formation of wave breaks by pressing the damper to the drill string [12];
- modulation of the drilling tool noise with a rotating body. During the turbodrill shaft revolution a change of a sound wave direction occurs, resulting in a reflection of sound energy with a wave break formation [13].

An acoustic filter is an acoustic resonator made in a form of a vessel open to the external environment through a small hole or a tube. A characteristic feature is an ability to perform low-frequency self-oscillations, the wavelength of which is much longer than the size of the resonator. According to the Helmholtz and Rayleigh theory, an acoustic resonator is regarded as an oscillation system with one degree of freedom.

The natural frequency of an acoustic resonator does not depend on the shape of the vessel and the cross-sectional shape of the pipe and is expressed by the formula [14]:

$$f = \frac{c}{2\pi}\sqrt{\frac{F}{Vh}},$$

where: c is a sound velocity in a medium (drilling mud), m/s;
V is a resonator volume, m^3;
F and h are a cross-section square, m^2 and a pipe length, m.

The quarter-wave resonator is a lateral branch line of the tubular cross-section with a closed end. At the frequency when a quarter of the muffled sound wavelength fits the length of the pipe, the sound wave, getting into the branch line, is reflected from the closed end back into the pipeline, but with the opposite phase in relation to the incident wave. The interference of the two waves of the same frequency, but opposite by phase causes an attenuation of the oscillating process. The resonant frequency at which maximum efficiency of the adjusted pipe occurs is determined by the ratio:

$$f = \frac{c}{4l}(2n+1)$$

where: f is resonant frequency, Hz;
c is sound velocity in a medium (drilling liquid), m/s;
n is the positive integers;
l is resonator length, m;

A property of the acoustic resonator, i.e. reflection and absorption of energy with given frequency from a sound vibration spectrum can be used:

- to reduce a noise, if the noise frequency is in the frequency range of the information signal, thereby to increase the signal/noise relationship; e.g. using acoustic resonator for absorbing the drill pump noise in an injection line;
- to control the turbodrill shaft revolution rate by setting the resonator in the internal cavity of a drill pipe.

The efficiency of the resonator operation is based on the phenomenon of a resonant interaction of chambers having a wave nature and associated with a formation of secondary waves in the resonators connecting pipes after the primary wave travel to the second reflection chamber, return to the first chamber and the secondary reflection from it. If the secondary and subsequent waves are added to the primary wave in phase there is a noticeable strengthening of the latter, and the efficiency of sound absorption decreases. An addition in antiphase leads to an increase in efficiency of the composite resonator.

The development of a sound vibration modulator to control turbodrill revolutions can be given as an example of use of an acoustic resonator [15], [16], [17].

Figure 1 shows a basic diagram of a sound vibration modulator to control turbodrill revolutions; Figure 2 shows a spectrogram of drill string vibrations.

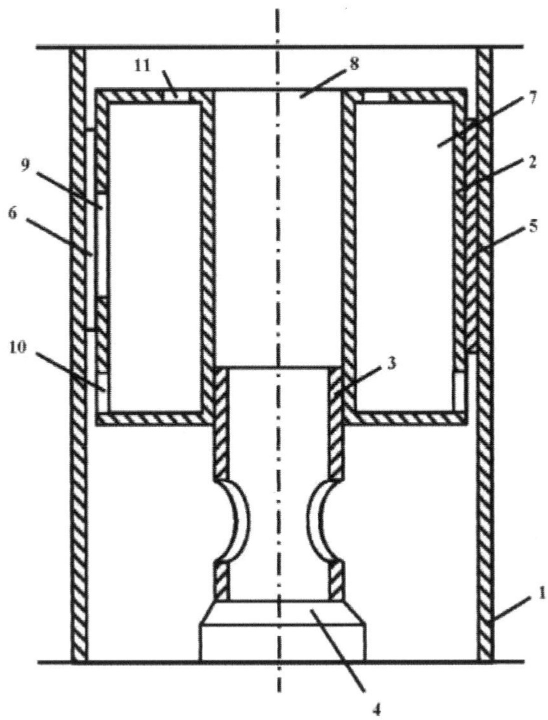

Figure 1. Sound vibration modulator.

A sound vibration modulator to control turbodrill operation comprises a casing-sub (1), a resonant chamber (2), a perforated pipe (3) associated with the turbodrill shaft (4), a bush (5) with a through longitudinal groove (6) installed between the casing-sub (1) and resonant chamber (2). The resonant chamber is made in the form of a hollow cylinder with two cavities: an internal closed annular cavity (7) and a central cylindrical cavity (8). On the outside of the cylinder there is a slotted perforation (9) and flow-through outlet holes (10). The flow-through inlet holes (11) are on the upper end face of the resonance chamber. The central cylindrical cavity is connected to the upper part of the perforated pipe.

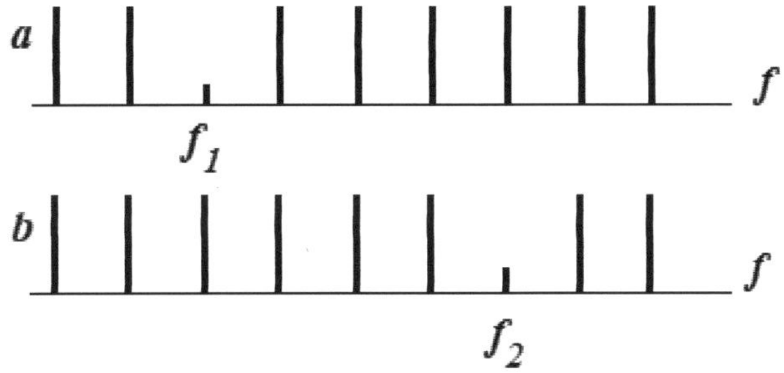

Figure 2. Spectrogram of drill string vibrations.

The sound vibration modulator to control turbodrill revolutions operates as follows:

Drilling fluid under pressure is supplied into the turbodrill through the central cylindrical cavity (8) and at the same time into the flow-through inlet holes (11) and then to the flow-through outlet holes (10).

When revolving the turbodrill shaft (4) rotates the perforated pipe (3) and reactive resonant chamber (2). With each shaft revolution the slotted perforation (9) is aligned with the through longitudinal groove (6) and the bush (5). In this case absorption of sound energy occurs at a frequency of f_2 (Figure 1) equal to the resonant chamber frequency from the drill string polyharmonic vibrations spectrum. When slotted perforation is blocked by the bush energy absorption via the flow-through inlet and outlet holes occurs and the spectrogram shows absorption of frequency f_1. Maintenance of constant internal volume of the closed annular cavity (7) is provided due to self-cleaning from sludge particles. Solid particles settled down on the bottom of the chamber (2) are removed by the drilling fluid under the centrifugal force flowing through the outlet holes.

4. Conclusion

The proposed sound vibration modulator to control the turbodrill revolutions does not require electrical energy because it uses the noise energy occurring in the process of well drilling. The simplicity of transmitting and receiving devices, and absence of any additional communication line built in the drilling tool provides increased reliability and performance of the hydraulic communication channel used to transmit acoustic impulses. In this regard, we can conclude that the way to control the number of turbodrill revolutions is safe. The absence of tuning and adjusting devices in the proposed modulator design provides an ease of operation and does not require high qualification.

References

[1] Savinykh Yu A and Bastiane E E 1981 AC (Author's certificate) 821688 USSR, IPC3 E21B 47/12 *A Way of Obtaining Bottomhole Information via the Drill String During Well Drilling* (USSR) 2192447/22-03; applied 17.11.75; issued in BI № 14

[2] 1980 *Monitoring of Bottomhole Conditions During Drilling Oil-Gas Wells* Survey information, Ser. Drilling Gas and Gas-Condensate Wells (VNIIEgazprom) 5 p 59

[3] Roberts W F Jr. and Johnson H A 1978 System available for measuring hole direction *Oil and Gas J* 22. **76** pp 68-76

[4] 1955 Js the O S Lagging behind Russia in turbodrill development *Oil and Gas J* 27 **54** pp 79-94

[5] Dozorov T A 1975 *Development and Study of an Acoustic Method for Evaluation of the Solid Mass Properties and the State of Drilling Tools* Author's abstract in support of candidature for a technical degree (Moscow)

[6] Dozorov T A and Kutuzov B N 1975 Study of the oscillations spectra occurring in the process of the roller cone bit. *Ser. RNTS Automation and Telemechanization of the Petroleum Industry* (Moscow: VNIIOENG) 7 pp 19-23

[7] Mikhailovsky V N and Ivanov S K 1983 *Measurement of Geophysical and Technological Parameters in the Process of Drilling* (Moscow: Nedra) p 192

[8] Shumilovsky N N and Mikhailovsky V N 1953 Revisiting a choice of a communication channel and an agent when measuring downhole parameters in an oil well *Scientific notes of IMSA of the Ukr.SSR AS* **1**

[9] Yamaliev V U 1989 *Spectral Representation of the Drilling Fluid Pressure Fluctuations When Evaluating the Turbodrill and Drill Bit Technical Condition* Author's abstract in support of candidature for a technical degree (Ufa)

[10] Koltsov O P 1973 Devices for measurement of drill string vibrations *RNTS Ser. Automation and Telemechanization of the Petroleum Industry* 2 pp 33-7

[11] Savinykh Yu A and Khanzhin V G 1985 AC 1154454 USSR, IPC3 E21B 47/12 *Method of Noise Modulation in the Drill String During Turbine Motor Drilling* (USSR) №369588/22-03; applied 19.01.84; issued in BI -№17
[12] Khanzhin V G and Savinykh Yu A 1995 AC 1218743 USSR, IPC3 E21B 47/12. *Method of Obtaining Downhole Information During Well Drilling* (USSR) № 3775825/03; applied 23.07.84; reg. 15 November 1995.
[13] Savinykh Yu A, Podbornov N V and Stolyarov O A 1995 Patent 2038472 RF IPC5 E21B 45/00 *Gauge to Control the Number of Revolutions of the Turbodrill Shaft* (RF) № 93013963/03; applied 17.05.93; issued in BI № 18
[14] Heckl M and Muller G D 1980 *Handbook of Engineering Acoustics* (Leningrad: Sudostroeniye) p 440
[15] Savinykh Yu A and Moskovkin V I 1988 AC 1406358 USSR, IPC4 E21B 47/12 *Sound Vibration Modulator to Control Turbodrill Revolutions* (USSR) №4137454/22-03; applied 20.10.86; issued in BI № 24
[16] Savinykh, Yu A, Savinykh R I, Ganyaev V P, Bogdanov V L and Muzipov Kh N *A Method of Preventing Paraffin Deposits in Oil Well* Patent of invention RUS 2263765 19.04.2004.
[17] Muzipov Kh N 2005 *Stimulation of Oil Production Using Sound Waves* Dissertation in support of candidature for a technical degree (Tyumen)

IOP Conf. Series: Earth and Environmental Science **181** (2018) 012021 doi:10.1088/1755-1315/181/1/012021

Operation of gas wells under conditions of intensive carrying out of mechanical impurities

A I Nasyrova, I I Nasyrov and A A Khairullin

Industrial University of Tyumen, 38 Volodarskogo st., Tyumen, 625000, Russia

E-mail: nasyrova_alexandra@mail.ru

Abstract: The presence of a sand plug at the bottom hole of a gas well has a significant impact on its production capabilities, which entails a reduction in the gas production capacity throughout the entire field. To solve this problem, it is necessary to carry out timely measures to clean the bottom hole of the sand, and also to optimize the operating mode of the well so that not only to keep the gas production at the proper level, but to reduce the carrying out of mechanical impurities. The work of a gas well under conditions of sand plug formation was considered. Based on the two-term law of non-linear gas filtration to the well, a mathematical model of accumulation of a sand plug at the bottom was developed, which was consistent with the field data.

1. Introduction

The West Siberian oil and gas region is the largest development center, rich in huge reserves of gas and gas condensate fields located in its depths.

The intensive nature of the development with high rates of reserves production over the past decades led the largest gas and gas condensate fields (Medvezhye, Urengoyskoye, Yamburgskoye, Vyngapurovskoye, Komsomolskoye and others) to depletion the resource base and, accordingly, the drop in production opportunities and attractiveness.

The total active well stock in the gas fields of Western Siberia compounds thousands of units, over a third of which require ongoing or major repairs, which is the result of inadequate control and the lack of timely optimization of the production wells.

The problems of most cases are directly related to an intensive drop in reservoir pressure, which leads to water flooding of the reservoir system, the bottom hole zone of the reservoir, the production, the deterioration of the productive characteristics of the reservoir and the subsequent destruction of its structure.

The natural carrying out of mechanical impurities on the surface with the fall of reservoir pressure is aggravated by the ever increasing formation of a sand plug at the bottom hole of the well, which, combined with water seepage, leads to a sharp increase in resistances, an intensive drop in gas production, up to a transition to hydrate operation and shutdown.

In this connection, there was a need to create effective mathematical models and methods for optimizing the technological mode of well operation.

2. Research

Content from this work may be used under the terms of the Creative Commons Attribution 3.0 licence. Any further distribution of this work must maintain attribution to the author(s) and the title of the work, journal citation and DOI.
Published under licence by IOP Publishing Ltd

- Analysis of the formation of the sand plug at the bottom hole of the well and the mathematical justification of its influence on the gas production process.
- Analysis and mathematical description of gas flow along the wellbore in the perforation interval, taking into account the formed sand plug.
- Analysis and description of the mathematical calculation of the profile of the gas inflow to a vertical well with the change in the size of the sand plug.
- Algorithmic presentation of the process of gas inflow to the well.

In order to more accurately describe the gas wells operation under the conditions of formation of a sand plug on the bottom hole, studies were carried out on the actual accumulation of sand in the control group of wells in a certain oil and gas condensate field in Western Siberia. The development of this field began in the 1980s, but due to forced exploitation in the mid-1990s it moved to the stage of the falling production.

At the current stage of the development of the Cenomanian gas deposit of the N field, the majority of wells operate under conditions of low reservoir pressure and reduced productivity [1], caused by the water flooding of gas-bearing interlayers and by intense sand ingress.

The decrease in the average daily gas production rate for a number of reasons led to a significant decrease in the total gas extraction (Figure 1).

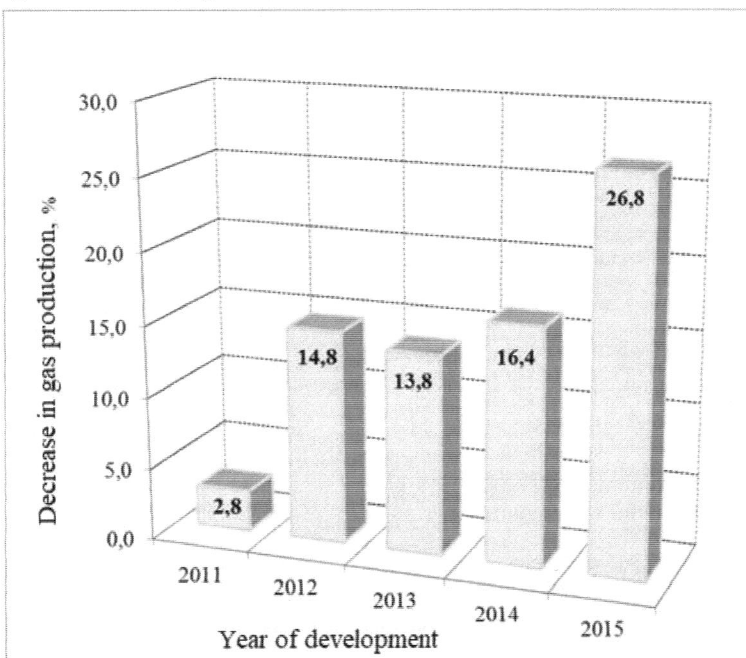

Figure 1. Dynamics of decrease in gas production.

One of the precautionary measures that will help maintain the gas production rate or reduce its fall the intensity to control the accumulation of sand on the bottom hole and determining the permissible well operating conditions to minimize its negative impact.

As of early 2016, the total Cenomanian well stock in the field was over 1200 units, of which slightly more than 1000 units refer to the producing well stock.

During the field operation, geophysical studies were carried out to determine the current total depth in 384 wells. The results of these studies showed that over 23% of the wells were transferred to low-yield well stock due to the presence of a sand plug on the bottom hole, which the size exceeded the perforation interval by more than 50%.

First of all, the presence of sand plugs at the bottom hole of the wells is due to the fact that, during their long operation in the bottom hole reservoir zone, the rock tends to disintegrate, resulting in a large number of mechanical impurities entering the wellbore together with the gas-liquid mixture.

Subsequently, at insufficient up-flow rate, sand particles settle along the wellbore and wall of the borehole to the bottom hole, as a result, conditions arise for the accumulation of a sand plug at the well bottom hole.

The depth of descent of the tubing also influences the process of sand accumulation. On the basis of conducted field research, it can be noted that more intense sand plugs accumulate in the wells, where the elevator column has a large diameter and is lowered to the upper holes of the perforation interval.

It should be noted that the accumulated information on measurements of the current total depth, which allows to determine the actual value of the sand plug, is very small. In this regard, it is impossible to give an accurate estimate of the degree and nature of the formation of sand plugs at the bottom holes of gas wells.

Selection of wells for analysis was carried out on the basis of active sand accumulation at the bottom hole and its influence on daily gas production. Over the past 10 years, gas production for this wells group has decreased significantly from 600-800 to 100-200 thousand m^3/day. In Figure 2 the dynamics of well production associated with the carrying out of mechanical impurities is shown.

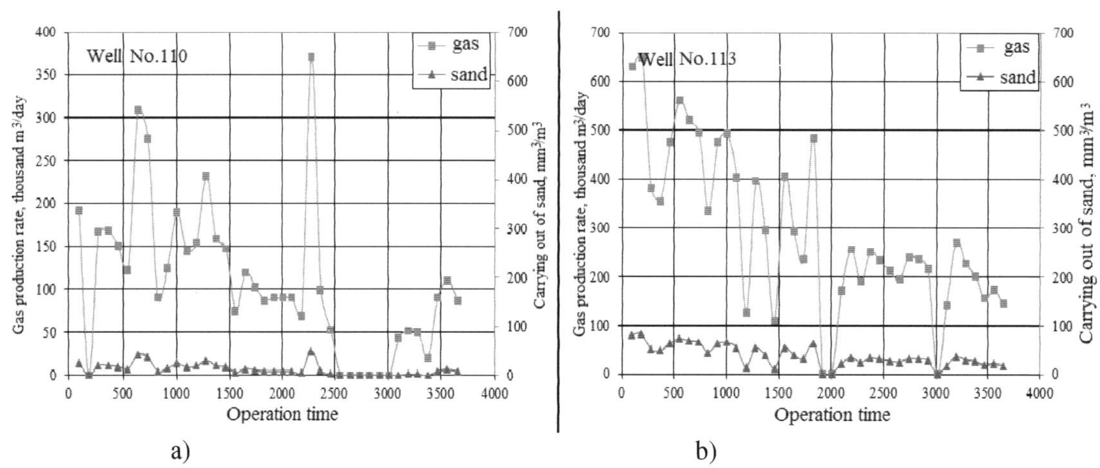

Figure 2. Dynamics of well production taking into account the carrying out of mechanical impurities.

As can be seen, the trend to the drop in production is monitored for all the wells under investigation, which, as indicated earlier, is associated with water flooding of gas-bearing interlayers and intense sand ingress. As it is known, the wells operation in high depression modes promotes a more intensive flow of sand material and water from the reservoir into the bottom hole space, where a dense sandy overlap of the perforation interval is formed depending on the particle size and conditions on the up-flow. Under these conditions, when the bottom hole cleaning suggests already stopping the well for the major workover service, re-launching may not ensure the previous gas production rates, which is due to the aggressive operating conditions of the wells and the lack of timely decisions to monitor the wells condition at the bottom hole.

In Figure 3 the dependence of the carrying out of mechanical impurities on the gas production rate is shown.

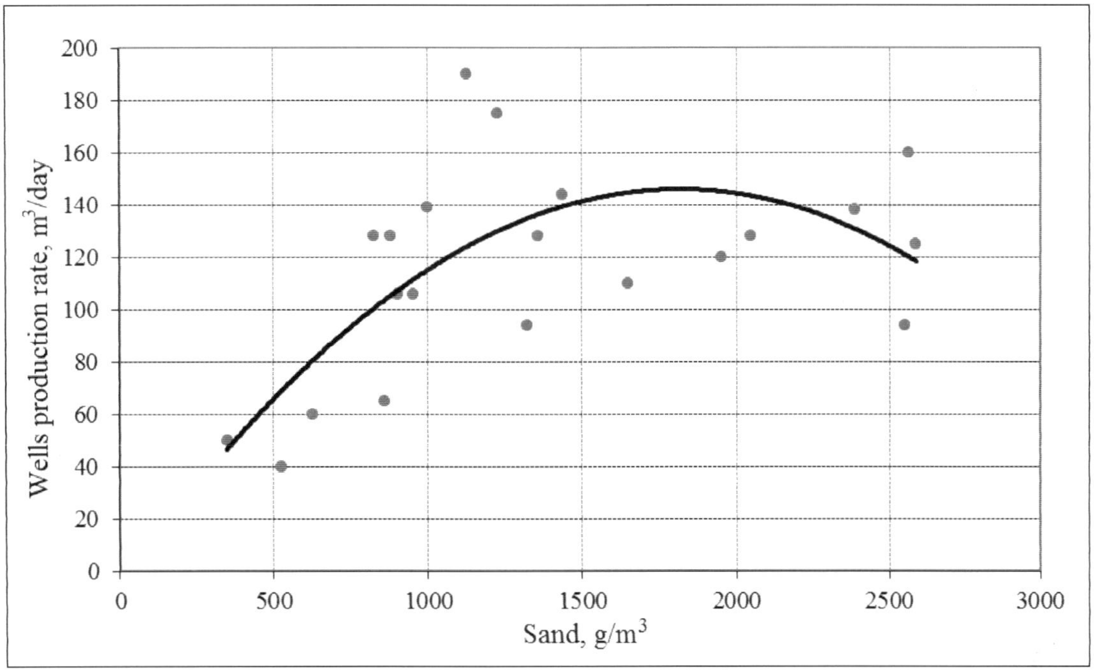

Figure 3. Dependence of gas production rate on sand content.

The presented dependence characterizes the increase in the volume of the extracted sand with the increase in the daily gas production.

The uncontrolled formation of sand plug, especially under conditions of high gas production rate, leads to an irreversible process of intensive growth of the sand plug. Under given conditions, there is a direct need for periodic implementation of measures to eliminate it.

The larger and denser the sand plug is, the more time and effort are needed to clean the bottom hole, which is accompanied by production loss and an increased risk of lowering the potential of the well [2].

Such periodicity of well operation is characterized by the concept of workover interval. In most cases, the planned workover interval assumes a bottom hole condition with a 50% overlap in the perforation interval. The increase in the workover interval only leads to additional gas production losses.

In Figure 4 a clearly expressed example of the volume of losses on the well as a result of idle time for the well workover is graphically shown.

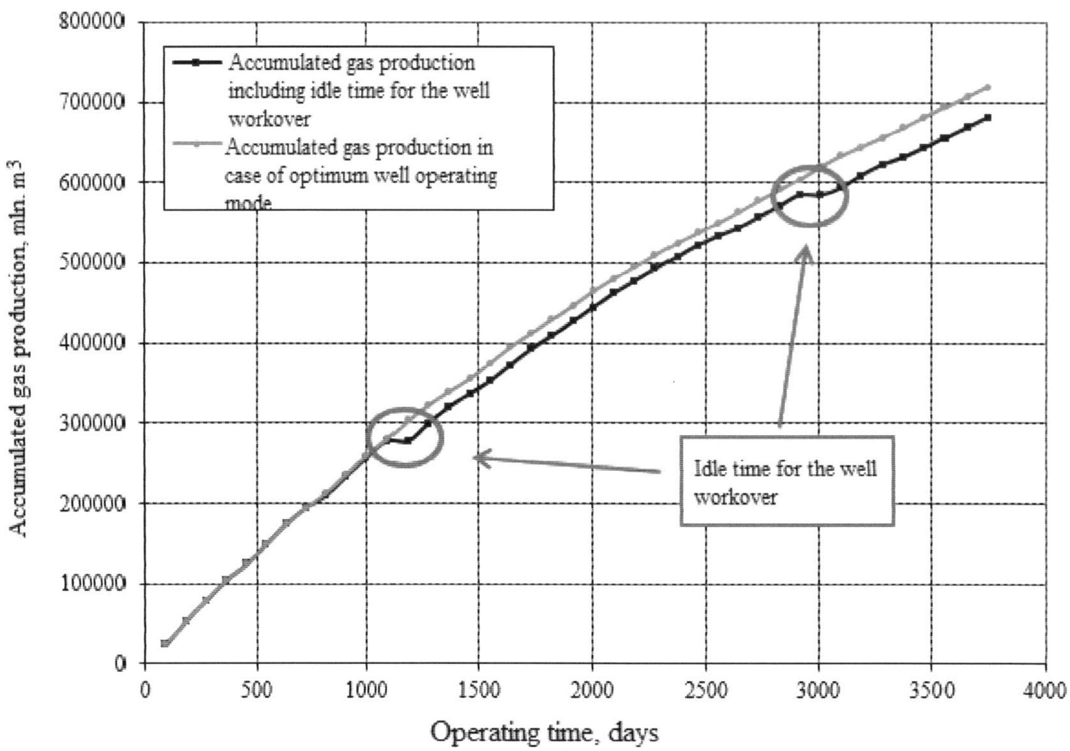

Figure 4. Influence of well workover on accumulated gas production.

According to the results of gas-dynamic studies of the given well, the identified stops were caused by the formation of a sand plug on the bottom hole. The decision to carry out repairs is often caused by a sharp drop in gas production at the well. Thus, according to the presented example, the estimation of the volume of losses was 38 mln. m^3 of gas. Such wells require an instrument for tracking and operative evaluation of the sand plug size without intervention on its technological mode.

Under conditions of a small amount of information, carrying out the major repairs to increase gas production is not always justified. Based on this, a method was developed to determine the sand plug size at the bottom hole of the well, which will solve the problem of conducting the repair work and determine the permissible conditions for the operating wells, which will enable not only to maintain the gas production rate, but also to reduce the impact of the sand plug on the well productivity (P):

$$\Delta P_i = e^{-\frac{2L\cdot g\cdot \cos(\alpha)}{zRT}}\left(P^2+\left(\frac{\mu}{k_{np}}+\frac{bkm_i}{\pi r_c^2 k}\right)\cdot\left(\frac{m_i zRT}{\pi r_c^2 k}\right)^2\frac{\pi r_c^2 k}{m_i g\cdot\cos(\alpha)}\right)- -\frac{\pi r_c^2 k}{m_i g\cdot\cos(\alpha)}\cdot\left(\frac{\mu}{k_{np}}+\frac{bkm_i}{\pi r_c^2 k}\right)\cdot\left(\frac{m_i zRT}{\pi r_c^2 k}\right)^2, \quad (1)$$

where ΔP_i - pressure drop at the i-th section of the perforation interval, Pa; R - individual gas constant, (J/kg K); m_i - the interval mass flow rate, kg/s; z - gas compressibility coefficient; k_{np} - coefficient of permeability m^2; bk - coefficient of hydraulic resistance during filtration in a porous medium according to the Minsky formula; r_c - well radius, m.

To check the efficiency of the developed method, three wells were selected (Table 1), taking into account the following conditions:

- conducting geophysical surveys on candidate wells;
- different amount of operating production rate;
- the presence of hydrodynamic studies with the determination of the coefficients *a* and *b*;
- different percentage of overlapping of the perforation interval with a sand plug.

In addition to these parameters, information was used on geophysical studies with the determination of the average permeability and porosity both for the entire thickness of the reservoir as a whole and for working interlayers in particular.

Table 1. Well data.

Criterion	Well No. 1	Well No. 2	Well No. 3
Perforation interval, m	62	73	73
Current production rate, thousand m^3/day	502	640	820
Reservoir pressure, MPa	4.821	4.858	4.900
Reservoir temperature, °C	25.4	25.4	25.4
Bottom hole pressure, MPa	4.676	4.709	4.780
Coefficient *a*	0.005	0.006	0.040
Coefficient *b*	0.00044	0.00600	0.00015
Sand plug size, m	31.6	25.2	14.6
Average permeability on the reservoir, mD	402	401.5	939

The work of the proposed method consists in determining the sand plug size according to the algorithm given below.

Further evaluation of the effectiveness of the developed algorithm is carried out by comparing the calculated and actual results obtained on the basis of field research.

Calculation of the sand plug size is performed using the average permeability at the effective thickness of the reservoir.

According to the investigated field under the laboratory conditions, the true value of the permeability of the gravel packing was not determined. During the study of sand formation, the empirical value of the permeability of gravel packing, which is 2000 D [3], was established.

In the process of modeling the current operating mode of each well, taking into account the presence of a sand plug at the bottom hole, an inflow profile and diagrams were obtained for the distribution of the main characteristics of the gas flow (flow rate and pressure loss).

In Figure 5 the gas inflow profile for the well No. 1, taking into account the current operating mode and the presence of a sand plug at the bottom hole is shown.

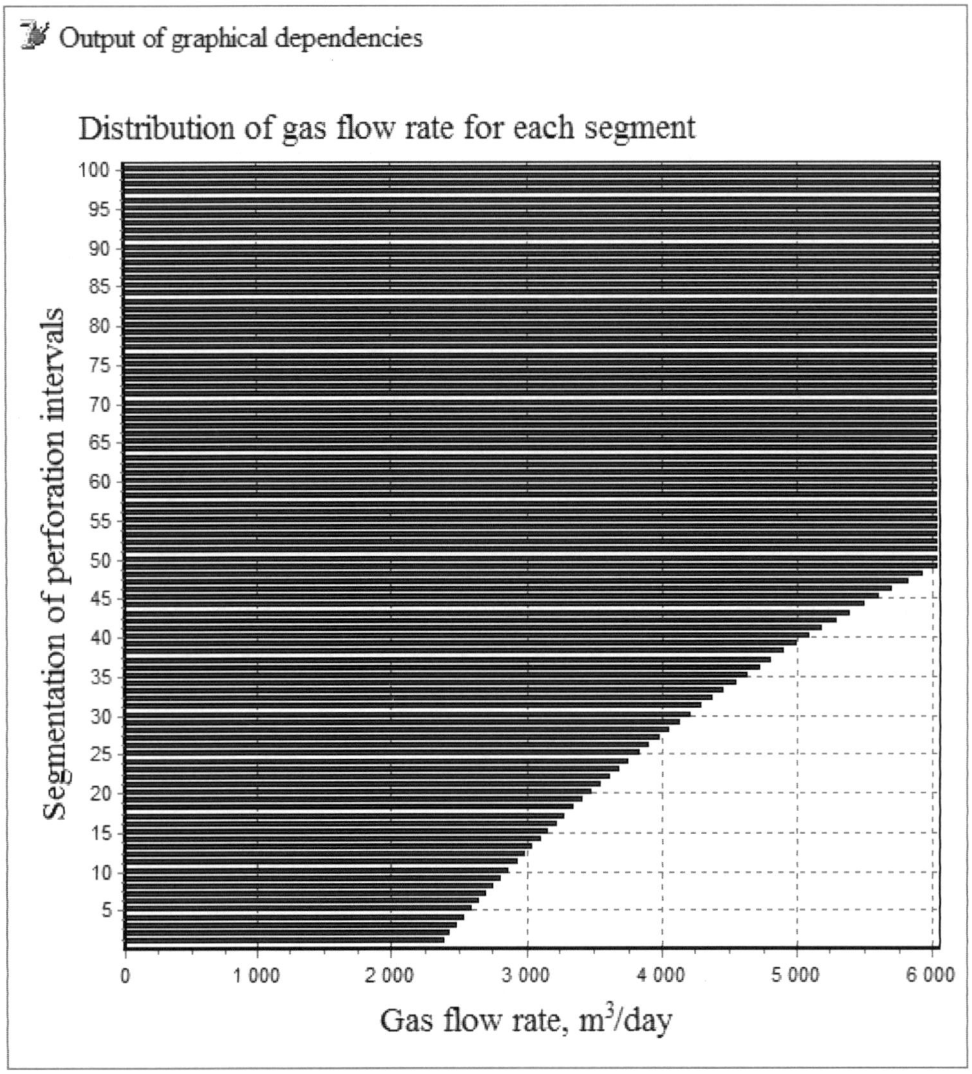

Figure 5. Profile of the gas inflow for the well No. 1.

According to the presented diagrams of the gas flow distribution, the following conclusions can be drawn:

1. This distribution shows that the maximum inflow of gas falls on the upper perforations, which are free of sand.

2. The change in the dynamics of the gas inflow for each section of the perforation free from the sand plug is of minimal value, due to insignificant pressure losses.

3. With a fixed permeability value throughout the working interval of the reservoir, the change in mass production starting from the bottom hole of the well depends on the calculated hydraulic pressure losses, especially in the perforation interval.

4. The profiles of the gas inflow in the section of the wellbore filled with a sand plug are described for each well by a parabolic dependence (Figure 6), which is mainly due to the gas filtration in the pore space (sand plug) by a quadratic law.

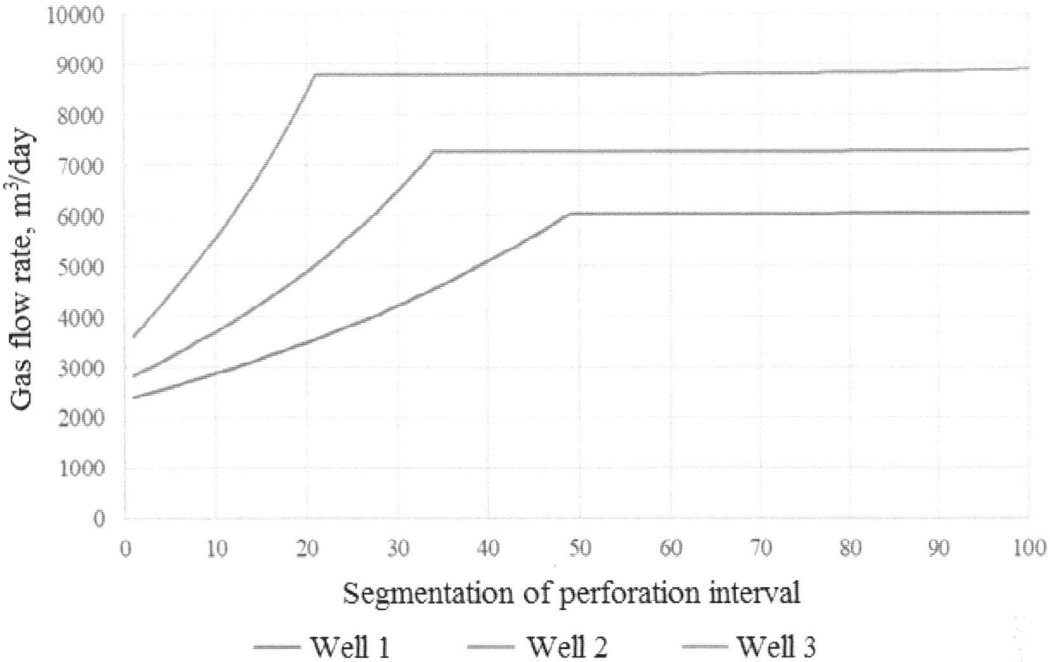

Figure 6. Profile of gas inflow for the investigated wells.

Calculation of the main distribution of pressure losses is confined directly to the estimated interval of the sand plug, as shown in Figure 7.

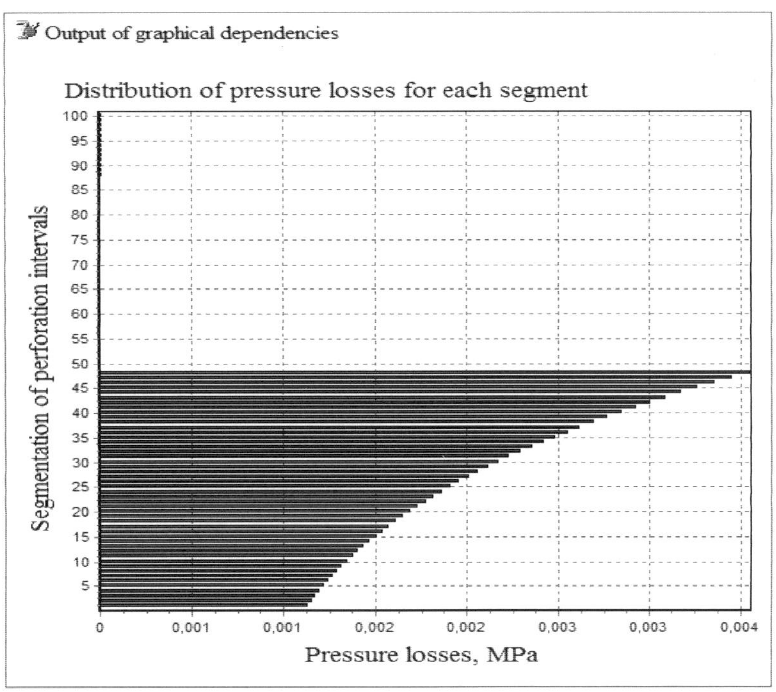

Figure 7. Distribution of pressure losses for the perforation interval of the well No. 1.

3. Results and discussion

According to the dynamics of the change in pressure losses for each well, the following conclusions can be drawn:

1. Losses increase from the bottom hole to the wellhead, which is due to the flow rate, since the gas flow rate in the tube changes from laminar to turbulent with the flow rate increasing (increase in the Reynolds number).

2. Under conditions of the increasing flow rate, the hydraulic pressure losses also directly depend on the presence of an aggressive medium (sand and water ingress), which leads to additional resistances during the gas filtration in each hole in the "reservoir – bottom hole - wellbore" system.

3. Under conditions of sustained permeability and productivity of each perforation hole, in the absence of additional resistance at the bottom hole of the well, a uniform decrease in the operating depression in the perforation interval is observed. When there is a sand plug, the main redistribution of pressure losses occurs at the overlapped intervals of the reservoir.

As a result of the work, the value of the sand plug for each well was estimated (Table 2).

Table 2. Estimation of the sand plug size for each well.

Permeability value	Well No. 1		Well No. 2		Well No. 3	
	Sand plug size, m	Percentage of discrepancy, %	Sand plug size, m	Percentage of discrepancy, %	Sand plug size, m	Percentage of discrepancy, %
Data obtained as a result of well logging	31.6		25.2		14.6	
Data obtained as a result of estimation	30.2	4.4	24.4	3.2	14.7	-0.7

Visualization of the comparison of calculated parameters is presented on the following histogram (Figure 8).

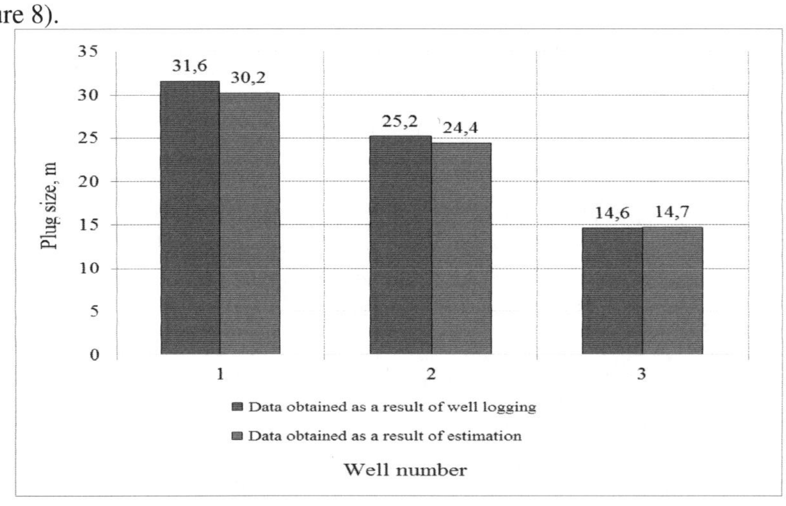

Figure 8. Comparison of estimated and actual values of the sand plug size.

IOP Conf. Series: Earth and Environmental Science **181** (2018) 012021 doi:10.1088/1755-1315/181/1/012021

As a result of the research, the following conclusions can be drawn:

1. Comparison of the calculated and actual values of the sand plug size shows that the maximum discrepancy on the average does not exceed 5%, which indicates the agreement of the results of field research and mathematical modeling of the well operation.

2. Based on the comparison of the estimations results, it can be argued that the use of uniform permeability for the reservoir during the simulation of the sand plug is optimal (Table 2). Since the estimation of the model depends on the amount of the given permeability, averaging its value for the reservoir minimizes the effect of highly productive interlayers, the inhomogeneity of their distribution in the perforation interval of the reservoir, and also reduces the sensitivity of the final calculation to its oscillations, which leads to simplification of the model and obtaining a more optimal result.

3. Using the adapted mathematical model of the well operation, the filtration coefficients a and b can be estimated. The dynamics of these coefficient changes describes the state of the bottom hole reservoir zone, in other words, the possible presence of sand and liquid plugs.

To verify the stability of the estimation of the studies performed and to assess the possibility of using the obtained results to determine the optimal gas production rate, it is necessary to compare the estimated and actual data on the basis of the indicator diagrams. This comparison will allow to identify the errors obtained during the research.

4. Conclusion

In the course of the work, the following provisions were developed and considered:

1. A mathematical model of the reservoir-well system was developed, complicated by the presence of a sand plug along the perforation interval.

2. Analysis of the formation of the sand plug at the bottom hole of the well and the mathematical justification of its influence on the gas production process.

3. Analysis and mathematical description of gas flow along the wellbore in the perforation interval, taking into account the formed sand plug.

4. Algorithmic presentation of the process of gas inflow to the well.

References

[1] Gasumov R A, Geikhman M G, Salikhov Z S and Mosienko V G 2004 *Carrying Out the Major Repairs in the Wells at the Fields of the Extreme North, Preventing the Formation of Sand Plugs and Destruction of the Bottom Hole Reservoir Zone* (Moscow: IRC Gazprom) p 108

[2] Sumen D, Ellis R and Snyder R 1986 *Handbook for the Monitoring and Control of Sand Ingresses in Wells* Trans. from English. and ed. by M A Tseiger (Moscow: Nedra) p 177

[3] Sokhoshko S K 2008 *Development of the filtration theory to flat and horizontal gas and oil wells and its application for solving applied problems* (Tyumen: Tyumen State Oil and Gas University) p 211

IOP Conf. Series: Earth and Environmental Science **181** (2018) 012022 doi:10.1088/1755-1315/181/1/012022

Modeling the task of optimal planning of wells repair and re-newal operations

N V Nazarova and A A Khairullin

Industrial University of Tyumen, 38 Volodarskogo st., Tyumen, 625000, Russia

E-mail: nelli-nazarova@mail.ru

Abstract: The article considers the task of optimal planning of repair and renewal operations (RRO). A graphical description of the change in well stock is given. The solution of the task is described as the Markov process; the well can only be in three possible states, while the time is continuous, since the transition from one state to another can occur at any time. The problem of selecting a well for repair is considered.

1. Introduction

The operating time of the well is divided into three periods: operation, waiting for repair and repair itself. Emergency shutdown of the well affects both the volume of oil production and the entire operation of the deposit. Therefore, the task of optimal planning of the RRO is relevant today.

2. Research

To improve the efficiency of the well, it is necessary to take into account both technical and economic factors:

- the number of new wells placed on production;
- the number of repair/drilling crews;
- operation life (quality of construction and repair of wells);
- period of well repair.

Then the task sounds as follows: for a given oil production plan, it is necessary to determine the amount of work of the drilling and repair crews at the maximum profit of the enterprise.

The problem of optimal planning of repairs at wells can be described as a linear programming task. [1]

It is required to find the maximum of the objective function

$$F = D - 3 \rightarrow \max \,, \tag{1}$$

under the following restrictions

$$\begin{cases} Q \geq Q_{pl} \\ K \leq K_{\max} \\ Z \leq Z_{\max} \\ K,\, Z > 0 \end{cases}, \tag{2}$$

,

Content from this work may be used under the terms of the Creative Commons Attribution 3.0 licence. Any further distribution of this work must maintain attribution to the author(s) and the title of the work, journal citation and DOI.
Published under licence by IOP Publishing Ltd

$$Q = q_1 \int_0^t x_1(t)dt + q_2 \int_0^t X_2\,(t)dt$$

where

D – enterprise total revenue;

$З$ –overall costs of the enterprise;

Q_{pl} – planned oil production;

Z_{max} – maximum number of drilling crews;

K_{max} – maximum number of repair crews.

The costs can be calculated using the following formula

$$З = З_1 + З_2 + З_3 + З_4, \quad (3)$$

where $З_1$, $З_2$, $З_3$, $З_4$ – expenses for maintenance and construction of wells, maintenance of repair and drilling crews.

Wherein

$$З_1 = C_r \cdot \frac{K}{T_r} \cdot T, \quad З_2 = C_m \cdot K \cdot T, \quad З_3 = C_c \cdot \frac{Z}{T_c} \cdot T, \quad З_4 = C_d \cdot Z \cdot T,$$

where

C_r – the average cost of repair of one well, rub.;

C_m – the average cost of maintaining one repair crew, rub./day;

C_c – the average cost of construction of one well, rub.;

C_d – the average cost of maintaining one drilling crew, rub./day.

The income can be calculated according to the following formula

$$D = C\left(Q - q_2 \int_0^{T} x_3(t)\,dt\right) \quad (4)$$

where

q_1 – the average volume of production of a working well, t/day;

q_2 – the average volume of production of the well waiting for repairs, t/day;

T – planning period (1 year).

3. Results and discussion

3.1. The task of optimal planning of drilling and repair works.

Since the number of working wells is constantly changing, the dynamics of changes in the well stock can be considered as the Markov process; the well can only be in three possible states, while the time is continuous, since the transition from one state to another can occur at any time.

In this case, we denote:

$X_1(t)$, $X_2(t)$ and $X_3(t)$ – the average number of wells that are in normal operation, stopped for queuing for repairs and under repair;

$T_М$, $T_о$ and $T_р$ – the average overhaul period of the well, the time in line for setting for repair and the time of the repair itself;

$\lambda_{12} = \frac{1}{T_М}$ – the intensity of the transition from the working state to the repair queue,

$\lambda_{31} = \dfrac{1}{T_r}$ – the intensity of the transition from repair to operational state of failure of wells and exit from repair.

Set the initial value of the operating well stock $N_0 = x_1(0) + x_2(0) + x_3(0)$. [2]

The number of crews in idle $K_m = K - X_3(t)$.

The change in the well stock is described in Figure 1.

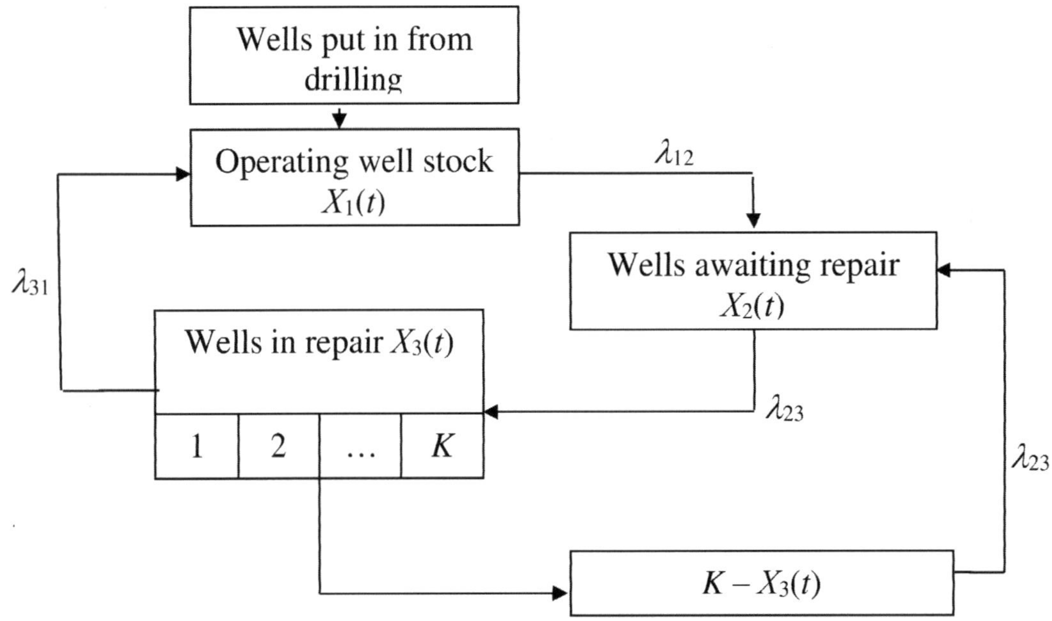

Figure 1. Change in the stock of wells and repair crews.

For Figure 1, we form the Kolmogorov equations with regard to the average numbers of states:

$$\left.\begin{aligned} \frac{dX_1}{dt} &= -\lambda_{12}X_1 + \lambda_{31}X_3 + \delta \\ \frac{dX_2}{dt} &= -\lambda_{23}X_2 + \lambda_{12}X_1 \\ \frac{dX_3}{dt} &= -\lambda_{31}X_3 + \lambda_{23}X_2 \end{aligned}\right\}. \qquad (5)$$

Instead of the last equation, we use the equation of material balance at time t:

$$N = N_0 + \delta \cdot t,$$

or

$$X_1(t) + X_2(t) + X_3(t) = N_0 + \delta \cdot t.$$

Thus, substituting X_3 into (5), we obtain a system of three equations with three unknowns:

$$\left.\begin{aligned} &\frac{dX_1}{dt} = -\lambda_{12}X_1 + \lambda_{31}X_3 + \delta \\ &\frac{dX_2}{dt} = -\lambda_{23}X_2 + \lambda_{12}X_1 \\ &X_3(t) = N_0 + \delta \cdot t - X_1(t) - X_2(t) \end{aligned}\right\}. \tag{6}$$

The obtained model of the well stock dynamics is modified taking into account the intensity of the flows λ_{12}, λ_{23} and λ_{31}, and the intensity of wells input from drilling $\delta = \frac{Z}{T_c}$. These flows depend on the overhaul period, the number of repair crews *K*, and the quality of well repair.

The objective function (1), taking into account the function of costs and income from the sale of extracted oil, or under constraints (2), completely determines the task of optimal planning of the repair works volume. In this case, the dynamics $X_1(t)$, $X_2(t)$ and $X_3(t)$, obtained as a result of the solution, as well as K_{opt} give the optimal trajectory of the oil and gas production management. Any changes in the number of $X_1(t)$, $X_2(t)$, $X_3(t)$ will deviate from the optimal trajectory, and therefore it is desirable to eliminate these deviations (in particular, for deviations $X_1(t)$ and $X_3(t)$).

3.2. The task of planning repair work at wells.
Having determined the volume of repair and drilling operations, it is necessary to solve one of the main tasks of efficient use of the well stock - the task of choosing a well for repair [3].

With a limited number of repair crews, there may be a situation where several wells need to be repaired simultaneously. Then the criterion for the order of repair can be, for example, the priority of high-output wells or the repair time, or the cost of repairs, etc. can be considered an important factor.

However, for the best planning, it is recommended to take into account several most important criteria at once, for example, the overhaul period, the time spent for repairs and the planned profit due to repair and renewal operations.

Overhaul period characterizes the duration of the well operation between the date of bringing the well on to stable production and the date of its shutdown, due to the failure of the deep pumping equipment, the underground part of the mechanized well stock facilities. The calculation is carried out according to the operating mechanized well stock [4].

Repair time is divided into two parts: repair and tripping. The most labor-intensive and time-consuming operation is tripping. Performing the tripping often takes between 50 and 80% of the total time. Now, it is technically impossible to reduce the time spent on tripping. Often, along with the RRO, various geophysical surveys are carried out inside the well; the time for their implementation, as well as the implementation of other measures, should also be taken into account [5].

When calculating profits, it is necessary to take into account high-output wells, whose production will increase insignificantly after repair, and even a reduction in profits is possible. Hence, in case of emergency on high-output wells, a sharp drop in the average daily production rate takes place. To account for such wells, we offer to conduct a double sampling. First, we sample, without additional conditions, and then we perform internal sorting taking into account only the current production rate. In this case, we avoid possible emergencies on high-output wells, whose downtimes will greatly reduce the average production rate.

In the aggregate of all factors, we calculate the priority criterion. Ranking the criterion in descending order, we get the schedule of the sequence of carrying out repair and renewal operations at the wells, taking into account priority maintenance of high-output wells in Figure 2.

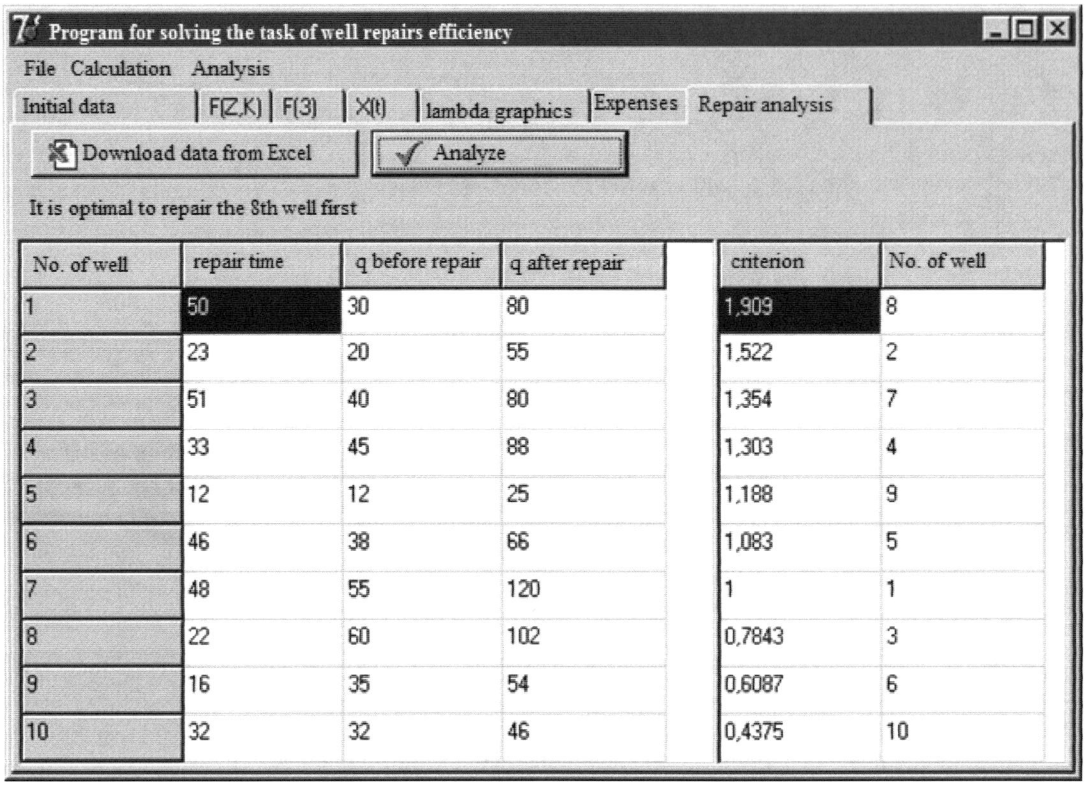

No. of well	repair time	q before repair	q after repair
1	50	30	80
2	23	20	55
3	51	40	80
4	33	45	88
5	12	12	25
6	46	38	66
7	48	55	120
8	22	60	102
9	16	35	54
10	32	32	46

criterion	No. of well
1,909	8
1,522	2
1,354	7
1,303	4
1,188	9
1,083	5
1	1
0,7843	3
0,6087	6
0,4375	10

Figure 2. Distribution of wells for repair and renewal operations.

Figure 2 shows that it is more profitable to repair well № 8 first. Its debit will increase by 1.7 times, while the debit of well № 7 will increase more than 2 times. However, considering that the repair time of these wells differ by almost 2 times, it is recommended to repair the well № 8 first.

4. Conclusion

Thus, a method of optimal planning of repair works was proposed, to stabilize the level of oil production and determine the dynamics of changes in the fund of production wells.

The tasks of effective use of a wells fund are formulated and solved. Criteria of optimality for the selection of wells for repair are determined. Using double sorting in the calculation of profits, high-output wells are taken into account.

References

[1] Papadimitriu H and Staiglitz N *Combinatorial Optimization. Algorithms and Complexity*

[2] Ukhalov K A, Kuchumov R Ya and Namestnikov S V 2003 Modeling of economic efficiency of maintenance and repair of wells *Coll. of Materials of the Sc.-Pr.Conf. "Modeling of Oil Production Technological Processes"* 4 (Tyumen: Vector Book) pp 291-8

[3] Mamchistova E I and Kuchumov R R 2005 Optimum organization of repair work at wells using the Dijkstra algorithm *Coll. of Materials of the Sc.-Pr.Conf. "Modeling of Oil Production Technological Processes"* 6 (Tyumen: Vector Book)

[4] Antoniadi D G, Zakharchenko E I, Streltsova Yu G, Pustovoy P A and Bondarenko V A 2011 Planning and management of well repair *Oilfield Engineering* 6 pp 59-62

[5] Antoniadi D G, Gaponenko A M, Vartumyan G T and Streltsova Yu G 2011 *Modern Technologies of Intensification of High-Viscosity Oil Extraction and an Estimation of their Application Efficiency* (The Publishing House-South) p 419

Research of the foaming agents' influence on the process of reservoir and condensate fluids recovery from Cenomanian gas wells on closing stage of development

A Yu Yushkov, V A Ogay, A F Khabibullin, V O Dovbysh and M N Rogov

Industrial University of Tyumen, 38 Volodarskogo st., Tyumen, 625000, Russia

E-mail: ogayvlad@mail.ru

Abstract: Large gas and gas-condensate fields of Russia are getting depleted and gradually coming to the closing stage of development and that means that an increased number of "drowned" and "self-killed" gas wells appear. The urgency of this problem is approved by the gas-recovery ratio of such unique gas fields of Russia as Urengoyskoye, Medvezhye and Yamburgskoye. A significant share of gas is produced there and current gas recovery factor of the Cenomanian reservoir is about 80%. Insufficient flow velocity of the gas-liquid mixture in production casing and tubing is the reason of liquid accumulation. One of the most efficient methods of solving this problem is pumping foaming compositions downhole. Nowadays both liquid foaming solutions and solid foaming agents have been used to solve this problem. Many foaming compositions have been tested at unique West-Siberian gas fields. Before applying surfactants at production fields, laboratory research should be performed to assess their efficiency and influence on the processes of foam formation and liquid removal from the wellbore. This research is often performed under dynamic conditions using an installation simulating processes taking place in a gas well. The results of the laboratory research of different surfactants' influence on the processes of foam formation and liquid removal from Cenomanian gas wells are presented in this article and can be used by gas-producing companies.

1. Introduction

A lot of large gas and gas-condensate fields of Russia are getting depleted and gradually coming to the closing stage of development and that means that an increased number of "drowned" and "self-killed" gas wells appear. Accumulation of liquid inside gas and gas-condensate wells leads to their gradual killing by a liquid column, i.e. bottomhole pressure is getting balanced by hydrostatic pressure and gas discharge stops. [1]

High flow velocity of the gas-liquid mixture in production casing and tubing is necessary for well self-purification. So, insufficient flow velocity of the gas-liquid mixture is the reason of liquid accumulation. When the flow rate and water-gas-liquid ratio are both high, flow velocity can be sufficient (more than 3-5 m/s). Nevertheless, some problems may occur when:

(a) liquid part (mainly reservoir water) from the recoverable fluid is increasing. The greater volume of the fluid phase heading to the bottomhole has no time to come out to the surface with the same gas flow velocity in the wellbore. In addition, the gas flow rate and gas flow velocity are decreasing, too due to the reduction in gas production (phase permeability of the

Content from this work may be used under the terms of the Creative Commons Attribution 3.0 licence. Any further distribution of this work must maintain attribution to the author(s) and the title of the work, journal citation and DOI.
Published under licence by IOP Publishing Ltd

reservoir for gas is decreasing while the fluid part in the reservoir is increasing). This situation is commonly known as "gas well drowning".

(b) flow rates (flow velocities) are decreasing gradually due to the impossibility of subsequent reduction of bottomhole pressure after natural reservoir pressure decrease. The continuation of wellhead and bottomhole pressures reduction is impossible mainly due to the compressor equipment. Even though the liquid part in gas is low (for example, condensation water only) there is no liquid carryover so it gradually accumulates and this leads to flow rate reduction and well shut-off. This situation is known as "self-kill of gas well".

Both reasons might take place separately or at the same time. If the (a) cause occurs it is possible to perform well service and water shut-off. The (b) case tends to be more urgent for gas-condensate wells. In both cases further well operation is possible using the following methods:

1. Re-equipment of a compressor equipment to obtain lower wellhead pressures.
2. Periodical blow-down of a well.
3. Replacement of tubing with a lower diameter one.
4. Blow-down of wells with high-pressure gas taken from "donor wells" without gas loss.
5. Application of concentric tubing.
6. Usage of bottomhole assemblies designed to pump out the liquid.
7. Application of plunger lift.
8. Pumping of foaming compositions downhole (surface active substances, surfactants). [2]

For many years "drowned" and "self-killed" wells have been the subject of study of such well-known scientists as James F. Lea, Henry V. Nickens, Mike R. Wells, A.S. Yepryntsev, P.S. Korotov, A.V. Nurmakin, A.N. Kiselev, A.A. Tochigin, G.E. Odishariya, N.V. Mikhaylov, V.L. Slivnev, L.S. Chugunov., A.V. Kusteshyv, K.I. Basniyev, L.F. Dementyev, S.N. Yermilov, Z.S. Aliyev, V.N. Shmyglya and others.

If gas extraction from a gas reservoir is about 80%, the amount of "self-killed" gas wells can reach one third of the total well stock. [3] The urgency of this problem is approved by the gas-recovery ratio of such unique gas fields of the Russian federation as Urengoyskoye, Medvezhye and Yamburgskoye. A significant share of gas is produced there and current gas-recovery factor of the Cenomanian reservoir is about 80%. [4]

2. Research

One of the most efficient methods of solving this problem is pumping foaming compositions downhole. After injection of surfactants into the wellbore they are getting mixed with the wellbore fluid by an upward gas flow which creates foam. Density of the gas-liquid mixture and surface tension at the gas-water interface are both decreasing. These processes lead to a reduction in critical gas velocity (subsequent velocity of gas to carry liquid out of the well), gas-liquid mixture carryover and stable well performance.

Nowadays, both liquid foaming solutions and solid foaming agents have been experienced in application. Among the big amount of compositions tested at the Medvezhye field such samples as "FA-4275", "Foamatron V-625" and "BT-Former" are worth noting. The results of application of various surfactants in different time periods are given in Table 1.

Table 1. Results of application of various surfactants at the Medvezhye field.

Surfactant	Experience of application	Results
"Morpen"	Medvezhye field – 2011	Increase in the flow rate is about 15 thousand m^3/day. Increase in liquid carryover.
"BT-Former"	Medvezhye field – 2015	Increase in the flow rate is about 15 thousand m^3/day. Increase in liquid carryover. Absence of negative influence on the processes of gas gathering and treatment.

"FA-4275"	Medvezhye field – 2013	Increase in the flow rate is about 30 thousand m^3/day. Increase in carryover volume of liquid. Absence of negative influence on the processes of gas gathering and treatment.
"Foamatron V-625"	Medvezhye field – 2015	Increase in the flow rate is about 27 thousand m^3/day.

It is worth noting that it is important to take into account the composition of the wellbore fluid, gas flow velocity in production casing and tubing, bottomhole and wellbore temperatures while choosing a surfactant solution (its composition and concentration). Also the cost of the surfactant is a significant factor at determining its optimal concentration.

Before applying surfactants at production fields, laboratory research should be performed to assess their efficiency and influence on the processes of foam formation and liquid removal from the wellbore.This research is often performed under dynamic conditions using an installation simulating processes taking place in a gas well. A certain amount of wellbore fluid (or its simulated solution) is placed in a small-diameter tube and then gas or air is pumped through it with different velocities. The main measuring parameters are stability, foam expansion ratio and share of liquid carryover.

Since the issue of the Cenomanian gas wells drowning is urgent for the domestic industry, the decision was made to conduct laboratory research to determine the influence of surfactants on the process of foam formation and liquid carryover, which is simulated to be a liquid from Cenomanian gas wells. The results of the component composition analysis of reservoir and condensate fluids taken from the Urengoyskoye field received during the systematic monitoring and hydrochemical analysis of well fluid samples were taken as a basis. [7]

Compositions of the prepared simulated solutions are presented in Table 2.

Table 2. Chemical compositions of the simulated reservoir and condensate fluids from the Urengoyskoye field.

Stage	Reservoir fluid	Condensate fluid
	Cenomanian	Cenomanian
Density, g/cm^3	1.011	0.998
Chloride ion, mg/dm^3	$\frac{9700–16462}{10795}$	53–720
Hydrocarbonate ion, mg/dm^3	$\frac{164.7–356.2}{242}$	43–190
Sulfate ion, mg/dm^3	$\frac{2.67–12.51}{6.49}$	1.85–6.48
Bromide ion, mg/dm^3	$\frac{40.88–62.25}{49.7}$	0.47–2.45
Calcium, mg/dm^3	$\frac{175–441}{272}$	3.17–72.82
Magnesium, mg/dm^3	$\frac{84.6–137.1}{99.69}$	0.78–14.06
Sodium, mg/dm^3	$\frac{6054–10507}{6729}$	7.01–276.53

The research was conducted using the installation simulating a section of a gas well wellbore, in the pipe with the inner diameter of 29 mm and height of 1.9 m. During the research, the simulated wellbore fluid with the volume of 50 ml was placed into the pipe with a certain volume of the surfactant and then air was pumped through the valve with a certain flow rate. The most important aims of this research were the conduction of tests with different values of the surfactant's concentration, the ratio of reservoir/condensate fluids and the air flow velocities, as well as recording and measurement of the quantity of liquid carryover.

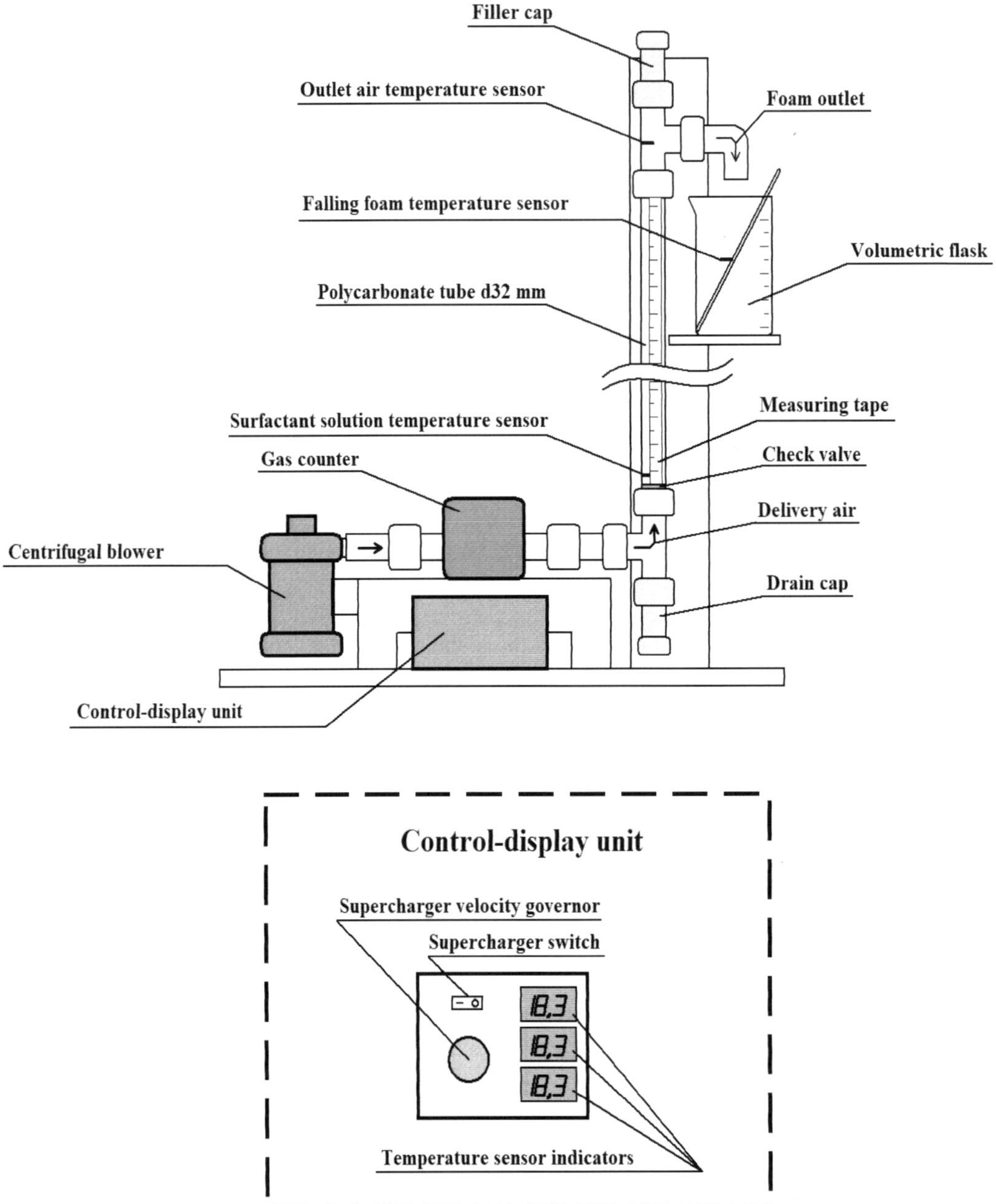

Figure 1. Laboratory installation scheme.

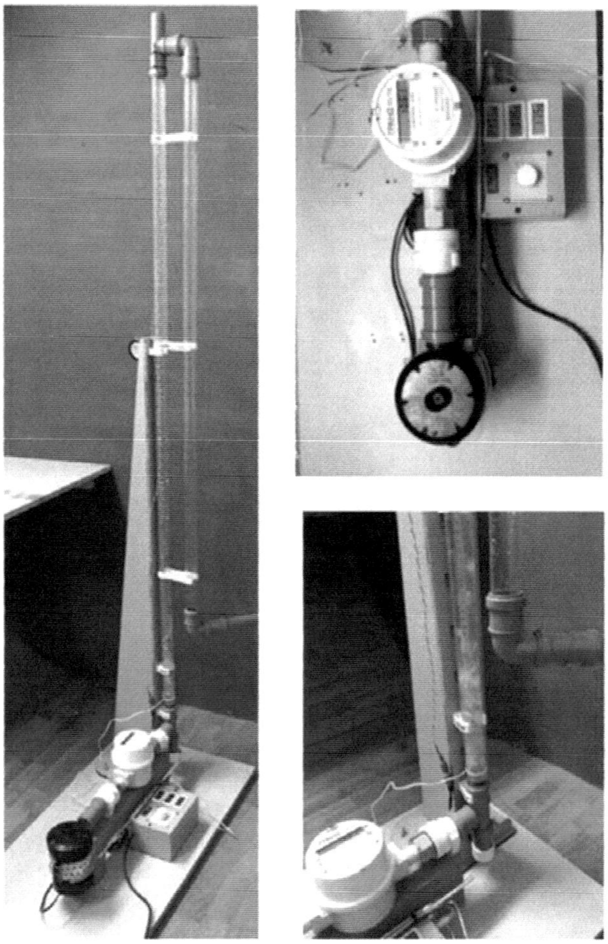

Figure 2. Photo of the laboratory installation.

The most common and available surfactants, such as anion-active surfactant sulphanole and non-ionic surfactant “OP-10”, were chosen as foaming agents. Based on the previous field experience of surfactant application, the surfactant’s concentration was changed from 0.5 g/l to 5 g/l. [7] Time of air delivery was 20 min, temperatures at the “bottomhole” and in the “wellbore” of the installation as well as the ambient temperature were changed from 18 to 27 °C.

3. Results and discussion

The research results are summarized in Table 3.

Table 3. Results of the research

Composition of simulated fluid	Quantitative content of components, g/l		Air velocity, m/s	Liquid carryover, ml
	Sulphanole	“OP-10”		
Cenomanian reservoir fluid	-	0.5	0.5	0
	-	2.5		26
	-	5.0		27.9
	0.5	-		0.8
	2.5	-		19.9
	5.0	-		26.5
	-	0.5	1.3	0

	-	2.5		26
	-	5.0		27.9
	0.5	-		0
	2.5	-		19.7
	5.0	-		26.8
	-	0.5	3	0
	-	2.5		4.1
	-	5.0		26.4
	0.5	-		0
	2.5	-		3.4
	5.0	-		19.3
Cenomanian condensate fluid	-	0.5	0.5	1.3
	-	2.5		23.8
	-	5.0		28.1
	0.5	-		1.6
	2.5	-		15.8
	5.0	-		22.1
	-	0.5	1.3	1.3
	-	2.5		23.8
	-	5.0		28.1
	0.5	-		0
	2.5	-		16.5
	5.0	-		24.3
	-	0.5	3	0
	-	2.5		6.3
	-	5.0		17
	0.5	-		0
	2.5	-		4.8
	5.0	-		13.2
50% Cenomanian reservoir fluid - 50% Cenomanian condensate fluid mixture	-	0.5	0.5	0
	-	2.5		9.1
	-	5.0		15
	0.5	-		0
	2.5	-		13.9
	5.0	-		20.8
	-	0.5	1.3	0
	-	2.5		10.2
	-	5.0		15
	0.5	-		0
	2.5	-		9.2
	5.0	-		12.4
	-	0.5	3	0
	-	2.5		9.1
	-	5.0		22.8
	0.5	-		0
	2.5	-		7.5
	5.0	-		17.6

The analysis of the research shows that the best carryover is observed with lower air flow velocities and higher surfactant's concentration. When the concentration of a surfactant was 0.5 g/l there was practically no liquid carryover. The reason is that if surfactant's concentration is low there is absence of such surface effects as surface tension reduction, resiliency of a bubble film, restoration of destroyed bubbles. [7] When the air flow velocity is increasing the degree of air usage (a ratio of foam flow rate to air flow rate) is decreasing due to partial air slippage and the reason is that the part of gas has no time to disperse and foaming agent disrupts. The lower the surfactant's concentration, the sooner the foaming agent will disrupt. So, the higher the surfactant's concentration the bigger the share of liquid carryover is. It is worth noting that the non-ionic surfactant "OP-10" proved to be more efficient during the research. In general, non-ionic surfactants are practically not exposed to the chemical effect from mineralized reservoir water that is why these surfactants are recommended to use when mineralization of the well is unknown.

4. Conclusion

The research results were compared to the results of the research conducted by SevCavNIPIgaz Research Institute. [5] Their research results showed higher share of liquid carryover and the reason might be the difference between test conditions: time of air delivery was 30 min and liquid volume was 250 ml. Under such conditions, the foam quantity in the installation's "wellbore" many times exceeds the foam quantity obtained during our research and that allows the foam to come out of the wellbore to the surface in a solid column without gaps. Time spent on the experiment also increases the amount of liquid carryover to the surface. Moreover, SevCavNIPIgaz used liquids of other compositions.

There was no research conducted using the mentioned surfactants under the conditions of the closing stage of development of Cenomanian gas reservoirs at the West Siberian fields. The results obtained can be taken into account by gas producing enterprises when choosing surfactants to solve the problem of liquid accumulation in gas wells under corresponding conditions.

References

[1] Andreev O P, Minlikaev V Z, Arno O B, Merkulov A V and Kirsanov S A 2014 Efficiency of natural gas fields exploitation on the closing stage of development under conditions of seasonal variation of selections *Gazovaya Promyshlennost* 707 **6** pp 31-6

[2] Zaboeva M I, Mitrofanov D A and Ogay V A 2015 Operations currently using for maintainance of operation mode of "self-killing" wells *Proc. of the Neft i Gaz Zapadnoy Sibiri* ed Yevtin P V (Tyumen: Industrial University of Tyumen) pp 294-8

[3] Yepryntsev A S, Krotov P S, Nurmakin A V and Kiselev A N 2011 Exploitation problems of liquid loaded wells of gas fields on declining production *Vestnik Orenburgskogo Gosudarstvennogo Universiteta* 135 **16** pp 41-5

[4] Sarancha A V, Ogay V A, Mitrofanov D.A and Khayrullina D V 2015 Estimation of low-pressure Cenomanian gas volumes at fields of Yamalo-Nenets Autonomous Okrug *Proc. of the Innovatsionnye Tekhnologii v Neftegazovoy Otrasli* (Tyumen: Industrial University of Tyumen) pp 84-8

[5] Gasumov R A 2015 Research of usage efficiency of foaming agents for removal of water from wells on the closing stage of field development *Vestnik Severo-Kavazskogo Federalnogo Universiteta* 45 **6** pp 27-37

[6] Lea James, Nickens Henry and Wells Michael 2008 *Gas Well Deliquification* Second Edition (Burlington: Gulf Professional Publishing)

[7] Koshelev A V, Li G S and Kataeva M A 2014 Operative hydrochemical control of encroachment of productive formation by reservoir water at the Urengoyskoye oil-gas-condensate fiel *Vesti Gazovoy Nauki* 19 **3** pp 106-15

[8] Kolmakov A V, Krotov P S and Kononov A V 2012 Development Technologies for Cenomanian Fields of Low-Pressure Gas (Saint Petersburg: Nedra)

[9] Kleshchenko I I 2002 *Research Report: Development of Water Shut-off Technologies in the Wells of the Fields in the North of Tyumen Oblast* (Tyumen: TyumeNIIgiprogaz)

[10] Chupova I M and Dmitruk V V 2001 Control results of gas-water contact advance at the medvezhie field *IRTs Gazprom* 4 pp 82-7
[11] Menshikov S N, Varyagov S A, Melnikov I V, Kharitonov A N and Arkhipov Yu A (2011) Features of gas wells exploitation at the Medvezhye field *Nauka i TEK* 3 **3**
[12] Khodzhaev V V 2008 *Technological Improvement of Gas Wells Exploitation with Presence of Liquid Phase at the Bottomhole: the Urengoyskoe Field Example* PhD Thesis (Ufa: Institut Problem Transporta Energoresursov)

Creation of the installation for studying the impact of current pulse excitation on the bottomhole formation zone

N M Paklinov, A N Shepelevich and A V Strekalov

Industrial University of Tyumen, 38 Volodarskogo st., Tyumen, 625000, Russia

E-mail:033253@mail.ru

Abstract: The article describes the laboratory research installation for studying the effect of plasma impulse excitation on the bottomhole formation zone with the aim of increasing oil recovery. The results of studies carried out using this installation form the basis of the author's dissertation paper. This installation allows conducting studies on the impact on the core of short-time pulses, including plasma-type ones, to assess the feasibility of its field use for bottomhole treatment.

1. Introduction

At present, there is a need to apply non-standard methods of stimulating the reservoir in order to increase oil and gas recovery which are based on increased controllability, energy efficiency, energy conservation and environmental friendliness, both in the late stages of field development and in the early stages with physically challenged hard-to-recover reserves. A controlled physical impact on filtration processes, for reserves categorised as hard-to-recover, can address the affected areas with residual reserves.

2. Research

Plasma impulse excitation (PIE) is one of the methods of intensifying oil production based on the use of resonant properties of the formation. The essence of this method is that when a specially created electric discharge is formed inside the volume of the fluid around its formation zone, ultrahigh hydraulic pressures occur that can perform useful mechanical work and are accompanied by a complex of physical and chemical phenomena [8,9] (7, 14).

The technology is based on an increase in the hydraulic and hydrodynamic effects and the amplitude of the shock action during a pulsed electric discharge in a conducting liquid [1].

At the same time, the following processes occur in the formation:

- heating of the near-well zone;
- acceleration (up to 1000 times) of gravity aggregation of oil and gas;
- increase in the relative phase permeabilities for oil to a greater extent than for water;
- increase in speed and completeness of capillary displacement of oil and water:
- emergence of seismoacoustic emission in reservoir rocks accompanied by the formation of microcracks;
- change in the stressed state of reservoir rocks and a related change in the structure of the pore space (dilatancy).

Content from this work may be used under the terms of the Creative Commons Attribution 3.0 licence. Any further distribution of this work must maintain attribution to the author(s) and the title of the work, journal citation and DOI.
Published under licence by IOP Publishing Ltd

High-voltage current passes through the spark gap electrodes into the working interval inside the well. An electric arc characterized by a high degree of decomposition of the molecules leads to the formation of an instantaneous increase in temperature. Due to this, within a few microseconds, high pressure develops. An instantaneous shock wave with a speed above the speed of sound is transmitted to the surrounding fluid in the well, which leads to the formation of a compression shock wave.

On average, the energy is small, but due to an extremely short discharge time, the maximum energy reaches 20 MW. Instantaneous expansion creates a shock wave, and subsequent cooling and compression cause backflow into the well through the perforations in the casing, which at the initial stage of the well treatment facilitates the removal of colmatizing substances into the wellbore.

When the plasma impulse excitation is applied, the permeability of the bottomhole zone increases, the connectivity of the oil formation with the bottomhole is increased by clearing old filtration channels and creating new ones, the pore space is cleared and new microcracks in the bottomhole zone of the well and the filtration channels of the formation are formed.

During the impact, a shock wave occurs which, passing through the perforations into the zone of penetration into the elastic medium, causes its movement, rapidly damps, becoming a series of successive oscillations propagating at the speed of elastic waves.

3. Results and discussion

In order to study the impact of a pulsed electric discharge on the bottomhole formation zone, on the basis of the Industrial University of Tyumen (IUT), a laboratory research installation for pulsed discharge of electric current is being assembled and improved. Figure 1 shows the installation diagram.

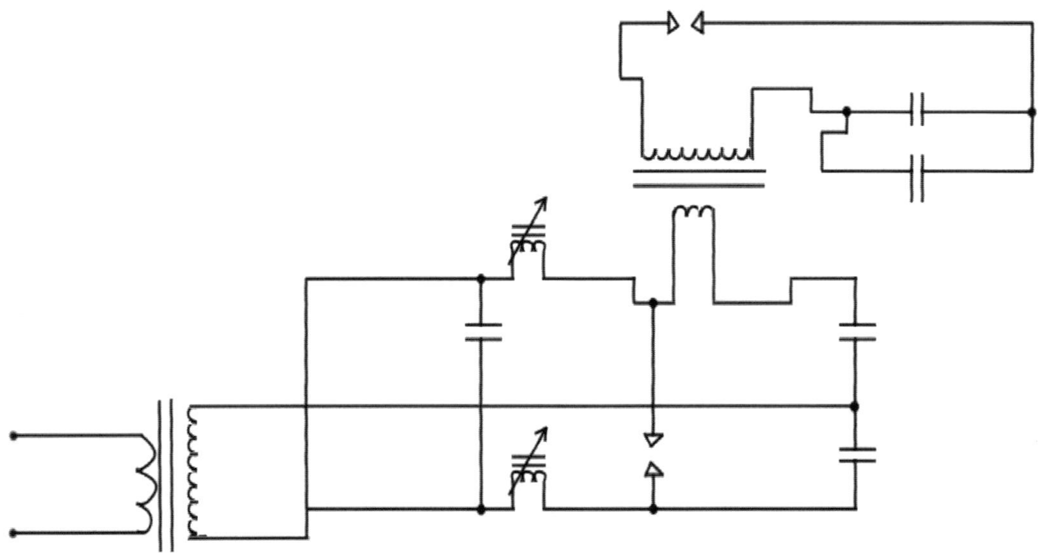

Figure 1. Installation diagram.

The principle of the installation operation: the voltage supplied to the step-up transformer is regulated by a laboratory autotransformer (LATR). After increasing the voltage, the current moves to the converter of electrical energy (conversion of the input electric current of an alternating direction into a unidirectional current) and then to n, near which the spark gap (SG) is located. The main purpose of the SG is overvoltage control. In the case when a certain voltage is reached at the electrodes, the conductivity of the SG increases sharply, and the electric breakdown occurring in it leads to a decrease in the voltage, preventing damage to the electrical equipment. When the energy in the capacitors accumulates to the desired value, it moves to the second step-up transformer, then to the high-capacity storage capacitors. When the voltage in the capacitors rises to a predetermined value, at which the breakdown and all the energy stored in the capacitor immediately flows to the working interval in the liquid (discharge, in the hydrodynamic part), where it is released as a short electric pulse of high power. The impact is carried out on rock samples located in the hydrodynamic part of the installation. This procedure is repeated a predetermined number of times. The effect is shown in Figures 2-3. [5-7, 10-14]

Figure 2. Effect of the installation.

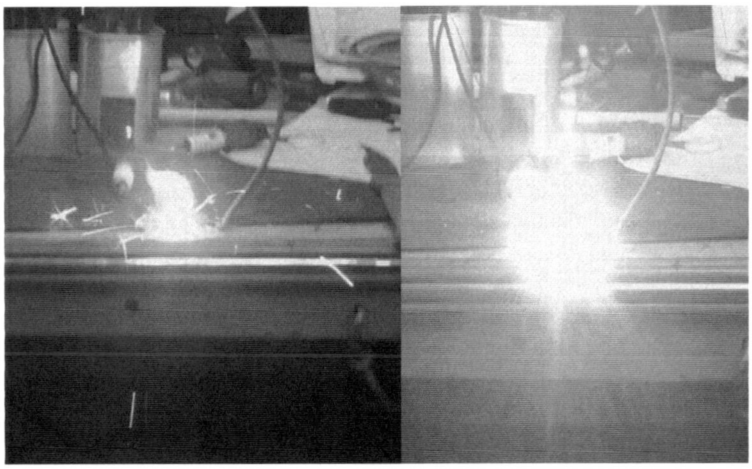

Figure 3. Effect of the installation.

The hydrodynamic part of the installation consists of metal pipes with a wall thickness of 8-10 mm, capable of withstanding the high pressure produced during a pulsed current discharge. The diagram of the hydrodynamic part is shown in Figure 4.

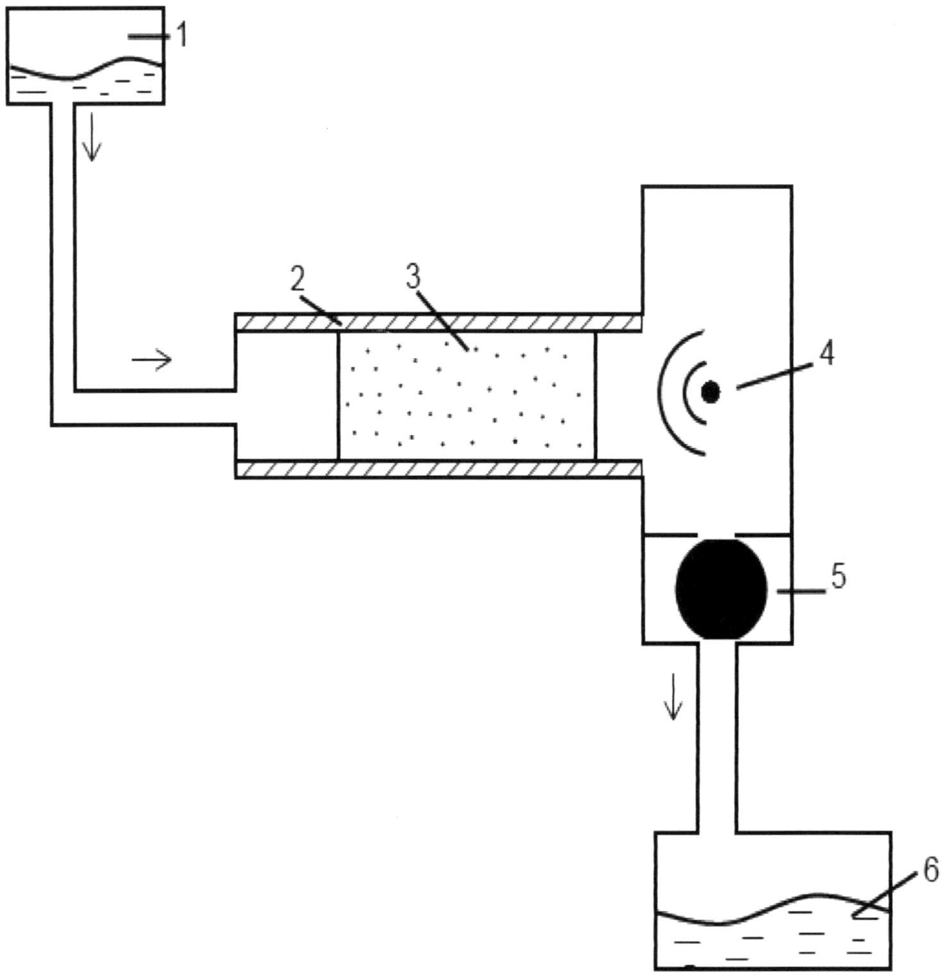

Figure 4. Diagram of the hydrodynamic part of the installation.

The operating principle of the hydrodynamic part of the installation: water enters from the vessel (1) into the compartment with the rock sample (3), then into the chamber, where the pulsed discharge (4) occurs and further into the vessel (6). Samples of rock are preliminarily fixed in the compartment (2) near the chamber of pulsed current discharges. This pulse-discharge chamber can be divided into two compartments. The first compartment is designed to discharge the voltage, instantly arriving at the working interval in the liquid, where it is released as a short electric pulse of high power, affecting the sample. The second compartment serves as a valve (5). At the time of the electric pulse, pressure is generated which acts on the valve closing it, thereby preventing the energy from moving further into the vessel. At the end of the action, pressure in the discharge chamber drops, allowing the valve to open, after which the liquid continues its movement to the vessel (6).

The process of controlling the voltage of the installation is based on pulse width modulation (PWM), by changing the pulse relative duration at a constant frequency. In PWM, transistors are used as key elements (other semiconductor devices can also be used) not in a linear but in a key mode, that is, the transistor is either all the time open (disengaged) or closed (in a saturation state). In the first case, the transistor has almost infinite resistance, so the current in the circuit is very small, and although all the supply voltage drops in the transistor, the power released in the transistor is practically zero. In the second case, the resistance of the transistor is extremely small, and, therefore, the voltage drop in it is close to zero - the power released is also small. In transition states (the transition of a key from a conducting state to a nonconducting state and vice versa), the power released in the key is significant, but since the duration of transition states is extremely small, with respect to the modulation period, the average loss power for switching is negligible.

Figure 5 exemplifies one of the two-level PWM methods using an analogue comparator. To one of the inputs of the comparator a sawtooth voltage is delivered from the auxiliary generator, and to the other input - a modulating voltage. The output state of the comparator is PW-modulation. [2,3,4].

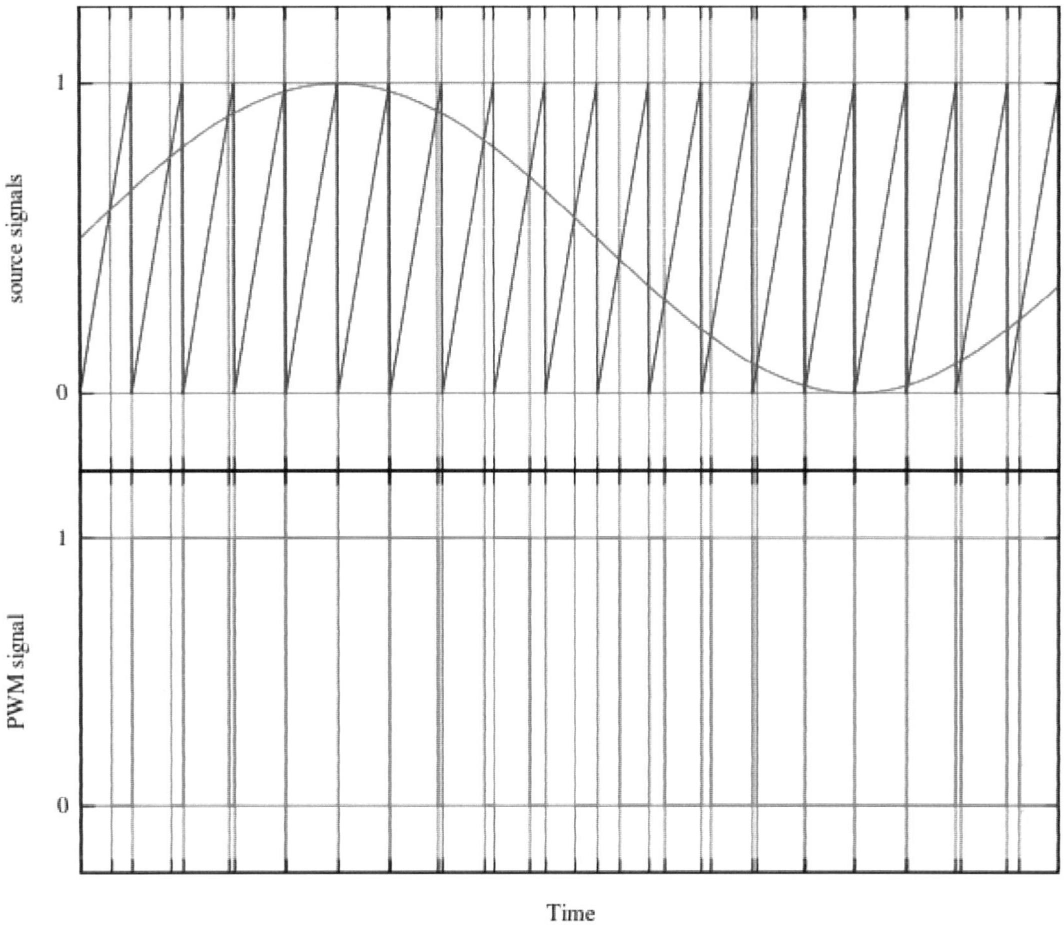

Figure 5. Above - a sawtooth signal and a modulating voltage, below - a PWM result.

4. Conclusion

Currently, the above-described method of stimulating the oil and gas bearing strata is poorly understood, represents a great prospect and requires further study in this area.

References

[1] Yutkin L A 1986 *Electrohydraulic Effect and its Application in the Industry* (Leningrad: Mechanical Engineering) 12-13
[2] MCU SPACE Time of MCU Technology PWM [Electronic resource] URL http: //mcus.ru/posts/pwm-shim
[3] MSP430 LaunchPad. Lesson 21: The Tricks of the Timer PWM [Electronic resource] URL: http://www.shelezyakin.ru/?p=1135

[4] AVR. Training Course. Using PWM [Electronic resource] URL: http://easyelectronics.ru/avr-uchebnyj-kurs-ispolzovanie-shim.html
[5] Maksiutin A V 2009 Experimental studies of rheological properties of high-viscosity oil under elastic wave action. Automation, telemechanization and communication in the oil industry 5 4-8
[6] Paklinov N M, Baryshnikov A A and Vedmenskiy A M 2015 Stimulation of the oil-bearing strata by physical fields to increase oil recovery *Modern Problems of Sc. and Edu.* 2(2) URL: https://www.science-education.ru/ru/article/view?id=22167
[7] Molchanov A A 2007 *Intensification of High-Viscosity Oil Inflow with the Application of Borehole Elastic Stimulation of Productive Strata* (Kazan: FEN) 417-20
[8] Yutkin L A 1955 *Electrohydraulic Effect* (Moscow: Mashgiz) 55
[9] Yutkin L A and Goltsova L I *A Method for Obtaining High and Ultrahigh Pressures* AC 105011 (USSR) applied 15.04.50, No. 416898 issued 1957, No. 1
[10] Ametov I M, Baydikov Yu N, Ruzin L M and Spiridonov Y A 1985 Extraction of Heavy and Highly Viscous Oils (Moscow: Nedra) 205 p
[11] Ashchepkov M Yu 2003 *Dilatative-Wave Stimulation of Oil Reservoirs* PhD Thesis (Ufa) 140 p
[12] Gazizov A Sh and Gazizov A A 2007 Scientific and technical foundations of energy-saving technologies for enhanced oil recovery *Oil Industry* 3 p 60-64
[13] Dyblenko V P 2008 *Wave Methods of Stimulation of Oil Reservoirs with Hard-to-Recover Reserves. Review and Classification* (Moscow: VNIIOENG) 80 p
[14] Kitsis S I, Belousov P L and Ulyanov M V 1988 Prospects of application of the method of electric stimulation of the productive oil and gas reservoir for the intensification of oil inflows to the wells. *Coll. of Sc. Papers "Problems of Development of Rnergy Resources of the West-Siberian Pil Complex"* (TSU TII named after Leniniskiy Komsomol) pp100-4
[15] Mikhailov N N 1996 *Information-Technological Geodynamics of Near-Well Zones* (Moscow Nedra)
[16] Baryshnikov A A, Strekalov A V, Vedmensky A M, Kustyshev D A and Dolgushin V A 2013 Methods of electrical reservoir stimulation in order to increase oil recovery *Oilfield Engineering* 7
[17] Baryshnikov A A, Strekalov A V, Vedmensky A M, Kustyshev D A and Dolgushin V A 2013 Application of electromagnetic and plasma-pulse stimulation for enhanced oil recovery *Oilfield Engineering* 6

Atmospheric pollution of the Ural Federal District subjects (2016): regional features

A I Petrov[1] and D A Petrova[2]

[1]Industrial University of Tyumen, Tyumen
[2]Ural Federal University named after the first President of Russia B.N. Yeltsin, Yekaterinburg

E-mail: ArtIgPetrov@yandex.ru

Abstract: The article deals with regional characteristics of the atmosphere ecology in the Ural Federal District (UFD). The official Rosstat data is given on total emissions of pollutants into the atmosphere of the Urals. Analytical work was carried out to assess the specific (for 1 person/year and for 1 km^2/year) total air emissions of pollutants and their distribution by sources of pollution - stationary and mobile. On the basis of data analytics, conclusions are drawn about regional peculiarities of atmospheric pollution in UFD subjects. It is established that atmospheric pollution in the UFD in terms of specific (for 1 person/year and for 1 km^2/year) indicators significantly exceeds the all-Russian level by almost 2 times. In the subjects of the Ural Federal District, a significant spatial uneven distribution of air pollution is formed. The fact of neglecting the necessity of trapping and neutralizing pollutants is revealed in the Khanty-Mansiysk Autonomous District (Yugra) and the Yamalo-Nenets Autonomous District.

1. Introduction

According to the World Health Organization (WHO), the problem of atmospheric pollution is one of the most acute in the field of ecology. According to WHO [1], air pollution caused about 3.7 million deaths in urban and rural areas of the world in 2012. According to Rosstat [2] in 2016, the total emissions of pollutants into the atmosphere in the Russian Federation amounted to 31617.1 thousand tons. Taking into account the population of Russia (on 1.01.2017 - 146804 thousand people [3]), an average of 215.3 kg of air pollutants was accounted for by each Russian in 2016. About 55% of them are products of stationary sources of pollution and 45% are produced by vehicles [2].

In the Ural Federal District (UFD) of Russia - one of the most important industrial clusters of the country - there is an even more threatening ecological situation. Specific (taking into account the population of the district - 12345.8 thousand people as of 01.01.2017 [3]) 416.0 kg of air pollutants was accounted for by 1 person in 2016 in the Ural Federal District [2]. The distribution of emissions between stationary and mobile sources was distributed approximately in a ratio of 75% and 25% [2]. These data illustrate not only the fact of the significant presence of the third and fourth technological industry in the region, but also the relative imbalance, even in comparison with the all-Russian picture, between a fairly new fleet of vehicles and relatively outdated industrial facilities. Within the framework of this article, the problems of spatial unevenness in the distribution of atmospheric air pollution within the boundaries of the Ural Federal District are being studied. The study focused on a

Content from this work may be used under the terms of the Creative Commons Attribution 3.0 licence. Any further distribution of this work must maintain attribution to the author(s) and the title of the work, journal citation and DOI.
Published under licence by IOP Publishing Ltd

comparative assessment of specific (for 1 person/year and for 1 km²/year) emissions of pollutants into the atmosphere in 6 UFD subjects.

2. Research

Based on the official data of Rosstat [2], it is necessary to determine a number of indicators that are not reporting. In particular, for the correct assessment of the level of environmental problems in different UFD subjects, the authors calculated the indicators "Specific emissions of pollutants for 1 person/year and for 1 km²/year" produced by all sources of pollution, as well as separately stationary and mobile sources.

3. Results and discussion

3.1. Estimation of the total pollution of atmospheric air (without taking into account its source)

Table 1 shows the official information of Rosstat [2] on the total (from stationary and mobile sources) emissions of pollutants into the atmosphere in 2016 in Russia, in general for the UFD and for 6 subjects of the UFD differentially.

Table 1. Total (from stationary and mobile sources) emissions of pollutants into the atmosphere in Russia and UFD in 2016

Russia, UFD and their subjects	Emissions of pollutants into the atmosphere, total, thous. t.	including	
		from stationary sources, thous. t.	from mobile sources, thous. t.
The Russian Federation (RF)	31617.1	17349.3	14267.8
Ural Federal District (UFD)	5136.5	3837.2	1299.3
Kurgan region	103.6	41.7	61.9
Sverdlovsk region	1336.8	906.4	430.4
Tyumen region	2790.4	2291.7	498.7
including:			
Khanty-Mansi Autonomous District - Yugra	1658.5	1428.0	230.5
Yamalo-Nenets Autonomous District	834.1	749.3	84.8
Tyumen Region without AD	297.7	114.3	183.4
Chelyabinsk region	905.8	597.5	308.3

Based on official statistics [2; 3; 4], the values of the indicator "Specific emissions of pollutants into the atmosphere (for 1 person/year - Table 2, for 1 km²/year - Table 3) were calculated for Russia, UFD and 6 subjects of UFD" irrespective of the source of pollution.

Table 2. Specific emissions (per 1 person/year) into the atmosphere of pollutants (in total, without taking into account the source) in Russia and the subjects of UFD in 2016.

Russia, UFD and their subjects	Emissions of pollutants into the atmosphere, total, thous. t. [2]	Population of Russia, UFB and their subjects, total, thousand people. [3]	Specific emissions of pollutants into the atmosphere without taking into account the source, kg/person/year
The Russian Federation (RF)	31617.1	146804.0	215.4
Ural Federal District (UFD)	5136.5	12345.8	416.0
Kurgan region	103.6	854.1	121.3
Sverdlovsk region	1336.8	4329.3	308.7

Russia, UFD and their subjects	Emissions of pollutants into the atmosphere, total, thous. t. [2]	Population of Russia, UFB and their subjects, total, thousand people. [3]	Specific emissions of pollutants into the atmosphere without taking into account the source, kg/person/year
Tyumen region	2790.4	3660.0	762.3
including:			
Khanty-Mansi Autonomous District - Yugra	1658.5	1646.1	1007.5
Yamalo-Nenets Autonomous District	834.1	536.0	1556.1
Tyumen Region without AD	297.7	1477.9	201.4
Chelyabinsk region	905.8	3502.3	258.6

The value of the indicator "Specific emissions of pollutants into the atmosphere (per person per year)" (in total, without taking into account the source) differs by a factor of 7 times in different regions of the UFD (from 121.3 kg/person/year in Kurgan region to 1556.1 kg/person/year in the Yamal-Nenets Autonomous District).

Table 3. Specific emissions (per 1 km2/year) of pollutants into the atmosphere (in total, without taking into account the source) in Russia and the subjects of the UFD in 2016.

Russia, UFD and their subjects	Emissions of pollutants into the atmosphere, total, thous. t. [2]	The area of the territory of Russia, UFD and their subjects, total, km^2. [4]	Specific emissions of pollutants into the atmosphere without taking into account the source per 1 km^2 of the territory, kg/km^2/year
The Russian Federation (RF)	31617.1	17125191	1846
Ural Federal District (UFD)	5136.5	1818497	2824
Kurgan region	103.6	71488	1449
Sverdlovsk region	1336.8	194307	6879
Tyumen region	2790.4	1464173	1905
including:			
Khanty-Mansi Autonomous District - Yugra	1658.5	534801	3101
Yamalo-Nenets Autonomous District	834.1	769250	1084
Tyumen Region without AD	297.7	160122	1859
Chelyabinsk region	905.8	88529	10231

The value of the indicator "Specific emissions of pollutants into the atmosphere (per 1 km^2/year)" (in total, excluding the source) differs by 7 times in different regions of the UFD (from 1449 kg/km^2/year in the Kurgan region to 10231 kg/km^2/year in the Chelyabinsk region).

3.2. Assessment of atmospheric air pollution by stationary sources

Based on the data of Table 1 [2], data on population [3] and area of regions [4], the specific emissions of pollutants from stationary sources into the atmosphere in 2016 (Table 4 and Table 5) were determined for Russia, UFD and 6 UFD subjects.

IOP Conf. Series: Earth and Environmental Science **181** (2018) 012025 doi:10.1088/1755-1315/181/1/012025

Table 4. Specific emissions (per 1 person/year) into the atmosphere of pollutants from stationary sources in Russia and subjects of the UFD in 2016.

Russia, UFD and their subjects	Index			
	Emissions of pollutants into the atmosphere from stationary sources, total, thous. t. [2]	Population of Russia, UFD and their subjects, total, thousand people [3]	Specific emissions of pollutants into the atmosphere from stationary sources, kg/person/year	Specific weight of emissions from stationary sources in total emissions, %
The Russian Federation (RF)	17349.3	146804.0	118.2	54.9
Ural Federal District (UFD)	3837.2	12345.8	310.8	74.7
Kurgan region	41.7	854.1	48.8	40.3
Sverdlovsk region	906.4	4329.3	209.3	67.8
Tyumen region	2291.7	3660.0	626.1	82.1
including:				
Khanty-Mansi Autonomous District - Yugra	1428.0	1646.1	867.5	86.1
Yamalo-Nenets Autonomous District	749.3	536.0	1397.9	89.8
Tyumen Region without AD	114.3	1477.9	77.3	38.4
Chelyabinsk region	597.5	3502.3	170.6	66.0

The value of "Specific emissions of pollutants into the atmosphere (per person/year)" (from stationary sources) in different regions of the UFD differs by 28 times (from 48.8 kg/person/year in the Kurgan region to 1397.9 kg/person/year in Yamalo-Nenets Autonomous District).

Table 5. Specific emissions (per 1 km2/year) of pollutants into the atmosphere from stationary sources in Russia and the UFD subjects in 2016.

Russia, UFD and their subjects	Index			
	Emissions of pollutants into the atmosphere from stationary sources, total, thous. t. [2]	The area of the territory of Russia, UFD and their subjects, total, km^2. [4]	Specific emissions of pollutants into the atmosphere from stationary sources, kg/person/year	Specific weight of emissions from stationary sources in total emissions, %
The Russian Federation (RF)	17349.3	17125191	1013	54.9
Ural Federal District (UFD)	3837.2	1818497	2110	74.7
Kurgan region	41.7	71488	583	40.3
Sverdlovsk region	906.4	194307	4665	67.8
Tyumen region	2291.7	1464173	1565	82.1
including:				
Khanty-Mansi Autonomous District - Yugra	1428.0	534801	2670	86.1
Yamalo-Nenets Autonomous District	749.3	769250	974	89.8
Tyumen Region without AD	114.3	160122	714	38.4
Chelyabinsk region	597.5	88529	6749	66.0

The value of the indicator "Specific emissions of pollutants into the atmosphere (per 1 km^2/year)" (from stationary sources) in different regions of UFD varies by 11.5 times (from 583 kg/km^2/year in the Kurgan region to 6749 kg/km^2/year in the Chelyabinsk region).

3.3. Assessment of air pollution by mobile sources

Based on the data of Table 1 [2], the corresponding values of specific emissions into the atmosphere in 2016 of pollutants from mobile sources were defined (Table 6 and Table 7).

Table 6. Specific emissions (per 1 person/year) to the atmosphere of pollutants from mobile sources in Russia and the UFD subjects in 2016.

Russia, UFD and their subjects	Index			
	Emissions of pollutants into the atmosphere from mobile sources, total, thous. t. [2]	Population of Russia, UFD and their subjects, total, thousand people. [3]	Specific emissions of pollutants into the atmosphere from mobile sources, kg/person/year	Specific weight of emissions from mobile sources in total emissions, %
The Russian Federation (RF)	14267.8	146804.0	97.2	45.1
Ural Federal District (UFD)	1299.3	12345.8	105.2	25.3
Kurgan region	61.9	854.1	72.5	59.7
Sverdlovsk region	430.4	4329.3	99.4	32.2
Tyumen region	498.7	3660.0	136.2	17.9
including:				
Khanty-Mansi Autonomous District - Yugra	230.5	1646.1	140.0	13.9
Yamalo-Nenets Autonomous District	84.8	536.0	158.2	10.2
Tyumen Region without AD	183.4	1477.9	124.1	61.6
Chelyabinsk region	308.3	3502.3	88.0	34.0

The indicator "Specific emissions of pollutants into the atmosphere (per person/year)" (from mobile sources) in different regions of the UFD differs slightly more than 2-fold (from 72.5 kg/person/year in the Kurgan region to 158.2 kg/person/year in the Yamal-Nenets Autonomous District).

Table 7. Specific emissions (per 1 km^2/year) into the atmosphere of pollutants from mobile sources in Russia and UFD in 2016.

Russia, UFD and their subjects	Index			
	Emissions of pollutants into the atmosphere from mobile sources, total, thous. t. [2]	The area of the territory of Russia, UFD and their subjects, total, km^2. [4]	Specific emissions of pollutants into the atmosphere from mobile sources, kg/km^2/year	Specific weight of emissions from mobile sources in total emissions, %
The Russian Federation (RF)	14267.8	17125191	833	45.1
Ural Federal District (UFD)	1299.3	1818497	714	25.3
Kurgan region	61.9	71488	865	59.7
Sverdlovsk region	430.4	194307	2215	32.2
Tyumen region	498.7	1464173	341	17.9

Russia, UFD and their subjects	Index			
	Emissions of pollutants into the atmosphere from mobile sources, total, thous. t. [2]	The area of the territory of Russia, UFD and their subjects, total, km^2. [4]	Specific emissions of pollutants into the atmosphere from mobile sources, $kg/km^2/year$	Specific weight of emissions from mobile sources in total emissions, %
including:				
Khanty-Mansi Autonomous District - Yugra	230.5	534801	431	13.9
Yamalo-Nenets Autonomous District	84.8	769250	110	10.2
Tyumen Region without AD	183.4	160122	1145	61.6
Chelyabinsk region	308.3	88529	3482	34.0

The indicator "Specific emissions of pollutants into the atmosphere (per km^2/year)" (from mobile sources) in different regions of the UFD differs slightly by a factor of 31 (from 110 per 1 km^2/year in the Yamalo-Nenets Autonomous District to 3482 per 1 km^2/year kg/person/year in the Chelyabinsk region).

3.4. Summary data table

Table 8, Figures 1...4 present the data illustrating the distribution of air pollutants in the regions of the UFD by source type.

Table 8. Summary data table on the distribution of air pollutants emissions in Russia and the subjects of the UFD in 2016 by source type.

Russia, UFD and their subjects	Specific emissions of pollutants per 1 person/year, kg/person/year		Specific emissions of pollutants per 1 km^2/year, $kg/km^2/year$	
	Stationary	Mobile	Stationary	Mobile
The Russian Federation (RF)	118.2	97.2	1013	833
Ural Federal District (UFD)	310.8	105.2	2110	714
Kurgan region	48.8	72.5	583	865
Sverdlovsk region	209.3	99.4	4665	2215
Tyumen region	626.1	136.2	1565	341
including:				
Khanty-Mansi Autonomous District - Yugra	867.5	140.0	2670	431
Yamalo-Nenets Autonomous District	1397.9	158.2	974	110
Tyumen Region without AD	77.3	124.1	714	1145
Chelyabinsk region	170.6	88.0	6749	3482

Comparative analysis of the data of Table 1 and Table 8 shows that the ranks of the regions in terms of contribution to the atmospheric pollution of the Ural Federal District are not the same in absolute and specific dimensions. In absolute figures, the hierarchy of regions in terms of pollution is as follows: KhMAD, Sverdlovsk, Chelyabinsk regions, YNAD, Tyumen and Kurgan regions. For specific indicators (for 1 person/year) the hierarchy of regions is reconstructed as follows: YNAD, KhMAD, Sverdlovsk, Chelyabinsk, Tyumen and Kurgan regions. Specific analogue at the rate of 1 km^2/year makes Chelyabinsk and Sverdlovsk regions the leaders in atmospheric pollution.

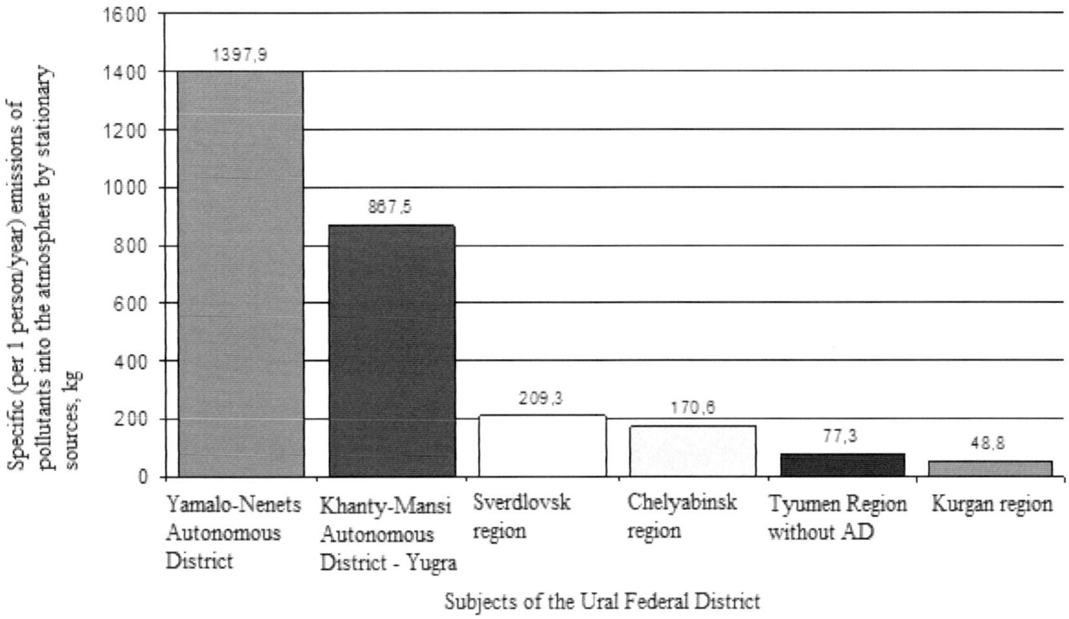

Figure 1. Comparative diagram of atmospheric pollution in UFD subjects at the rate of 1 person/year, formed by stationary sources.

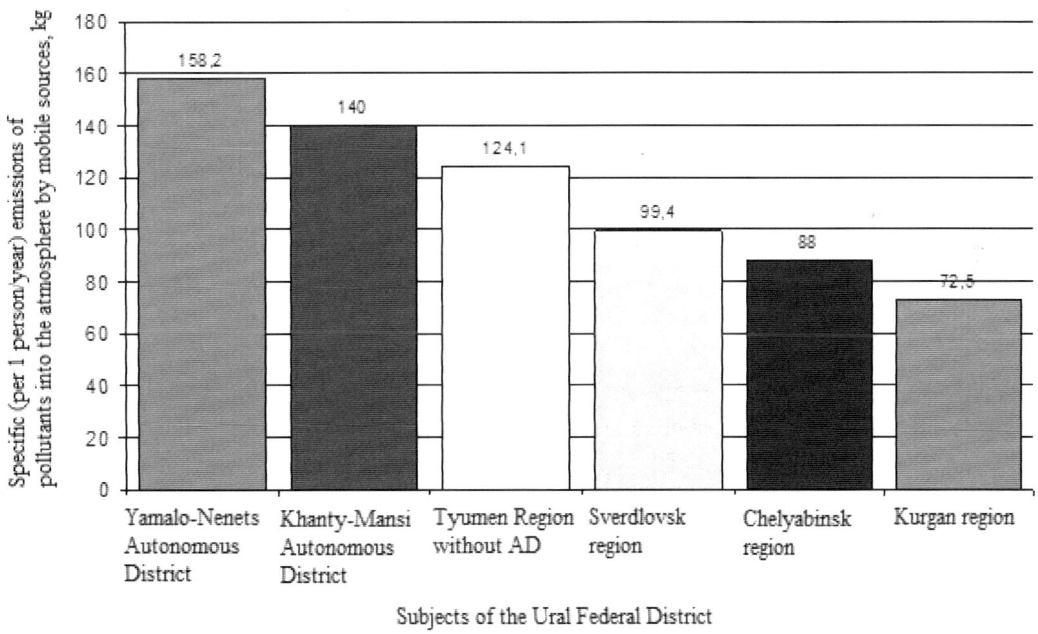

Figure 2. Comparative diagram of atmospheric pollution in UFD subjects at the rate of 1 person/year, formed by mobile sources.

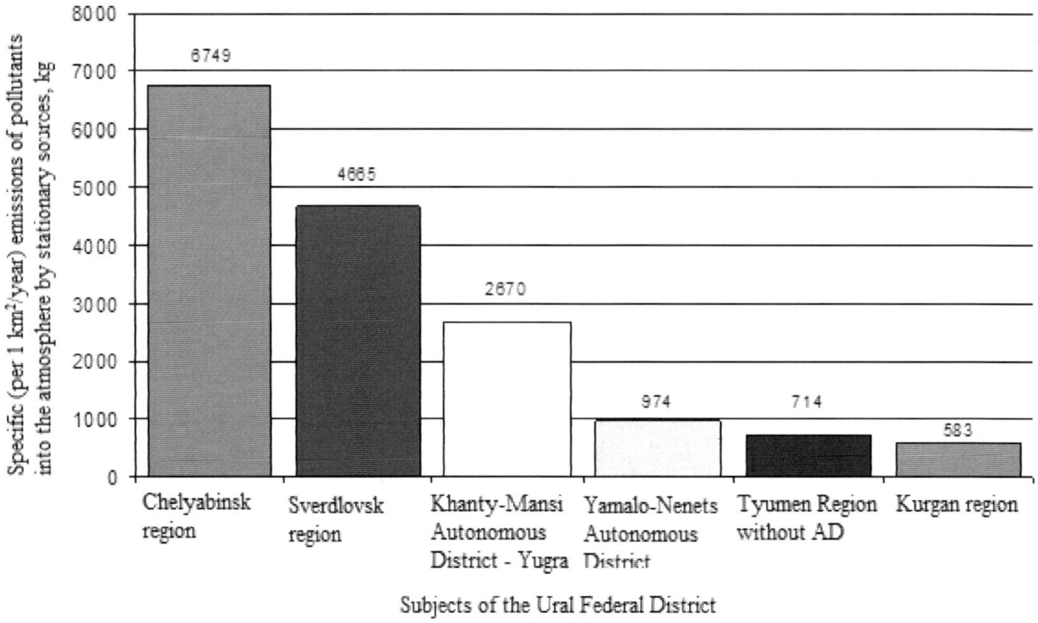

Figure 3. Comparative diagram of atmospheric pollution in the UFD subjects based on 1 km^2/year, generated by stationary sources.

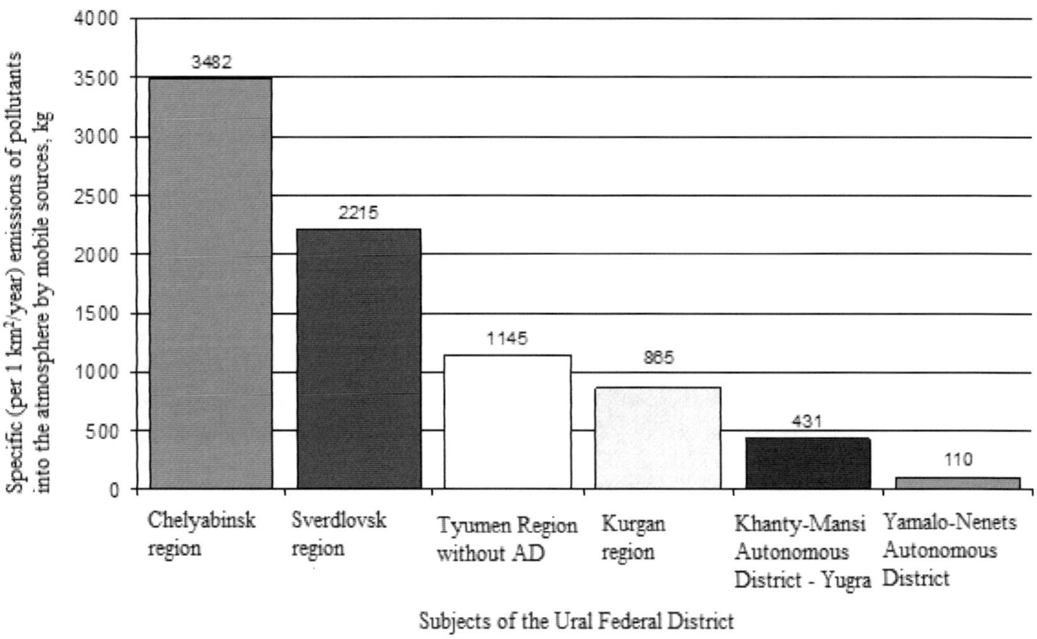

Figure 4. Comparative diagram of atmospheric pollution in the UFD subjects based on 1 km^2/year, generated by mobile sources.

3.5. Assessment of regional successes in the sphere of ecology of the atmosphere in the Ural Federal District subjects

If over the reduction of emissions into the atmosphere of pollutants by vehicles, both the producers and operators and control organizations work in the complex, serious work still remains to be done in the sphere of improving environmental supervision over industrial production. This is evidenced by the data [2], presented in Table 9. Figure 5 presents a diagram clearly illustrating the data of Table 9.

Table 9. Assessment of regional (UFD) successes in the field of capture and neutralization of atmospheric pollutants at stationary facilities in 2016. [2]

Russia, UFD and their subjects	Emissions of pollutants into the atmosphere from stationary sources, total, thous. t	Production of air pollutants, thous. t		Capture and neutralization of pollutants, thous. t.	
		total	in% of the total amount of waste pollutants	total	in% of the total amount of waste pollutants
The Russian Federation (RF)	17349.3	66586.2	100.0	49236.9	73.9
Ural Federal District (UFD)	3837.2	14751.7	100.0	10914.5	74.0
Kurgan region	41.7	111.3	100.0	69.6	62.5
Sverdlovsk region	906.4	8308.6	100.0	7402.2	89.1
Tyumen region	2291.7	2450.3	100.0	158.6	6.5
including:					
Khanty-Mansi Autonomous District - Yugra	1428.0	1434.6	100.0	6.6	0.5
Yamalo-Nenets Autonomous District	749.3	749.6	100.0	0.3	0.05
Tyumen Region without AD	114.3	266.0	100.0	151.7	57.0
Chelyabinsk region	597.5	3881.6	100.0	3284.1	84.6

The data of Table 9 allow us to state that if in industrial regions such as Sverdlovsk and Chelyabinsk, 85...89% of industrial waste are trapped and neutralized, then in KhMAD-Yugra and YaNAD, this indicator tends to zero. In the Tyumen and Kurgan regions, only about 60% of the waste pollutions are trapped and neutralized.

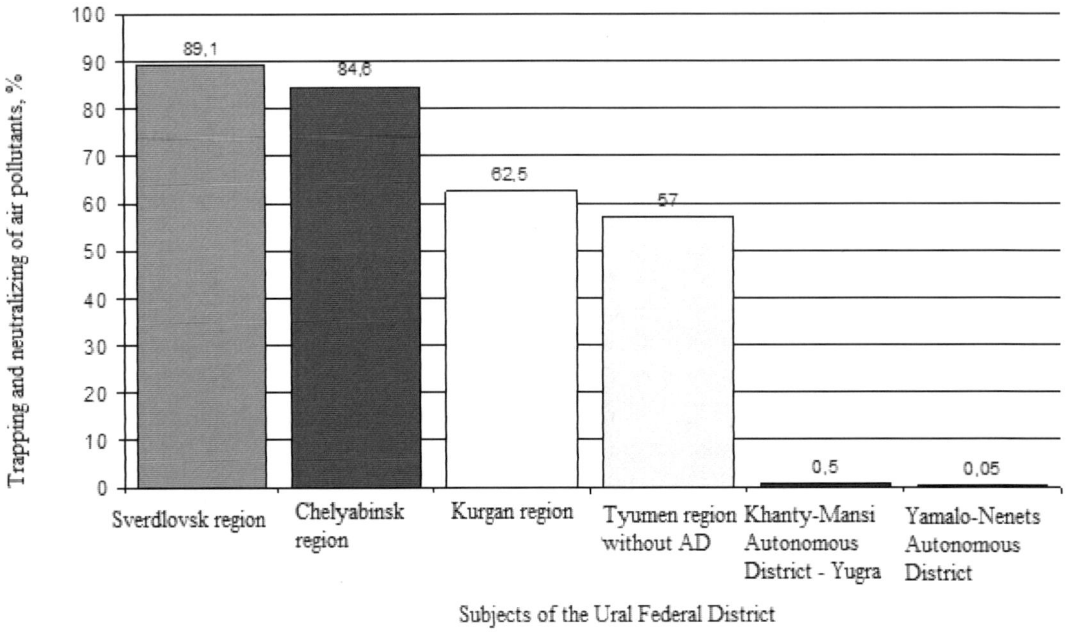

Figure 5. Comparative diagram of estimation for trapping and neutralizing air pollutants by regions of the UFD.

4. Conclusion

In 2014, Federal Law № 219-FZ "On Introducing Amendments to the Federal Law "On Environmental Protection" was issued and certain legislative acts of the Russian Federation. In [5], developed for the purpose of successful implementation of the Federal law, 160 pollutants are presented, which must be carried out with respect to the work of monitoring and gradual elimination of emissions. To be sure, this work is carried out with a certain measure of success. Thus, official data [2] and the corresponding calculations [6] show that year after year the air emissions of pollutants in Russian cities are gradually decreasing. Nevertheless, in some cities of Russia atmospheric pollution can be characterized as catastrophic [6].

The results of the assessment of atmospheric pollution of the UFD subjects can be briefly characterized as follows.

1. In the UFD in 2016, atmospheric pollution by specific indicators significantly exceeds the all-Russian level: at the rate of 1 person/year, the excess is 93.5%; at the rate of 1 km^2/year, the excess is at the level of 52.9%.
2. According to the absolute indicators of atmospheric pollution, the worst indicators in 2016 are characteristic for the Khanty-Mansiysk Autonomous District (1658.5 thousand tons); the best - in the Kurgan region (103.6 thousand tons).
3. Based on the calculation for 1 person/year, the leader among the subjects of the UFD in atmospheric pollution is the Yamal-Nenets Autonomous District; at the rate of 1 km2/year, the Chelyabinsk region is the leader in air pollution among the subjects of the UFD.
4. The most favorable is the environmental situation in the Kurgan region. Indicators of pollution of its atmosphere are much lower than in neighboring regions.
5. The analysis of data on disposal and neutralization of pollutants allows asserting that in the Sverdlovsk and Chelyabinsk regions there is a great deal of organizational work to improve the ecology of the atmosphere. In these regions, in 2016, 85...89% of industrial waste was captured and neutralized. In Khanty-Mansiysk and Yamalo-Nenets autonomous districts, on

the contrary, these works are practically not carried out (no more than 1% of industrial waste discharged into the air is trapped and neutralized).

6. Analysis of economic and geographical features of the distribution of atmospheric pollution in the Ural Federal District makes it possible to conclude that atmospheric pollution correlates with the level of industrial development of regions. In the Kurgan region, where the ecology of the atmosphere is maximally favorable, production is rather weak. On the contrary, in the industrialized regions of the Ural Federal District, the overall situation in the sphere of ecology is much worse. On the general background, the Sverdlovsk and Chelyabinsk regions stand out in terms of work on improving the ecological situation. In the Khanty-Mansiysk and Yamalo-Nenets Autonomous District, environmental issues need to be given much more effort than is done today.

References

[1] WHO: 7 million deaths per year are associated with air pollution (2014) [Electronic resource] Access mode: http://www.who.int/mediacentre /news/releases/2014/air-pollution/ru/

[2] Federal Service of State Statistics of the Russian Federation. Main indicators of environmental protection (2017) [Electronic resource] Access mode: http://www.gks.ru/wps/wcm/connect/rosstat_main/rosstat/ru/statistics/publications/catalog/doc_1140094699578

[3] Federal Service of State Statistics of the Russian Federation. Population of the Russian Federation (2017) [Electronic resource] Access mode: http://www.gks.ru/wps/wcm/connect/rosstat_main/rosstat/ru/statistics/population/demography/#

[4] State (national) report on the status and use of land in the Russian Federation (2017) [Electronic resource] Access mode: https://rosreestr.ru/site/activity/gosudarstvennoe-upravlenie-v-sfere-ispolzovaniya-i-okhrany-zemel/gosudarstvennyy-monitoring-zemel/sostoyanie-zemel-rossii/gosudarstvennyy-natsionalnyy-doklad-o-sostoyanii-i-ispolzovanii-zemel-v-rossiyskoy-federatsii/

[5] 2017 *Characteristics of pollutants from section "I. For atmospheric air" of "The list of polluting substances in respect of which state regulation measures are used in the field of environmental protection", approved by the order of the Government of the Russian Federation of 07/08/2015 № 1316-P. Directory* (Perm: FGBU UralNII "Ecology") 284 p

[6] Petrov A and Petrova D 2017 Atmospheric pollution in cities of Russia: statistics, causes and characteristics *IOP Conf. Ser.: Earth and Environmental Science* **72** 012007.

IOP Conf. Series: Earth and Environmental Science **181** (2018) 012026 doi:10.1088/1755-1315/181/1/012026

Research analysis of mechanical impurities of wells

I A Pogrebnaya and S V Mikhailova

Industrial University of Tyumen, 38 Volodarskogo st., Tyumen, 625000, Russia

E-mail: ya.irina0607@yandex.ru

Abstract: The article is devoted to the investigation of the mineralogical and granulometric composition of mechanical impurities which destroy the pipeline network and the pumping hydraulic system in the fields of Western Siberia. Premature failure of the productive work of the equipment is due to the aggressiveness of groundwater pumped into the reservoir. A physicochemical analysis of the compositions of solid deposits on the damaged parts of oilfield equipment was performed. Measures were proposed to prevent the processes of saline deposits.

1. Introduction

Since the development of Western Siberia, over sixty years have passed. Pipeline transportation of oil and gas industry is pretty dilapidated. In injection wells, reservoir pressure is maintained by pumping sewage into the reservoir, which is pumped through the pipeline. Fields in the Far North have changed considerably - the share of reservoir water with oil extracted from the fields is over 95%. This fluid has an aggressive environment, leading to the destruction of the pipeline network and the hydraulic pumping system.

The cause of aggressiveness, leading to the destruction of oilfield pipes and equipment, is the mineralization and granulometric composition of the pumped liquid.

2. Research

On the basis of the Nizhnevartovsk Research and Design Institute of the Oil Industry (NIPIneft), determination of the mineralogical and granulometric composition is carried out in the laboratory "Physics of Oil and Gas Reservoirs and Systems". Under the microscope, a complete analysis of the mechanical impurities of the disintegrated metal entering the exploration from the fields is determined. Microscopic analysis makes it possible to determine the shape, structure and characteristics of the destroyed surface, as well as optical properties, which makes it possible to decide on the expediency of a method of inhibitors of sediments and destruction of the pipes under investigation. To prepare an inhibitor for research under a microscope, less than 1 mg of the substance is required. Measurement of spherical particles under the microscope delivers reliable and accurate values.

Measurement of grains size is carried out using a petrographic microscope "Polam". In this case, the scale with the divisions or the grid located on the eyepiece is projected onto a sample to measure the granulometric composition of the cuttings. To determine the magnification of the microscope, an object micrometer is placed under the lens. To count the number of particles in an automatic mode using a photocell, the particles in the field of view are measured and recorded, and by the result they are assigned to the selected classes of a certain range of sizes. From these data, the frequency distribution of the number of particles is constructed. The granulometric composition is shown in the histograms of the report.

Content from this work may be used under the terms of the Creative Commons Attribution 3.0 licence. Any further distribution of this work must maintain attribution to the author(s) and the title of the work, journal citation and DOI.
Published under licence by IOP Publishing Ltd

3. Results and discussion

3.1. Sample No. 1

Reason for failure and defects:

In the control laboratory came in a detail of the pump - the tip of a dark color as a result of electrical breakdown in the cable entry clutch, while the extension cable heated.

Field	Well pad	Well	Reser-voir	Stop reason	Operating time, days.	Sampling date	Sampling point
		4738	AC10, AC12	Geological and technical actions-bottom-hole treatment	359	23.04.2016	Core sampling

Table 3.1. Mineral composition. Sample No. 1.

Mineral composition,%	Content,%
Carbonate	81
Barite	12
Iron hydroxides	7

The mineral composition of mechanical impurities is represented by microcrystalline carbonate, barite and iron hydroxides. The size of the micro-grain of the carbonate is smaller than 0.02 mm. Carbonate forms compressed lamellae, with irregular and isometric micrometers, with irregular jagged edges. Sphericity is 0.9. Roundness is 0.5. Barium admixture is noted, in the form of thin lamellar, acicular and scaly aggregates, no larger than 0.15 mm. Iron hydroxides colored in a rusty brown color are found in the form of colloidal and amorphous accumulations, up to 0.03 mm in size.

Table 3.2. Granulometric composition. Sample No. 1.

Siltstone fraction, %		Pelitic fraction, %	
Coarse-grained	0	Coarse-grained	85
Medium-grained	0	Medium-grained	0
Fine-grained	15	Fine-grained	0

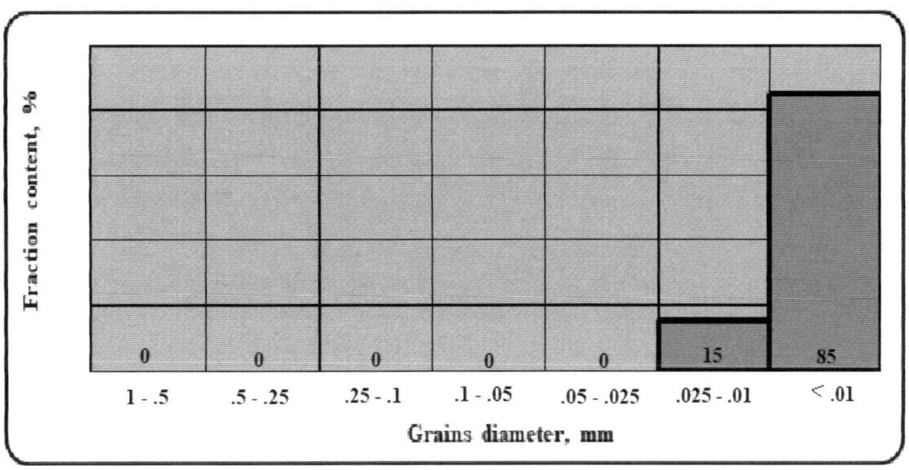

Figure 3.1. Histogram of granulometric composition. Sample No.1.

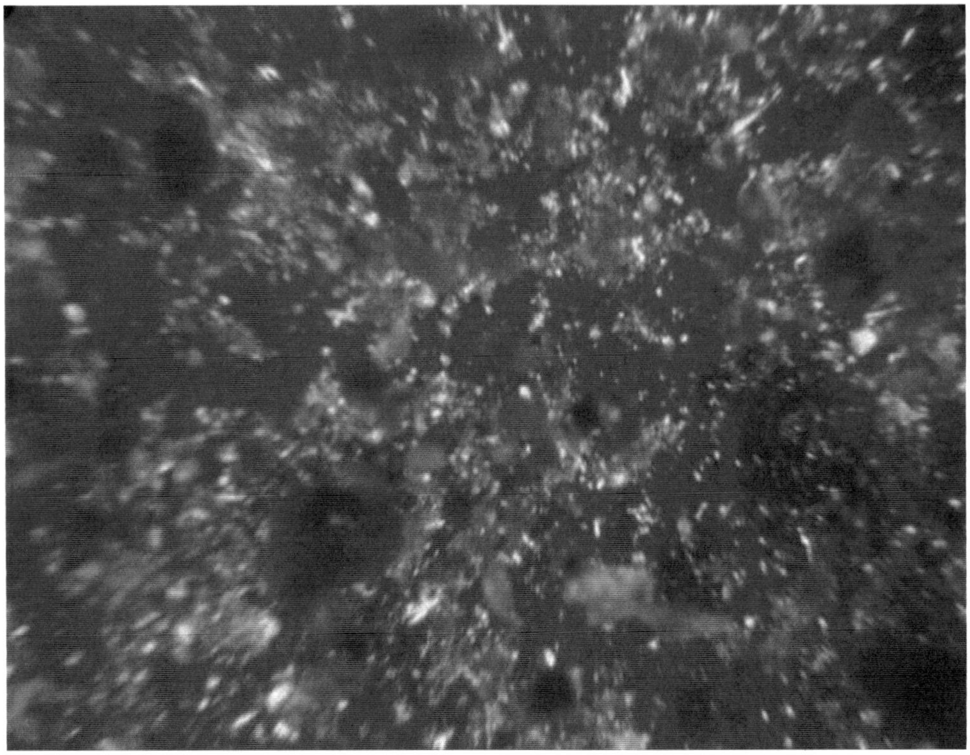

Figure 3.2. Photo of the sample No. 1.

The mineral composition of mechanical impurities is represented by microcrystalline carbonate, barite and iron hydroxides.

3.2. Sample No. 2

Reason for failure and defects:

1. The washers were completely worn out (51-100%);
2. The sleeves were worn;
3. Working wheels had average wear out (21-50%);
4. The guiding devices were completely worn out (51-100%)

Field	Well pad	Well	Reser-voir	Stop reason	Operating time, days.	Sampling date	Sampling point
		41332	AC12	Seismic prospecting	23	16.05.2016	Gas well, ESP

Table 3.3 Mineral composition. Sample No. 2.

Mineral composition	Content, %
Carbonate	89
Iron hydroxides	11
Coal substance	7

Mineral composition of mechanical impurities is represented by carbonate, iron hydroxides and carbonaceous matter. Micro-grained carbonate is distributed in the form of grains of irregular, tapered form, with uneven, sinuous edges, up to 0.03 mm in size. Iron hydroxides of amorphous and colloidal composition are distributed unevenly in the form of clusters, up to 0.05 mm in size with the sphericity

0.7 and roundness 0.7. The carbon bearing substance (in reflected light - black) is distributed by formless spreads, up to 0.05 mm in size.

Table 3.4. Granulometric composition. Sample No. 2.

Pelitic fraction,%	
Coarse-grained	100
Medium-grained	0
Fine-grained	0

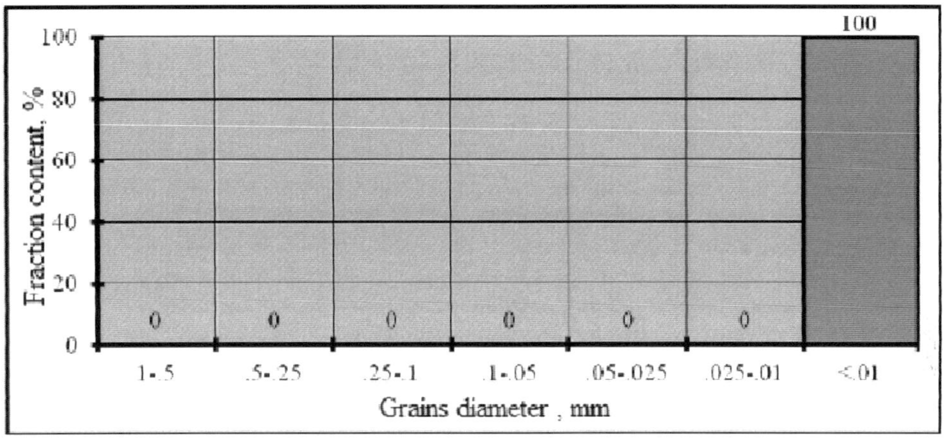

Figure 3.3. Histogram of the particle size distribution. Sample No. 2.

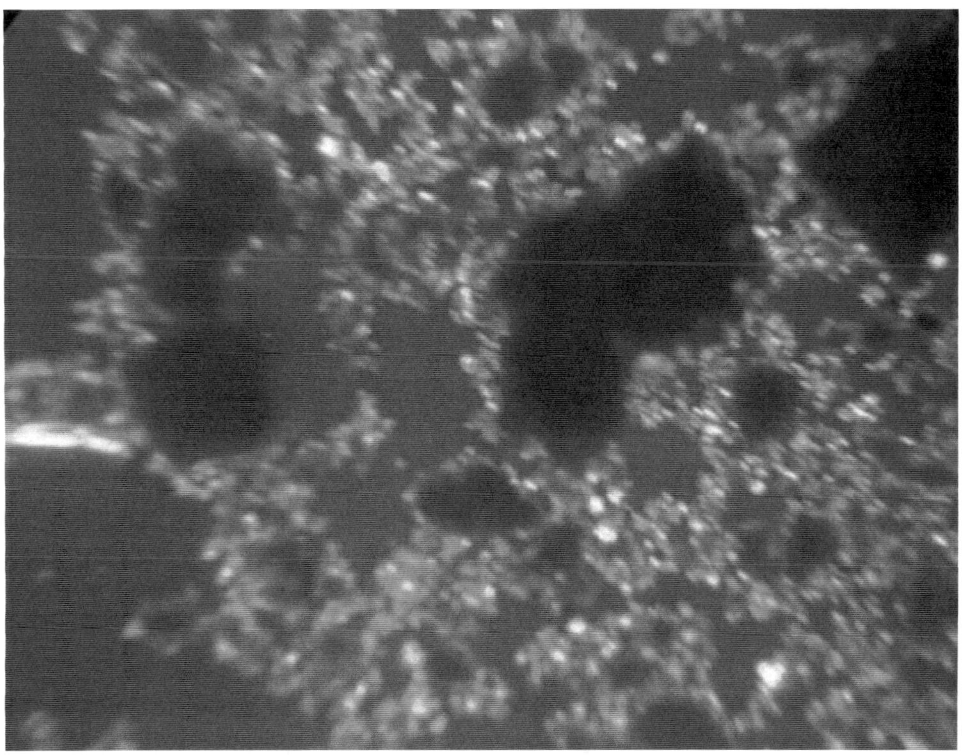

Figure 3.4. Photo of sample No. 2.

Mineral composition of mechanical impurities is represented by carbonate, iron hydroxides and carbonaceous matter.

The analysis of the studied mechanical impurities of wells in the mineralogical relation is represented by the presence of carbonate salts and a small amount of iron hydroxides admixed with carbonaceous matter. Of all the examined samples, the presence of clastic grains of quartz and rarely plagioclase was noted in the wells: Priobskoye, Ombinskoye and Malobalykskoe West Siberian fields, where the percentage ratio was 61% to 24%, respectively. The fragments of plagioclase were not more than 5%. The admixture of iron hydroxides did not exceed 3-32%, and the carbonaceous matter did not exceed 7-33%.

According to the granulometric composition, the studied mechanical impurities of oil were represented by the psammo-aleurite fraction, with a predominance of coarse-grained siltstone fraction, with grain sizes from 0.05 to 0.1 mm. The large-medium-fine-grained psammite fraction was noted in the range from 15 to 69%. The pelitic fraction was 21-100%.

Samples contained a mixed composition with a predominance of the psammitic and siltstone fraction in an equal ratio.

Deposits of calcium and magnesium carbonate, calcium sulfate, barium, strontium, chlorides and other salts in wells, equipment, etc. occur during the development and exploitation of fields. On the deposit, calcium carbonate $CaCO_3$ and, in rare cases, deposits carried up from the reservoir are noted. The sedimentation of inorganic salts occurs when all methods of well operation - fountain, pump, gas lift, but most when pumping one. Of the total number of wells with saline deposits, about 60% of the fields in Western Siberia are equipped with submersible ESP pumps.

Aggressiveness of groundwater is formed due to the presence in them of a certain chemical composition - dissolved components that enhance the dissolution of carbonate-clay cement, with subsequent removal of rocks. Aggressive waters affect the chemical composition and concentration of certain chemical components. Great influence is exerted by temperature and filtration rate of the reservoir solution.

4. Conclusion

Thus, the physics-chemical analysis of the compositions of saline deposits allows to draw the following conclusions:

firstly, the presence of samples with a clear advantage of a particular type of sediment indicates the corresponding process of formation of hardness deposition;

secondly, carbonate salts are present in all types of deposits, which indicates that the process of deposition of carbonate salts takes place at all points of sampling, but with different rates of reaction of their formation;

thirdly, in order to combat carbonate-type sediments, it is necessary to influence the production of wells in order to prevent their formation processes, to carry out salt deposition prevention measures on them, namely, inhibitors of saline deposits LEYSAN 3003 make 3, IPRODENT.

References

[1] Antipin Yu V 1987 *Prevention of Complications in the Production of Watercut Oil* (Ufa: Bashk. knizh. Izdat) p 168

[2] Latypov A R *Regulation of Saline Deposits in Oil and Gas Condensate Wells of OOO "RN-Purneftegaz"*

[3] Martos B N 1973 *Review of Foreign literature, Series Extraction* (Moscow: VNIIOENG) p 112

[4] Pogrebnaya I A, Malinovsky M I and Suleimanov I T 2015 Analysis of the application of gear hydraulic machines in the oil and gas industry *Experience, Current Problems and Prospects for Development of the Oil and Gas Industry, Proc. of the 5th Regional Sc.-Pr. Conf. of Higher Education and Post-Graduate Students and Scientists* (Tyumen: TyumGNGU) pp 343-347

[5] Pogrebnaya I A and Mikhaylova S V 2016 On the question of application of alternative materials for pipelines in the fields of Western Siberia *In the World of Scientific Discoveries,*

Proc. of the Int. Conf. Scientific Creativity of the 21st Century **39** 6-2 (Krasnoyarsk: Scientific and Innovation Center) pp136-40
[6] *Collect. of Reports of Intern. Scientific-Practical Conf.* 2010 (Krasnodar: NPF Nitpo) pp109-17
[7] TU 2257-075-26161597-2007. *Polymer compositions "Geotherm-01", "Geotherm-02", "Geotherm-03". Specifications*
[8] TU 2458-001-45305665-2006 with rev. 1. *Technical characteristics of inhibitor of saline deposirs Iproden C-1. Technical conditions*
[9] Bourne H M 2000 *Effective Treatment of Subsea Wells with a Solid Scale Inhibitor System* SPE 60207
[10] Collins I R 2001 *Development of Emulsion-Based Production Chemical Deployment Systems* SPE 65026
[11] Nasr-El-Din H A, Al-Saiari H A, Al-Ruwaily A-A A and Al-Gamber S D 2006 *Field Application of Emulsified Scale Inhibitor Treatment to Mitigate Calcium Carbonate Scale in Horizontal Wells* SPE 100456-MS
[12] Fitzgerald S M, Stephens G and Terwilliger P L *Fitzgerald* S.M. *Warm Air Coking Sand Consolidation Field Results in Viscous Oil Sands*
[13] Fitzgerald A M and Cowie L G 2008 *History of Frac-Pack Scale-Inhibitor Deployment* (BP) SPE 112474- MS.
[14] Geehan T 1999 *Use of Resin-Coated Proppant to Control Shallow, Low-Temperature Unconsolidated Sand* SPE 54631 Western Regional Meeting, Anchorage (Alaska)
[15] Gupta D V S 2008 *Multiyear Scale Inhibition from a Solid Inhibitor Applied during Stimulation* SPE-115655-MS-P
[16] Romero C 2007 *Behavior of a Scale Inhibitor Water-in-Oil Emulsion in Porous Media* SPE-98275-PA-P

IOP Conf. Series: Earth and Environmental Science **181** (2018) 012027 doi:10.1088/1755-1315/181/1/012027

Analysis of the precision problem in numerical calculation of the reliability indices of technical systems by the topological method at application of the floating-point arithmetic

P A Rahman and E R Ishkinina

Department of Automated Technological and Informational Systems, Ufa State Petroleum Technological University, 2, October Avenue, Sterlitamak, 453118, Russia

E-mail: pavelar@yandex.ru

Abstract. This paper deals with the problem of precision error arising in numerical calculation of the reliability indices of technical systems by the topological method at application of the floating-point arithmetic. The topological method for calculation of the reliability indices in accordance with the reliability models based on the Markov chains is discussed. An example of correct calculation of the reliability indices on the basis of analytical derivation of calculating formulas and substitution of the initial reliability parameters into them is given. The floating-point types and arithmetic supported in hardware by the modern computers and examples of the precision loss in floating-point arithmetic operations are also observed. Finally, an example of incorrect calculation of the reliability indices, based on the numerical solving of the system of linear algebraic equations, using the floating-point arithmetic is also given.

1. Introduction

Nowadays for reliability analysis of technical systems there are widely used such key reliability indices as: availability factor, mean time to failure and mean time to repair [1, 2]. For calculation of these reliability indices the reliability models on the basis of Markov chains and an effective topological method are often applied [3, 4]. At the same time the topological method of reliability indices calculation is based on solving of the system of linear algebraic equations, which is, as well as the solving methods, affected by the precision error arising in the case of application of numerical methods and floating-point arithmetic [5]. As a result, in the event of difference in values of initial reliability parameters on several tens or hundreds orders it is impossible to provide correct calculations of the reliability indices on computer in the case of application of the floating-point arithmetic.

This scientific paper is dedicated to the analysis of the problem of precision error arising in numerical calculation of the reliability indices of technical systems by the topological method at application of the floating-point arithmetic.

2. Topological method of calculation of the reliability indices of technical systems

In general case, Markov chain with continuous time (figure 1) in reliability models of the technical systems contains the set of states E, consisting of a subset of up-states E_+ and a subset of down-states E_-. The subset E_+, in its turn, contains a subset H_+ of border up-states, which have direct transition links to the downstates. Similarly, the subset E_- contains a subset H_- of border downstates, which have direct transition links to upstates.

Content from this work may be used under the terms of the Creative Commons Attribution 3.0 licence. Any further distribution of this work must maintain attribution to the author(s) and the title of the work, journal citation and DOI.
Published under licence by IOP Publishing Ltd

Each state $i \in E$ (figure 2) can have as inbound transition links with the rates γ_{ri} from the subset of states $\{r \in R_i\}$, which have transition links to the state *i*, so outbound transition links with the rates γ_{is} into the subset of states $\{s \in S_i\}$, which have transition links from the state *i*.

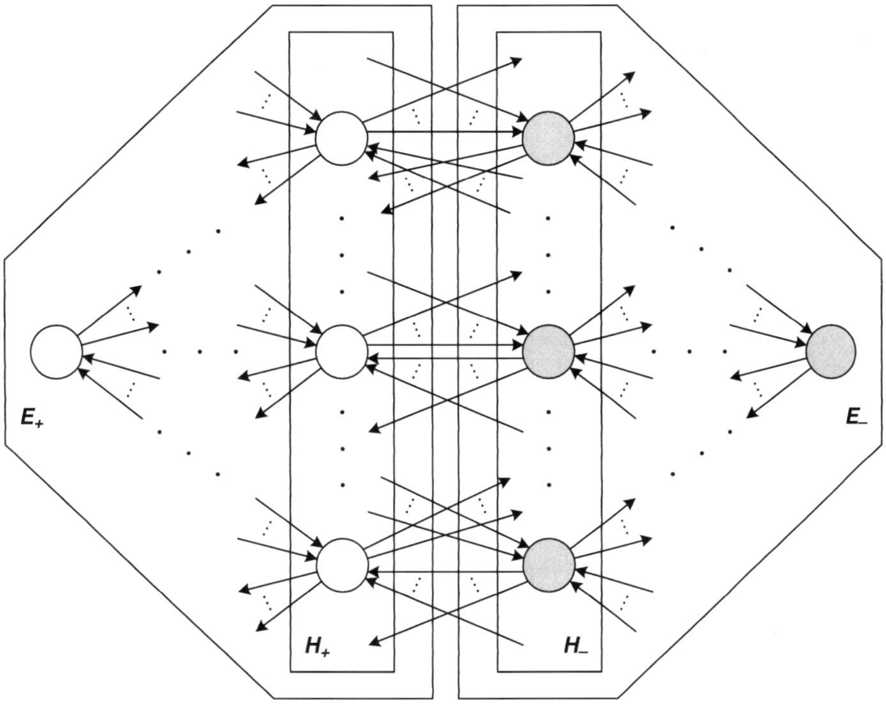

Figure 1. Generalized Markov chain in reliability models of the technical systems.

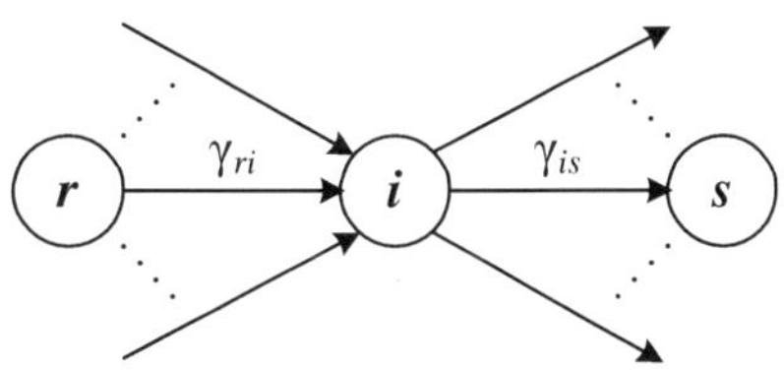

Figure 2. Inbound and outbound transitions for state *i*.

Mathematically the Markov chain model with continuous time represents the system of Kolmogorov-Chapman differential equations for each state $i \in E$. Having solved it, we can find the probabilities functions $P_i(t)$ at the given initial conditions $P_i(0)$ for all states of the technical system and, considering that the sum of probabilities of all states of the system is always equal to 1:

$$\begin{cases} \sum_{i \in E} P_i(t) = 1; \\ \forall i \in E: \dfrac{dP_i(t)}{dt} = \sum_{r \in R_i} (\gamma_{ri} P_r(t)) - P_i(t) \sum_{s \in S_i} \gamma_{is}. \end{cases} \tag{1}$$

At the stationary case $t \to \infty$, derivatives of probabilities functions tend to zero $\dfrac{dP_i(t)}{dt} \to 0$, and the system becomes the system of linear algebraic equations, having solved which it is possible to find stationary probabilities P_i for all states of the technical system:

$$\begin{cases} \sum\limits_{i \in E} P_i = 1; \\ \forall i \in E: \sum\limits_{r \in R_i} (\gamma_{ri} P_r) - P_i \sum\limits_{s \in S_i} \gamma_{is} = 0. \end{cases} \quad (2)$$

According to the topological method [3, 4], by using the stationary probabilities P_i we can calculate stationary availability factor K of the technical system as the sum of stationary probabilities of all upstates of the system.

The mean time to failure T_F of the technical system can be calculated as the ratio of the sum of stationary probabilities of all upstates to the weighed sum of stationary probabilities of border upstates. The sum of rates of transition links from border state to the all downstates is used as the weight for each of border upstates.

At last, it is also possible to calculate the mean time to repair T_R of the technical system using the similar formula. The only difference is that in numerator of the calculation formula it is used the sum of stationary probabilities of all downstates of the system.

$$K = \sum_{i \in E_+} P_i; \quad T_F = \frac{\sum\limits_{i \in E_+} P_i}{\sum\limits_{i \in H_+} P_i \sum\limits_{j \in E_-} \gamma_{ij}}; \quad T_R = \frac{\sum\limits_{i \in E_-} P_i}{\sum\limits_{i \in H_+} P_i \sum\limits_{j \in E_-} \gamma_{ij}}. \quad (3)$$

Example. A high-availability system with parallel structure consisting of 4 nodes with identical functionality and reliability parameters is given. The system is considered to be operable while at least one of nodes is in upstate. In figure 3 the Markov chain, which represents the reliability model of the high-availability system, is given.

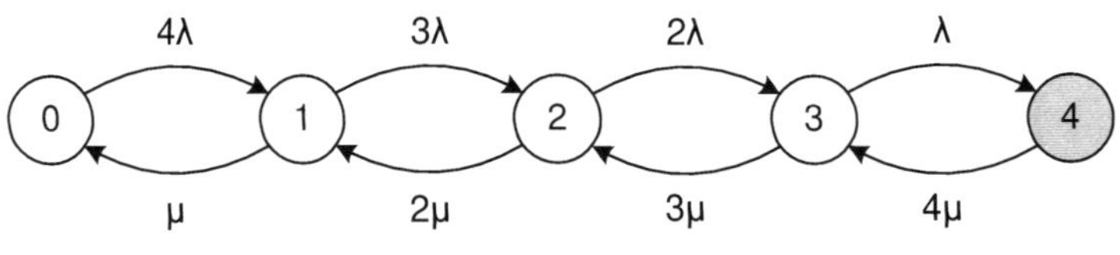

Figure 3. Markov chain for the system of 4 independent nodes.

State 0 – all nodes are in upstate, 1 – one node is failed, 2 – two nodes are failed, 3 – three nodes are failed, 4 – all nodes are in downstate. Accordingly, states 0, 1, 2 and 3 are considered to be operable for the system. Initial state of the system is 0 (all nodes are in upstate).

Nodes are independent from the viewpoint of failures and repairs, and represent repairable elements with exponential distribution laws of the failure and repair time.

Nodes failure rates are $\lambda = 10^{-20}$ hour^{-1}. Node repair rates are $\mu = 1$ hour^{-1}. Let us note that the values of the initial reliability parameters differ from each other on 20 orders.

Now, let us calculate the stationary availability factor, mean time to failure and mean time to repair of the given high-availability system by topological method. At first, we will solve the reliability task in analytic form with derivation of all the necessary calculating formulas.

In accordance with the given above Markov chain, we have the following stationary system of Kolmogorov-Chapman's equations:

$$\begin{cases} P_0 + P_1 + P_2 + P_3 + P_4 = 1; \\ -4\lambda P_0 + \mu P_1 = 0; \\ 4\lambda P_0 - (\mu + 3\lambda)P_1 + 2\mu P_2 = 0; \\ 3\lambda P_1 - (2\mu + 2\lambda)P_2 + 3\mu P_3 = 0; \\ 2\lambda P_2 - (3\mu + \lambda)P_3 + 4\mu P_4 = 0; \\ \lambda P_3 - 4\mu P_4 = 0. \end{cases} \tag{4}$$

By solving the system of equations in analytic form we obtain the formulas for calculation of the stationary probabilities for all states of the system:

$$P_0 = \frac{\mu^4}{(\mu+\lambda)^4}; \quad P_1 = \frac{4\mu^3\lambda}{(\mu+\lambda)^4}; \quad P_2 = \frac{6\mu^2\lambda^2}{(\mu+\lambda)^4}; \quad P_3 = \frac{4\lambda^3\mu}{(\mu+\lambda)^4}; \quad P_4 = \frac{\lambda^4}{(\mu+\lambda)^4}. \tag{5}$$

Considering that the system is operable only in the states 0-3, we obtain the following formula for the stationary availability factor:

$$K = P_0 + P_1 + P_2 + P_3 = \frac{\mu(\mu^3 + 4\mu^2\lambda + 6\mu\lambda^2 + 4\lambda^3)}{(\mu+\lambda)^4}. \tag{6}$$

Next, formula for the mean time to failure in accordance with the topological method and with taking into consideration the fact that the system has the only transition link with rate λ from upstate 3 into downstate 4:

$$T_F = \frac{P_0 + P_1 + P_2 + P_3}{\lambda P_3} = \frac{\mu^3 + 4\mu^2\lambda + 6\mu\lambda^2 + 4\lambda^3}{4\lambda^4}. \tag{7}$$

Finally, formula of the mean time to repair is:

$$T_R = \frac{P_4}{\lambda P_3} = \frac{1}{4\mu}. \tag{8}$$

Now, by substituting the given values of failures and repair rates into the obtained formulas, we obtain the following values of stationary probabilities for all states of the technical system:

$P_0 \approx 1$; $P_1 = 4\cdot10^{-20}$; $P_2 = 6\cdot10^{-40}$; $P_3 = 4\cdot10^{-60}$; $P_4 = 10^{-80}$.

Correspondingly, the stationary availability factor of the system: $K \approx 1$.

The mean time to failure of the system: $T_F \approx 2.5\cdot10^{79}$ hours.

The mean time to repair of the system: $T_R = 0.25$ hour.

It should be noted, that the key step in calculations of the reliability indices of the technical system by the topological method is solving of the system of linear algebraic equations for the corresponding Markov chain in reliability model.

In particular cases, for the small number of states it is possible to derive the analytical calculation formulas and obtain proper values for the reliability indices. However, for more difficult cases – tens and hundreds of system states, we obviously have to apply the numerical methods for solving the reliability task using the modern computers. In order to perform calculations on computer with providing high performance and simplicity of software implementation of the calculation method, it is usually used floating-point arithmetic supported by modern microprocessors in hardware. But here we can meet serious «hidden pitfall», which is discussed below.

3. Calculation of the reliability indices using the floating-point arithmetic

The modern computers in the majority are based on processors with Intel® x86™ architecture [5], containing as hardware processing unit for the integer arithmetic, so the hardware processing unit for the floating point arithmetic (beginning from Intel® Pentium™ processor).

The three types of real numbers with floating point are supported in hardware: 32-bit real numbers of single precision (single), 64- bit real numbers of double precision (double) and 80- bit real numbers of extended precision (extended). It stands to reason that 80-bit type provides the broadest range of the represented real numbers and the largest precision among the aforesaid three types.

The format of 80-bit real number is given below in figure 4.

79	78 · · 64	63	62 · · · 0
S	*E*	1	*M*

Figure 4. Format of 80-bit real number.

Where,

S (sign) – bit of sign, 0 – positive number, 1 – negative.

E (exponent) – 15-bit power of two, shifted on +16383.

M (mantissa) – 63-bit mantissa of a fractional part of number

The actual value of the floating-point number encoded in the 80-bit format can be calculated by the next formula:

$$X = (-1)^S (2^{E-16383})(1 + M/2^{63}). \qquad (9)$$

The range of the nonzero numbers is the following: $3.36 \cdot 10^{-4932} \leq |X| \leq 1.19 \cdot 10^{+4932}$. However, despite such broad range of the represented numbers, precision of numbers is quite restricted as far as mantissa contains only 63 binary bits, and actually 80-bit type of numbers in the best case can provide only 19 valid decimal digits.

Now, let us discuss the following example of 80-bit floating-point arithmetic. Let there be given a floating-point number $X = +1.0 = (-1)^0(2^{16383-16383})(1+0/2^{63})$ (figure 5).

79	78 · · 64	63	62 · · · 0
0	01 … 1	1	0 … 00

Figure 5. Initial floating-point number +1.0 in 80-bit format.

Let us add a small amendment $+10^{-19}$ to the initial value +1.0. In the 80-bit floating point arithmetic we receive the following result, which shows that 63-bit mantissa is still enough for representation of the sum value (figure 6):

79	78 · · 64	63	62 · · · 0
0	01 … 1	1	0 … 01

Figure 6. Result of calculation $(1.0 + 10^{-19})$ in 80-bit format.

However, it is necessary to mention, that even in this case we have precision error, as far as the number given above actually represents value $(1.0 + 2^{-63}) \approx (1.0 + 1.0842 \cdot 10^{-19})$, and it is obviously differing from the true sum value $(1.0 + 10^{-19})$.

Moreover, if we try to add more small amendment $+10^{-20}$ to the initial value +1.0, then 63-bit mantissa will not be able to provide representation of sum value, and the result in 80-bit format will look as if nothing at all was added, so actually it will represent the initial value +1.0 (figure 5). So, in the 80-bit floating point arithmetic the result of addition of +1.0 and $+10^{-20}$ is equal to +1.0.

Now, returning to the topological method of reliability indices calculation discussed above, we should note that the key stage of calculations is the solving of the linear algebraic equations system which involves a large number of arithmetical operations including operations of addition and subtraction. It is easy to see, that if initial data, as well as the intermediate data will differ from each other on the significant number of orders (20 and above), we can face incorrect results because of precision errors in floating-point arithmetic at the stage of solving of the equation system.

The authors programmed the numerical calculation of the reliability indices for the given above technical system of 4 independent elements in accordance with appropriate Markov chain (figure 3) and matrix of the transition rates W_{ij} from one state to another by using of the built-in function for solving the linear algebraic equation system in the Microsoft® Excel™, based on the inverse matrix method, and obtained the following values of probabilities of all states of the system P_j (figure 7).

It is rather obvious that only values of probabilities P_0 and P_1 are correct and equal to the calculated above values, and the values of other probabilities P_2, P_3 and P_4 are incorrect and absurd in point of fact, as the probabilities cannot be the negative by definition.

Respectively, the negative probability P_3 also results in absurd negative value of mean time to failure of $T_F = -2.5 \cdot 10^{59}$ hours after application of the corresponding topological formula (7).

		W_{ij}			P_j
0	4.00E-20	0	0	0	1
1	0	3.00E-20	0	0	4E-20
0	2	0	2.00E-20	0	-8.2E-56
0	0	3	0	1.00E-20	-4E-40
0	0	0	4	0	-3E-40

Figure 7. Example of incorrect results of numerical calculation of the reliability indices in Microsoft® Excel™ at application of the floating-point arithmetic.

4. Conclusion

Thus, the numerical calculation of the reliability indices of technical systems by the topological method, based on the solving of the system of linear algebraic equations, is affected by the precision error, if the floating-point arithmetic is used. In case of «bad» initial reliability parameters (difference of initial values among themselves on several tens orders and more), they can lead to incorrect calculation results or even impossibility of solving of the reliability task. Respectively, in such cases it is necessary to use the methods of error-free computation [6] based on using of the multi-precision rational numbers and arithmetic [7, 8], where each real value is represented by a multi-precision rational number as a relation of the two multi-precision integers, stored in memory as arrays of bytes.

Within the research work in the field of reliability models of the technical systems on basis of the Markov chains, the authors developed a software realization of the topological method for calculation of the reliability indices using the multi-precision rational numbers and arithmetic and obtained correct calculation results for a wide range of values of the initial reliability parameters for the several reliability models [9, 10]. In figure 8 is shown an example of correct results of numerical calculation of the reliability indices for the discussed above technical system of 4 independent elements in the author's calculation software at application of the multi-precision rational numbers and arithmetic.

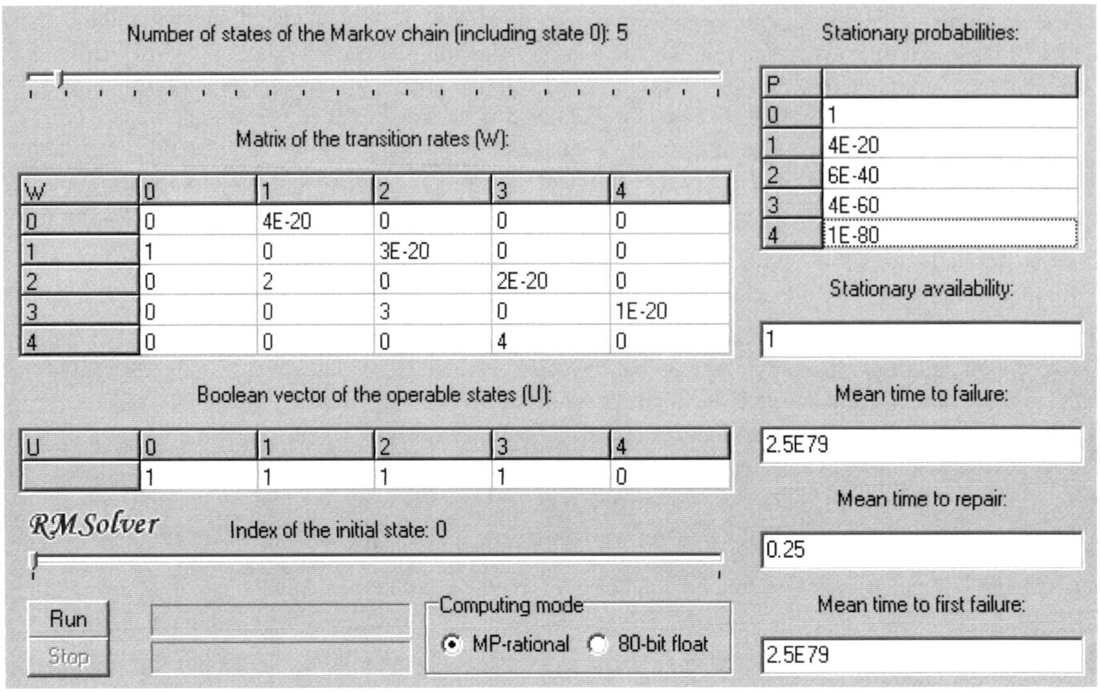

Figure 8. Example of correct results of numerical calculation of the reliability indices in the author's software at application of the multiple precision rational numbers and arithmetic.

References

[1] Shooman M L 2002 *Reliability of computer systems and networks* (John Wiley & Sons, Inc.)
[2] Koren I and Krishna C M 2007 *Fault-Tolerant Systems* (Morgan Kaufmann Publishers)
[3] Cherkesov G N 2005 *Reliability of Hardware and Software Systems* (Saint-Petersburg: Piter)
[4] Polovko A M and Gurov S V 2006 *Basis of Reliability Theory* (Saint-Petersburg: BHV-Petersburg)
[5] Barry B. Brey 2003 *The Intel Microprocessors. Architecture, Programming and Interfacing* (Pearson Education Inc.)
[6] R. T. Gregory and E. V. Krishnamurthy 1984 *Methods and Applications of Error-free Computation* (Springer)
[7] Donald E. Knuth 1998 *The Art of Computer Programming. Seminumerical Algorithms* (Addison-Wesley Longman Inc.)
[8] Tom St Denis 2006 *BigNum Math: Implementing Cryptographic Multiple Precision Arithmetic* (Syngress Publishing)
[9] Rahman P A and Bobkova E Yu 2017 *J. Phys.: Conf. Ser.* **803** 012124
[10] Rahman P A and Bobkova E Yu 2017 *J. Phys.: Conf. Ser.* **803** 012125

Acknowledgements

The authors are grateful to professor I. I. Ladygin from Moscow Power Engineering Institute for scientific support and theoretical base in the field of reliability models of technical systems.

IOP Conf. Series: Earth and Environmental Science **181** (2018) 012028 doi:10.1088/1755-1315/181/1/012028

Reliability model of the symmetric dual-node data processing system with functional redundancy

P A Rahman and V D Kondrateva

Department of Automated Technological and Informational Systems, Ufa State Petroleum Technological University, 2, October Avenue, Sterlitamak, 453118, Russia

E-mail: pavelar@yandex.ru

Abstract. This scientific paper deals with the reliability model of the symmetric system based on two data processing nodes with the functional redundancy. The simplified reliability model of the system and the advanced model offered by the authors considering nonzero time of activation of the node primary functions, nonzero times for node failure detection and switching to the function backing-up mode are presented. The formulas for calculation of stationary availability factor, mean time to failure and mean time to repair are also discussed. Finally, the examples of reliability indices calculations according to the simplified and advanced reliability models are also given.

1. Introduction

Nowadays not only the specifications concerning computational performance are important for the distributed data processing systems [1], but there are also important the reliability aspects, on which the safety and stability of information processes of the modern enterprises directly depends. Respectively, development of reliability models of data processing systems is rather urgent task.

There is a set of generalized reliability models and specialized reliability models of data processing systems which are discussed in the literature [2-5] based on the Markov chain models [6, 7]. However, these models do not take into account nonzero time of activation of the primary and backup functions of the data processing nodes in the specialized systems with the functional redundancy [8] used for information systems in the modern enterprises.

Within the research work the authors developed reliability models for the data processing systems with structural redundancy taking into account nonzero time of node activation [9, 10]. Respectively, within this scientific paper the authors developed the reliability model of symmetric dual-node data processing system with the functional redundancy taking into account time of activation of the primary and backup functions of nodes.

2. Simplified reliability model of symmetric dual-node system with functional redundancy

Let us overview a symmetric data processing system (Figure 1) with functional redundancy based on two identical nodes. The system interacts with the outside world via local area network obtaining source data from it and delivering processed data into it.

Data processing nodes have identical computational power and reliability parameters. In order to provide bi-directional control, the nodes also exchange the service data about their state using the internal heartbeat network.

Content from this work may be used under the terms of the Creative Commons Attribution 3.0 licence. Any further distribution of this work must maintain attribution to the author(s) and the title of the work, journal citation and DOI.
Published under licence by IOP Publishing Ltd

Each node by default performs only the primary function, but in case of failure of the neighbor node also performs the function of failed node in addition. After recovery of the failed node, the node performs only its primary function again. The functions have different meaning and purpose, but for simplification we will assume that from the reliability viewpoint they create identical load on data processing nodes. Data processing system is considered to be operable only in the case of availability of the both functions.

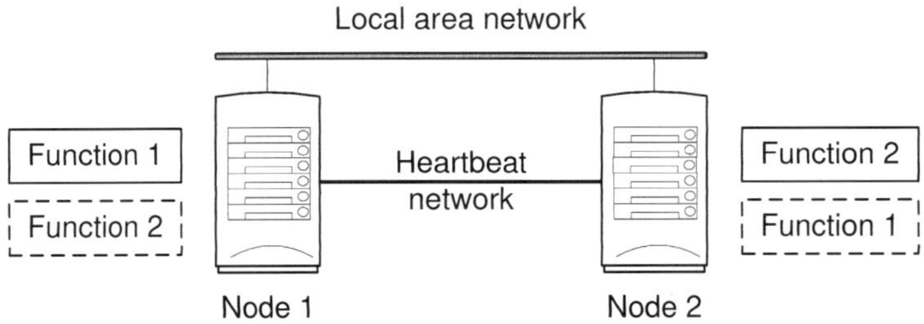

Figure 1. Symmetric system with two data processing nodes and functional redundancy.

In the simplified reliability model we will not take into account time necessary for the primary function activation, time of down-state detection and time of the neighbor node recovery, as well as entrance of the node into the backing-up mode and time of exit from the backing-up of the neighbor node function.

Each of nodes by default performs only its primary function, and in this mode its failure rate is λ_A. In case of failure of the other node it can also perform function of the neighbor node, and in this mode its failure rate is λ_B.

The nodes can be recovered independently (unlimited repair). Node repair rate is μ_N.

Then, taking into account symmetry of the system, let us consider the following set of system states and graphs of transitions from one state to another (Figure 2):

- State AA (Active / Active) – both nodes are in upstate, and every node performs only its own function.
- State FB (Failed / Backup) – one node is in downstate; the other node performs as its own function, so the function of the failed node.
- State FF (Failed / Failed) – both nodes are in downstate.

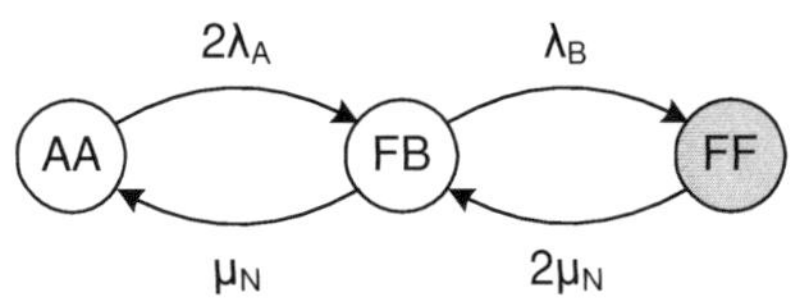

Figure 2. Simplified reliability model of the symmetric system of the dual-node system with functional redundancy.

In accordance, the stationary system of the Kolmogorov-Chapman equation is as follows:

$$\begin{cases} P_{AA} + P_{FB} + P_{FF} = 1; \\ 0 = -2\lambda_A P_{AA} + \mu_N P_{FB}; \\ 0 = 2\lambda_A P_{AA} - (\mu_N + \lambda_B) P_{FB} + 2\mu_N P_{FF}; \\ 0 = \lambda_B P_{FB} - 2\mu_N P_{FF}. \end{cases} \quad (1)$$

Then, with due regard that only in the states AA and FB the system is operable and provides the both functions, its stationary availability factor can be calculated as the sum of probabilities of the states AA and FB:

$$K_{\mathrm{SYS}} = P_{\mathrm{AA}} + P_{\mathrm{FB}} = \frac{\mu_{\mathrm{N}}(\mu_{\mathrm{N}} + 2\lambda_{\mathrm{A}})}{\mu_{\mathrm{N}}(\mu_{\mathrm{N}} + 2\lambda_{\mathrm{A}}) + \lambda_{\mathrm{A}}\lambda_{\mathrm{B}}}. \tag{2}$$

Hereafter, the mean time to failure of the system can be calculated as ratio of the sum of probabilities of up-states AA and FB to the probability of border up-state FB, multiplied by the rate λ_{B} of transition into down-state FF:

$$T_{\mathrm{FSYS}} = \frac{P_{\mathrm{AA}} + P_{\mathrm{FB}}}{\lambda_{\mathrm{B}} P_{\mathrm{FB}}} = \frac{\mu_{\mathrm{N}} + 2\lambda_{\mathrm{A}}}{2\lambda_{\mathrm{A}}\lambda_{\mathrm{B}}}. \tag{3}$$

At last, the mean time to repair of the system can be calculated from a formula of the reliability theory, which binds the stationary availability factor with mean time to failure and mean time to repair via simple equation $K = T_{\mathrm{F}}/(T_{\mathrm{F}} + T_{\mathrm{R}})$:

$$T_{\mathrm{RSYS}} = \frac{1 - K_{\mathrm{SYS}}}{K_{\mathrm{SYS}}} T_{\mathrm{FSYS}} = \frac{1}{2\mu_{\mathrm{N}}}. \tag{4}$$

3. Advanced reliability model of symmetric dual-node system with functional redundancy

In the advanced reliability model, we will consider data processing nodes as a complex of hardware-software systems which require nonzero time for activation (transition from passive to active state after completion of the software initialization) of its primary function.

Also, we will take into consideration the nonzero time of detection of neighbor node failure and entrance into the backing-up mode of its function, as well as nonzero time of detection of neighbor node recovery and exit from the backing-up mode of its function.

Now let us consider the following set of states (Figure 3) for a node as a part of the dual-node data processing system with functional redundancy:

- State P (Passive) – node is in upstate but it is still passive, does not perform any function and waiting for completion of the software initialization.
- State A (Active) – node is in upstate and performs its primary function.
- State B (Backing-up) – node is in upstate and performs as its primary function, so the function of the failed neighbor node.
- State F (Failed) – node is in downstate.

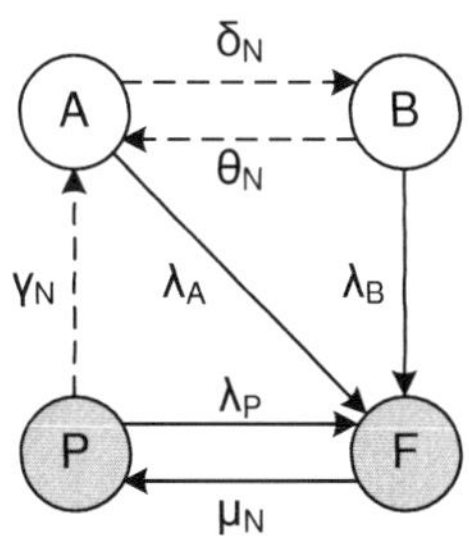

Figure 3. Reliability model of the data processing node within the symmetric dual-node system with functional redundancy.

Where,

λ_P – node failure rate in passive state;

λ_A – node failure rate in active state;

λ_B – node failure rate in backing-up state;

μ_N – node repair rate;

γ_N – rate of node primary function activation;

δ_N – rate of entrance into the mode of backing-up of the neighbor node function;

θ_N – rate of exit from the mode of backing-up of the neighbor node function.

It should be noted, that node cannot be in state B, if another node is in states A or B, thus the nodes cannot perform any of the functions simultaneously, and from this viewpoint they are interdependent. Now taking into account this interdependence of the nodes and symmetry of the system (identical nodes from the reliability viewpoint), we consider the following set of states of the dual-node data processing system with functional redundancy (Figure 4):

- State PP (Passive / Passive) – the both nodes are in up-state, but still passive, do not perform their functions and are waiting for completion of the software initialization.
- State AP (Active / Passive) – the both nodes are in upstate, one of them is still passive and does not perform its function, the other one is active and performs its own function.
- State PF (Passive / Failed) – one node is in downstate; the other one is in upstate, but passive and does not perform its function.
- State AF (Active / Failed) – one node is in downstate; the other one is in upstate, active and performs its own function.
- State FF (Failed / Failed) – the both nodes are in downstate.
- State PB (Passive / Backup) – both nodes are in upstate, one of the nodes is passive, the other one is in backing-up mode and performs as its own function, so the function of the passive neighbor node.
- State AA (Active / Active) – both nodes are in upstate and each of them performs only its own function.
- State FB (Failed / Backup) – one of the nodes is in downstate; the other one performs its own function and function of the failed neighbor node.

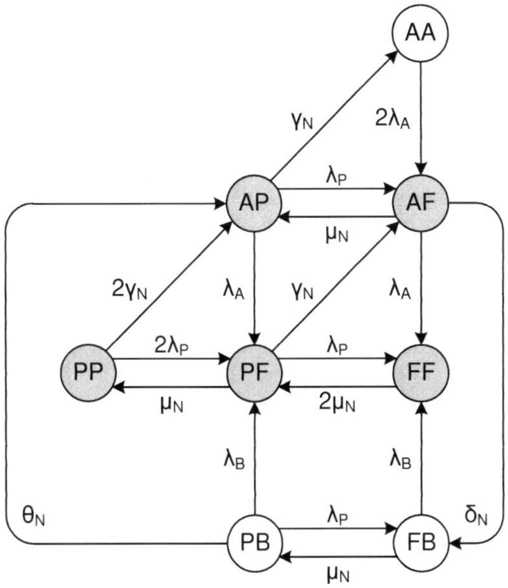

Figure 4. Advanced reliability model of the symmetric system with two nodes of data processing with functional redundancy.

Accordingly, the stationary system of the Kolmogorov-Chapman equation for the aforesaid Markov chain is as follows:

$$\begin{cases} P_{PP} + P_{AP} + P_{PF} + P_{AF} + P_{FF} + P_{AB} + P_{FB} + P_{AA} + P_{PB} = 1; \\ 0 = -(2\lambda_P + 2\gamma_N)P_{PP} + \mu_N P_{PF}; \\ 0 = 2\gamma_N P_{PP} - (\lambda_P + \lambda_A + \gamma_N)P_{AP} + \mu_N P_{AF} + \theta_N P_{PB}; \\ 0 = 2\lambda_P P_{PP} + \lambda_A P_{AP} - (\mu_N + \lambda_P + \gamma_N)P_{PF} + 2\mu_N P_{FF} + \lambda_B P_{PB}; \\ 0 = \lambda_P P_{AP} + \gamma_N P_{PF} - (\mu_N + \lambda_A + \delta_N)P_{AF} + 2\lambda_A P_{AA}; \\ 0 = \lambda_P P_{PF} + \lambda_A P_{AF} - 2\mu_N P_{FF} + \lambda_B P_{FB}; \\ 0 = -(\lambda_P + \lambda_B + \theta_N)P_{PB} + \mu_N P_{FB}; \\ 0 = \gamma_N P_{AP} - 2\lambda_A P_{AA}; \\ 0 = \delta_N P_{AF} - (\lambda_B + \mu_N)P_{FB} + \lambda_P P_{PB}. \end{cases} \quad (5)$$

The authors obtained the analytical solution of the system of equations (5). Taking into account that only in states PB, AA and FB the system is operable, when each node performs its own function or one node performs the both functions (its own and neighbor's), the following analytical formulas for calculation of stationary rate of the system availability, mean time to failure and mean time to repair were obtained by the authors:

$$K_{SYS} = P_{PB} + P_{AA} + P_{FB} = \frac{M_5 + M_6 + M_7}{1 + M_1 + M_2 + M_3 + M_4 + M_5 + M_6 + M_7}. \quad (6)$$

$$T_{FSYS} = \frac{P_{PB} + P_{AA} + P_{FB}}{(\lambda_B + \theta_N)P_{PB} + 2\lambda_A P_{AA} + \lambda_B P_{FB}} = \frac{M_5 + M_6 + M_7}{(\lambda_B + \theta_N)M_5 + 2\lambda_A M_6 + \lambda_B M_7}. \quad (7)$$

$$T_{\text{RSYS}} = \frac{1-K_{\text{SYS}}}{K_{\text{SYS}}} T_{\text{FSYS}} = \frac{1+M_1+M_2+M_3+M_4}{(\lambda_{\text{B}}+\theta_{\text{N}})M_5+2\lambda_{\text{A}}M_6+\lambda_{\text{B}}M_7}. \quad (8)$$

Where,

$$M_1 = \frac{2\gamma_{\text{N}}}{\lambda_{\text{A}}+\lambda_{\text{P}}+\gamma_{\text{N}}}\left(1+\frac{1}{\Psi}\left(\begin{array}{c}(\lambda_{\text{P}}+\gamma_{\text{N}})(\mu_{\text{N}}+\lambda_{\text{A}}+\lambda_{\text{P}}+\gamma_{\text{N}})\times \\ \times(\lambda_{\text{B}}\lambda_{\text{P}}+(\lambda_{\text{B}}+\mu_{\text{N}})(\lambda_{\text{B}}+\theta_{\text{N}})+\theta_{\text{N}}\delta_{\text{N}})\end{array}\right)\right);$$

$$M_2 = \frac{2\lambda_{\text{P}}+2\gamma_{\text{N}}}{\mu_{\text{N}}};\ M_3 = \frac{2\gamma_{\text{N}}(\lambda_{\text{P}}+\gamma_{\text{N}})(\mu_{\text{N}}+\lambda_{\text{A}}+\lambda_{\text{P}}+\gamma_{\text{N}})(\lambda_{\text{B}}\lambda_{\text{P}}+(\lambda_{\text{B}}+\mu_{\text{N}})(\lambda_{\text{B}}+\theta_{\text{N}}))}{\mu_{\text{N}}\Psi};$$

$$M_4 = \frac{\lambda_{\text{P}}(\lambda_{\text{P}}+\gamma_{\text{N}})}{\mu_{\text{N}}^2}\left(1+\frac{\gamma_{\text{N}}}{\lambda_{\text{P}}\Psi}\left(\begin{array}{c}(\mu_{\text{N}}+\lambda_{\text{A}}+\lambda_{\text{P}}+\gamma_{\text{N}})\times \\ \times\left(\begin{array}{c}\lambda_{\text{A}}(\lambda_{\text{B}}\lambda_{\text{P}}+(\lambda_{\text{B}}+\mu_{\text{N}})(\lambda_{\text{B}}+\theta_{\text{N}}))+ \\ +\lambda_{\text{B}}\delta_{\text{N}}(\lambda_{\text{P}}+\lambda_{\text{B}}+\theta_{\text{N}})\end{array}\right)\end{array}\right)\right);$$

$$M_5 = \frac{2\gamma_{\text{N}}\delta_{\text{N}}(\lambda_{\text{P}}+\gamma_{\text{N}})(\mu_{\text{N}}+\lambda_{\text{A}}+\lambda_{\text{P}}+\gamma_{\text{N}})}{\Psi};$$

$$M_6 = \frac{\gamma_{\text{N}}^2}{\lambda_{\text{A}}(\lambda_{\text{A}}+\lambda_{\text{P}}+\gamma_{\text{N}})}\left(1+\frac{1}{\Psi}\left(\begin{array}{c}(\lambda_{\text{P}}+\gamma_{\text{N}})(\mu_{\text{N}}+\lambda_{\text{A}}+\lambda_{\text{P}}+\gamma_{\text{N}})\times \\ \times(\lambda_{\text{B}}\lambda_{\text{P}}+(\lambda_{\text{B}}+\mu_{\text{N}})(\lambda_{\text{B}}+\theta_{\text{N}})+\theta_{\text{N}}\delta_{\text{N}})\end{array}\right)\right);$$

$$M_7 = \frac{2\gamma_{\text{N}}\delta_{\text{N}}(\lambda_{\text{P}}+\gamma_{\text{N}})(\mu_{\text{N}}+\lambda_{\text{A}}+\lambda_{\text{P}}+\gamma_{\text{N}})(\lambda_{\text{P}}+\lambda_{\text{B}}+\theta_{\text{N}})}{\mu_{\text{N}}\Psi};$$

$$\Psi = \left(\begin{array}{c}\lambda_{\text{A}}(\mu_{\text{N}}+\lambda_{\text{A}}+\lambda_{\text{P}}+\gamma_{\text{N}}+\delta_{\text{N}})(\lambda_{\text{B}}\lambda_{\text{P}}+(\lambda_{\text{B}}+\mu_{\text{N}})(\lambda_{\text{B}}+\theta_{\text{N}}))+ \\ +\delta_{\text{N}}\lambda_{\text{B}}(\lambda_{\text{P}}+\gamma_{\text{N}})(\lambda_{\text{P}}+\lambda_{\text{B}}+\mu_{\text{N}}+\theta_{\text{N}})\end{array}\right).$$

4. Examples of reliability indices calculation

A symmetric system with functional redundancy based on two identical data processing nodes with the following initial reliability parameters is given:

- Node failure rate in passive state: $\lambda_{\text{P}} = 1/8760$ hour^{-1} (on average one failure per year).
- Node failure rate in active state: $\lambda_{\text{A}} = 2/8760$ hour^{-1} (on average 2 failures per year).
- Node failure rate in backing-up state: $\lambda_{\text{B}} = 3/8760$ hour^{-1} (on average 3 failures per year).
- Node repair rate: $\mu_{\text{N}} = 1/12$ hour^{-1} (on average one recovery per 12 hours).
- Node activation rate: $\gamma_{\text{N}} = 20$ hour^{-1} (on average one activation per 3 minutes).
- Rate of entrance into the mode of backing-up of the neighbor node function: $\delta_{\text{N}} = 20$ hour^{-1} (on average one entrance into the backing-up mode per 3 minutes).
- Rate of exit from the mode of backing-up of the neighbor node function: $\theta_{\text{N}} = 20$ hour^{-1} (on average one exit from the backing-up mode per 3 minutes).

In order to estimate reliability indices in accordance with the simplified reliability model, we calculate stationary availability factor, mean time to failure and mean time to repair of the system using the formulas 2, 3 and 4. To estimate reliability indices in accordance with the advanced reliability model, we use the formulas 6, 7 and 8, correspondingly.

Results of the reliability indices calculations are given in table 1.

Table 1. Results of reliability indices calculations for the symmetric dual-node data processing system.

Reliability model	K_{SYS}	T_{FSYS} (hours)	T_{RSYS} (hours)
Simplified	0.999988802	535820	6
Advanced	0.999943315	1101	0.0624

5. Conclusion

Results of reliability indices calculations of dual-node data processing system with functional redundancy show that though the values of availability factor, obtained from the both reliability models, are close to each other, the value obtained from the advanced model is lower as far as it takes into consideration nonzero time of activation of the nodes primary function, detection of downstate of neighbor node and entrance into the backing-up mode of its function.

As for values of mean time to failure and mean time to repair, there are essential distinctions between the values obtained from the simplified and advanced reliability models.

The simplified model shows that the fault-tolerant system with two nodes possesses very long time before failure and it corresponds more to the model of ideal duplex system in which it is considered that system failure comes only after a failure of the both nodes.

The advanced reliability model shows that its mean time to failure is very short and high value of the availability factor is reached due to the short mean time to repair. So, this model more corresponds to real dual-node system with functional redundancy. The real system during most of the time is in the state when the both nodes are active, and each of them performs only its primary function. Failure of any of them results in short unavailability time of one of the system functions, and consequently in short downstate time of the system in general. However, due to the quite fast failure detection of the neighbor node and switching of its function to the remained operable node, the availability factor of the system remains to be rather high.

References

[1] Andrew S. Tanenbaum and Maarten van Steen 2002 *Distributed Systems* (Prentice Hall Inc.)
[2] Ostreykovsky V A 2003 *Reliability Theory* (Moscow: Vyshaya Shkola)
[3] Ushakov I A 2008 *Reliability Theory Course* (Moscow: Drofa)
[4] Rausand M and Holyand A 2009 *System Reliability Theory* (John Wiley & Sons)
[5] Abd-El-Barr M 2007 *Design and Analysis of Reliable and Fault-Tolerant Computer Systems* (London: Imperial College Press)
[6] Kijma M 1997 *Markov Processes for Stochastic Modeling* (London: Chapman & Hall)
[7] Anderson W J 1991 *Continuous-Time Markov Chains* (Springer-Verlag)
[8] Chandra Kopparapu 2002 *Load Balancing Servers, Firewalls and Caches* (John Wiley & Sons, Inc.)
[9] Rahman P A and Bobkova E Yu 2017 *J. Phys.: Conf. Ser.* **803** 012124
[10] Rahman P A and Bobkova E Yu 2017 *J. Phys.: Conf. Ser.* **803** 012125

Acknowledgements

The authors are grateful to professor I. I. Ladygin from Moscow Power Engineering Institute for scientific support and theoretical base in the field of reliability models of technical systems.

IOP Conf. Series: Earth and Environmental Science **181** (2018) 012029 doi:10.1088/1755-1315/181/1/012029

Creation of an automation system for engineering calculation of preparation for the production at high-technology enterprises of mechanical engineering

N N Savelyeva

Industrial University of Tyumen, Branch in the city of Nizhnevartovsk, Russia

E-mail: nnsavelieva@yandex.ru

Abstract: The article considers the creation of the automation system for labor-intensive engineering calculations, which is carried out by a technologist to prepare for work of metal-cutting equipment. This way allows to reduce the time for the process of technical preparation of production. The all calculations are performed using modeling in the program Borland Delphi 7. The main technical parameters of the technological preparation of production are marked. The information model of the cutting system is designed. The modeling algorithm for selecting optimal parameters for the operation of the equipment is made. A graphical interface has been created for sequential selection of equipment and production tooling for operation of the technological process. Consequently, the program allows to automate the creation and processing of technological documentation, and reduce the time for designing technological processes in enterprises. Also, the developed interface can be used in the educational process to study production tasks.

1. Introduction

A problem of automation of technological preparation of production exists in the conditions of the development of high-tech engineering production in Russia, especially designing technological processes, operation development, tooling selection and calculation of cutting modes for process operations. It is necessary for solving this problem to automate the creation of technology for manufacturing products and calculations for the organization of work of metal cutting equipment.

There are many programs created to automate the work of a technologist and designer. All of them have a common problem: there is no situation accounting in a particular production. Necessity arises for long training and adaptation of personnel, adjusting the program for the conditions of the enterprise. Therefore, it is important to create a software from scratch, and taking into account all the features of technological equipment and organization of the production process.

2. Research

The purpose of our research was the creation of a program for the automated design of technological processes and calculation of cutting regimes of technological process. Modern metal cutting machining centers were selected for the research.

Objectives of the research:

1. Automation of designing of technological process of product manufacturing.
2. Automation of technological calculations with optimization of cutting modes.

Content from this work may be used under the terms of the Creative Commons Attribution 3.0 licence. Any further distribution of this work must maintain attribution to the author(s) and the title of the work, journal citation and DOI.
Published under licence by IOP Publishing Ltd

IOP Conf. Series: Earth and Environmental Science **181** (2018) 012029 doi:10.1088/1755-1315/181/1/012029

3. Results and discussion

Research methods: analytic method and model analysis.

Description of the experiment. An experimental site equipped with high-tech equipment was created for the experiment [3]. Devices having computer numerical control of type WinNC: lathe with EMCO Turn 55 and milling Mill 155 became the technical means of carrying out the experiment. The following program software was used: Borland Delphi, CAD; systems: Kompas-3D, SolidWorks, Pro-Engineer, the program of monitoring and diagnostics, simulation and editing of processing of a detail, etc.

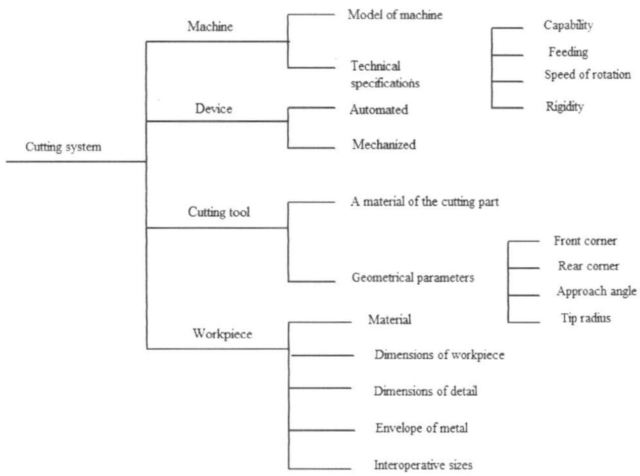

Figure 1. Information model of the cutting system.

Initially, we marked the main technical parameters that affect the cutting system: machine, device, cutting tool, workpiece. After that the information model of the cutting system was compiled that shown in Figure 1. Next, we compiled an algorithm for calculating the cutting modes for the operations of the technological process shown in Figure 2.

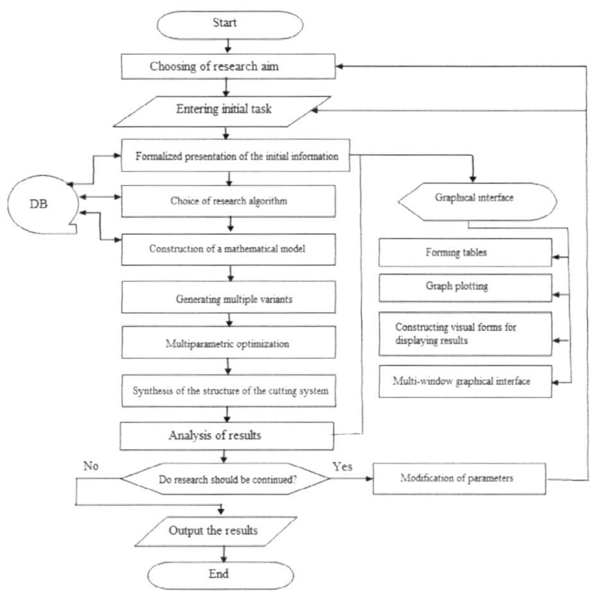

Figure 2. The structure of the modeling algorithm.

The algorithm includes the following components: the statement of the problem, nomination and testing of hypothesis, formalization of the initial information, generating of alternative variants, expert evaluation and screening out of rough variants, construction of a mathematical model, formation of a modeling algorithm, optimization of the solution of a mathematical model, visualization of the task solution results, the analysis of the results, the variation of the initial data and models in a series of successive iterations of the research process, the creation of a database and a number of other steps.

Next, we made a graphical interface for modeling and calculating the optimal parameters of the cutting process, that is shown in Figure 3. This interface was made in the program of Borland Delphi.

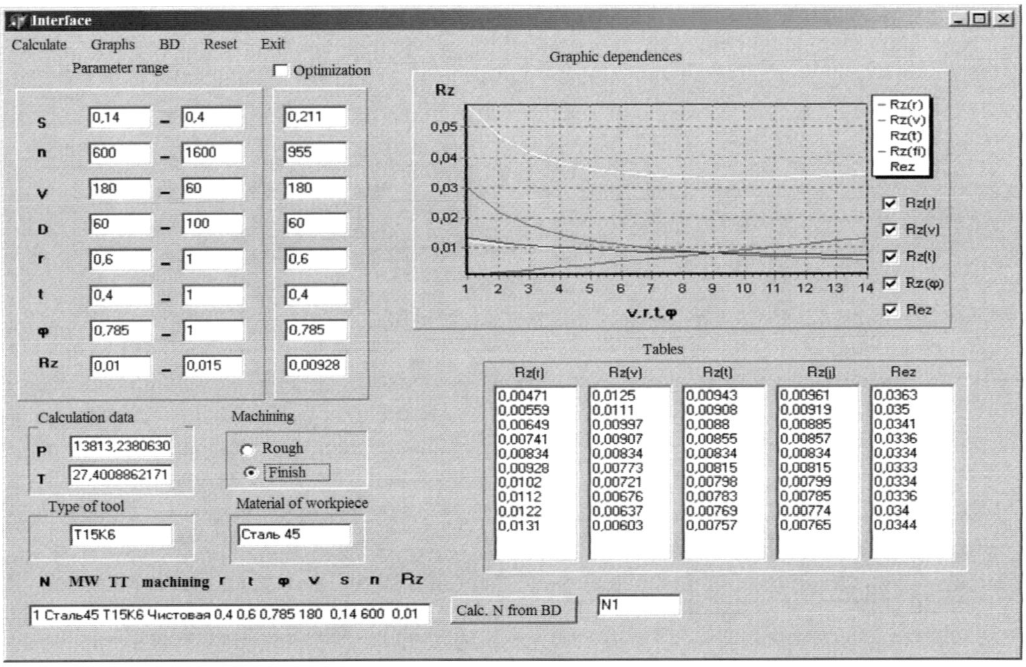

Figure 3. The interface for calculating the operating modes of equipment.

This is due to the fact that, the choice of optimal processing modes is a very laborious and complex task, because of the following reasons:

1) the efficiency of the manufacturing process of parts on metal cutting machines depends on many interrelated factors such as processing modes, tool geometric parameters and its wear, the properties of the workpiece material and tool material, the stiffness of the SPID system (the abbreviation in Russian that means machine, device, tool, part) etc. [4].

2) there is a wide variety of factors that affect the cutting process ambiguously, i.e. some of them are contradictory from the point of view of optimality of efficiency criteria such as, for example, roughness, accuracy, productivity, cost price, etc .;

3) most of the parameters of the cutting modes specified in the reference books and various normative materials have a certain range, which is not optimal, and has a recommendatory character. [5,6];

4) many dependencies of the cutting process are non-linear. Dependence of the wear rate of the tool on the cutting speed, which has a pronounced nonmonotonic character with a very narrow range of optimum cutting speeds, is an example for this.[2];

5) The tasks of optimizing the cutting process are multiparameter, multicriteria.

But, at the same time, among the set of acceptable variants of machining of parts, there is an optimal variant, estimated by singular criteria and as well as by additive criteria of efficiency [1,2]. Finding the optimal variant of the treatment process by experimental means is a complex and time-consuming task that requires a lot of money and time. To reduce the labour intensity and exclude expensive field studies, it is necessary to use computers and optimization methods that exclude a complete enumeration of the set of admissible variants.

A certain set of initial parameters is shown on the graphic form: cutting modes; the workpiece material; geometric parameters of tools; type of processing, etc. An individual composition of parameters for solving a specific tasks will be chosen from the set of initial parameters.

Output data include the following: calculation results of Rz, P, s and etc.; dependency tables Rz=f(r,v,t,φ), P=f(T,D,s,t,n); the panel of graphical plotting of diagrams according to the table data; optimal values of the parameters. The plotting of diagrams is accomplish in accordance with indicated their type and quantity.

The studying mathematical models are presented in the form of a set of multiparameter nonlinear dependencies and functions. These functions are stored in the database and used to form the optimization model and present them in the required format. A "coordinate descent method" for multidimensional functions [2], and heuristic algorithms are used to find the optimal value of the efficiency criterion. Limitations are valid parameter values. Optimization is accomplished out by variation these parameters in the specified boundary ranges. The optimal solution is given in the form of refined values of the parameters where the objective function takes the extremal value Fmin (or Fmax) for a given efficiency criterion.

Figure 3 shows the data of determining the optimal parameters of the cutting mode according to the roughness criteria Rz and the productivity P. We express the dependence of the roughness and productivity criteria on the cutting modes in the following form [4]:

$$R_z = \frac{\sqrt[0.4]{\left(8.25 r^{0.15} V \Big/ t^{0.3} \, sin\varphi^{0.4} \right)^2}}{8r}, \tag{1}$$

r – radius at the top of the lathe tool, mm; V – cutting speed, m/min; t – depth of cut, mm; φ - the main angle in the plan, deg.

$$P = \frac{\pi \cdot D \cdot n \cdot t \cdot s \cdot}{1 + \frac{\tau_{ch}}{\mathrm{T}}}, \tag{2}$$

D – workpiece diameter, mm; n – rotation speed, rpm.; t – depth of cut, mm.; s – feed, mm/rev; τ_{ch}– tool change time, min; T – withstandability of lathe tools, min.

The withstandability of lathe tools is determined by the following formula:

$$T = \frac{C_v \cdot k_v}{v \cdot t^x \cdot s^y}, \tag{3}$$

v – cutting speed, m/min; C_v – a constant value for determining the group of processed material; k_v – a coefficient, depending on the properties of the processed material and the cutting tool, on its geometry, wear, and also on the lubricating-cooling liquid; x,y – indexes of degree depending on tool properties and cutting conditions.

Substituting (3) into (2) and taking for the perturbing effect a change in the size of the allowance for a constant t, we obtain the objective function:

$$P = \frac{\pi \cdot \mathrm{D} \cdot \mathrm{n} \cdot \mathrm{t} \cdot \mathrm{s}}{1 + \frac{\tau_{\mathrm{ch}} \cdot t^x \cdot s^y \cdot v}{C_v \cdot k_v}}, \tag{4}$$

Then the databases for each parameter of the cutting system were created by us. Figure 4 shows a database for selecting and adding a tool. In this database you can select the tool by material, by typical size and by number. The database provides technical characteristics, markings and other parameters necessary for the technical calculation of the operation of the technical process.

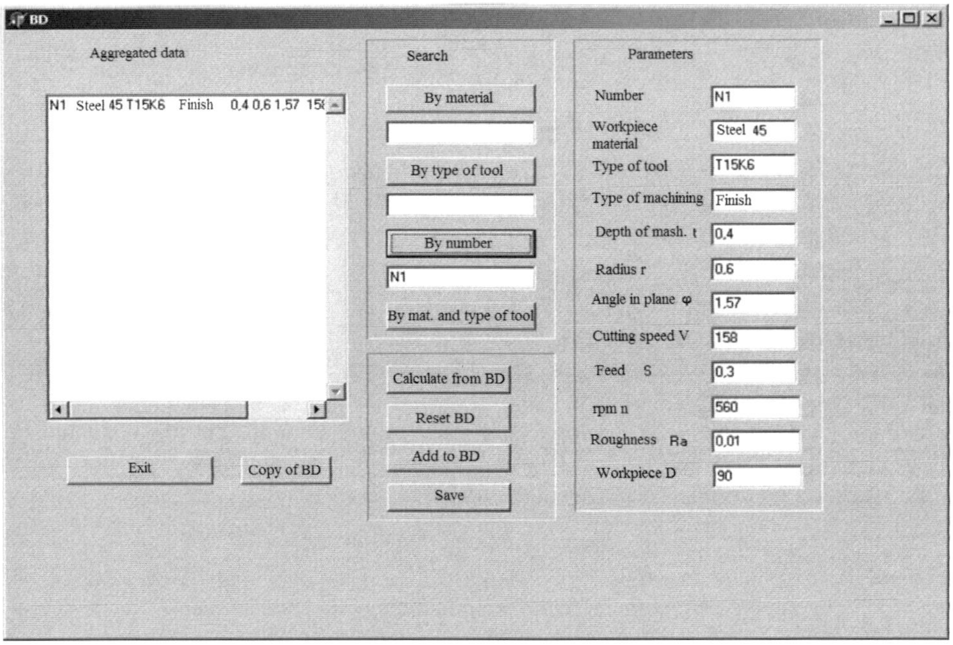

Figure 4. The database of cutting tools.

Figure 5 shows the database of workshop equipment.

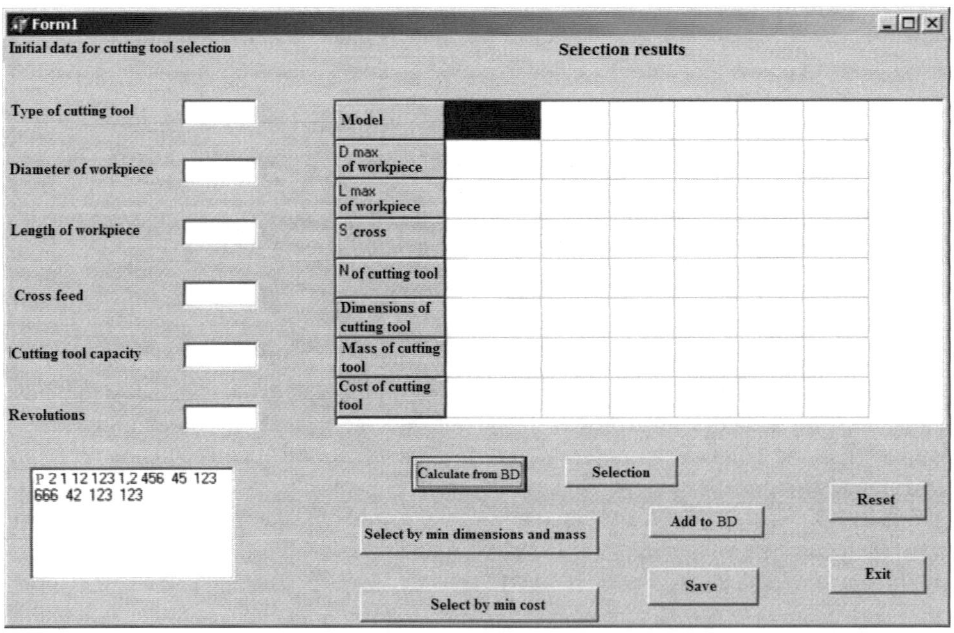

Figure 5. The database of metal cutting tools.

IOP Conf. Series: Earth and Environmental Science **181** (2018) 012029 doi:10.1088/1755-1315/181/1/012029

The database is intended for entering technical characteristics of metal-cutting machine tools in the database of the enterprise. It is possible in the database to constantly make changes. During purchasing new equipment, it is entered in the database. During it is written off or out of order, it is deleted from the database.

4. Conclusion

The created program allows doing the following:

1. Automating the execution of technological documentation and simplify the calculations for technological operations;
2. Reducing the design time of technological processes;
3. Using the developed interface in the learning process to study production tasks.

The result of our work is the developed interface for calculating cutting modes for metal cutting machines and a catalog in the form of a database of cutting tools and equipment passports.

References

[1] Grubiy S V 2014 Optimization of the Machining Process and Control of the Operating Parameters (Moscow: Mechanical engineering) 422 p

[2] Makarov A D 1976 Optimization of Cutting Processes (Moscow: Mechanical engineering) 278p

[3] Savelyeva N N 2017 *Schooling of Future Bachelors-Oil Workers for Professional Work in High-Tech Enterprises* (Tyumen) 122 p

[4] Savelyeva N N, Bogolubova M N and Proskuryakov P U 2012 Design-technological schooling of engineering students on the basis of electronic educational resources Fundamental Research 6-2 (Moscow) pp 388-91

[4] 1985 *Handbook of the Technologist-Mechanical Engineer* ed Kosilova A G and Mesheryakova R K (Moscow: Mechanical engineering)

[5] Wu B *Manufacturing Systems Design & Analysis: Context and Techniques* 2/e (UK: Chapman & Hall)

[6] Sandvik Coromant 2015 *Common Turning Processes* Digital Library 18 p

[7] Vinesh Raja and Kiran J Fernandes 2008 Reverse Engineering – an Industrial Perspective (London: Springer)

IOP Conf. Series: Earth and Environmental Science **181** (2018) 012030 doi:10.1088/1755-1315/181/1/012030

Mathematic modelling of filtering data of an oil reservoir based on successive random additions

Semenov B V, Svatov V P, Semenova S P

Industrial University of Tyumen, 38 Volodarskogo st., Tyumen, 625000, Russia

E-mail: semenov062@ya.ru

Abstract: The article presents a method of successive random additions for modelling intermediate values between wells, the data of which are obtained from deposits by special geographic probes. The main idea is to identify the relationship between the data obtained and further calculations in the form of pseudo-values that can be used for further research. Fractal data filtering allows modelling pseudo-random values based on correlated values of known data, simulating random motion. This approach will allow a more detailed study of the petrophysical properties of individual sections of the formation.

1. Introduction

The period of recovering "light" oil is almost over. Extraction is hampered by various complications [1]. Specialists in the oil industry are making great efforts to design new methods of developing hard-to-reach oil deposits [2]. But the extraction of hard-to-reach reserves does not always bring profit. One such example is oil shale deposits which are difficult to reach. Therefore, reservoir development technologies with shale reservoirs can be effectively implemented only after a comprehensive study of the filtration mechanisms. The relevance of this work is the possibility of applying the method for the study of shale deposits, the method of successive random additions makes it possible to compensate for the missing data and conduct a detailed study of the formation.

The purpose of this study is to apply mathematical modelling using computer technology and develop a general method and description of the formation that can be used for further research in the oil and gas industry. The development of the reservoir description technique was achieved due to the study of geostatistics, which allows modelling the phenomena of processes in a geographic space. The study of fractal geometry made it possible to apply the method of successive random additions to model intermediate values between wells, the data of which were obtained from deposits by special geographic probes. The main idea of this method is to identify the relationship between the data obtained and further calculations in the form of pseudo-values that can be used for further research. Fractal data filtering allows modelling pseudo-random values based on correlated values of known data, simulating random motion.

2. Research

For the purpose of further research, the application to the method is the use of such known methods as the Archie, Waxman-Smits model, which allow an evaluation of the petrophysical properties of the obtained pseudo-values between the wells [3], [4]. This solution allows building gamma-ray logs, investigating permeability, building a lateral section of layers of the formation at a given depth. Application of these methods will allow presenting the obtained data in the form of blocks into which

Content from this work may be used under the terms of the Creative Commons Attribution 3.0 licence. Any further distribution of this work must maintain attribution to the author(s) and the title of the work, journal citation and DOI.
Published under licence by IOP Publishing Ltd

IOP Conf. Series: Earth and Environmental Science **181** (2018) 012030 doi:10.1088/1755-1315/181/1/012030

the investigated oil-bearing layer is divided. This approach will allow a more detailed study of the petrophysical properties of individual sections of the formation.

A comprehensive application of numerous research methods allows for a detailed study of the reservoir and identification of potential porous rocks.

In recent years, geostatistical and fractal methods of geometry have made a significant contribution, which allows performing a qualitative description of oil reservoirs.

In this study, methods of geostatistics and fractal geometry were used, which are the main research methods. Application of these methods allowed developing a quantitative and computerized method of reservoir description. In this method, basic data are used to determine the lithology of the reservoir and the petrophysical properties. Fractal geometry concepts are used to estimate the fractal character (transmission coefficient, H) of logs (gamma logarithm, porosity of logarithm and resistance). The estimation of coefficients H allows generating pseudo (or fractal) distributions of these logs throughout the reservoir.

3. Results and discussion

The reliability of the developed algorithms was verified on the basis of data obtained from wells, the construction of fractal interpolation at a given depth between two wells.

Figure 1 shows fractal well interpolation, where Z1 and Z2 are the sources of data obtained from real wells, Z3 is a pseudo-well, the data were obtained using the method. Possible pseudo-wells are at points Z4, Z5.

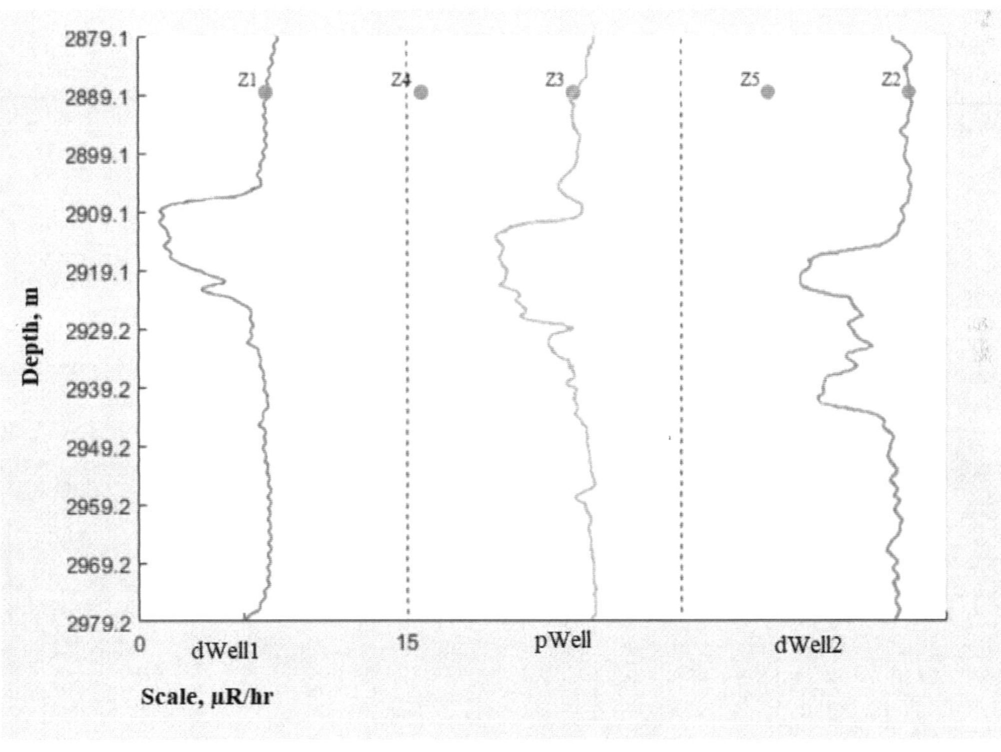

Figure 1. Fractal interpolation of wells (depth 2879.1-2979.2 m).

An additional implementation of the method is the construction of an image without loss of quality, the missing intermediate data is calculated, and a new image is formed from the obtained data (Figure 2).

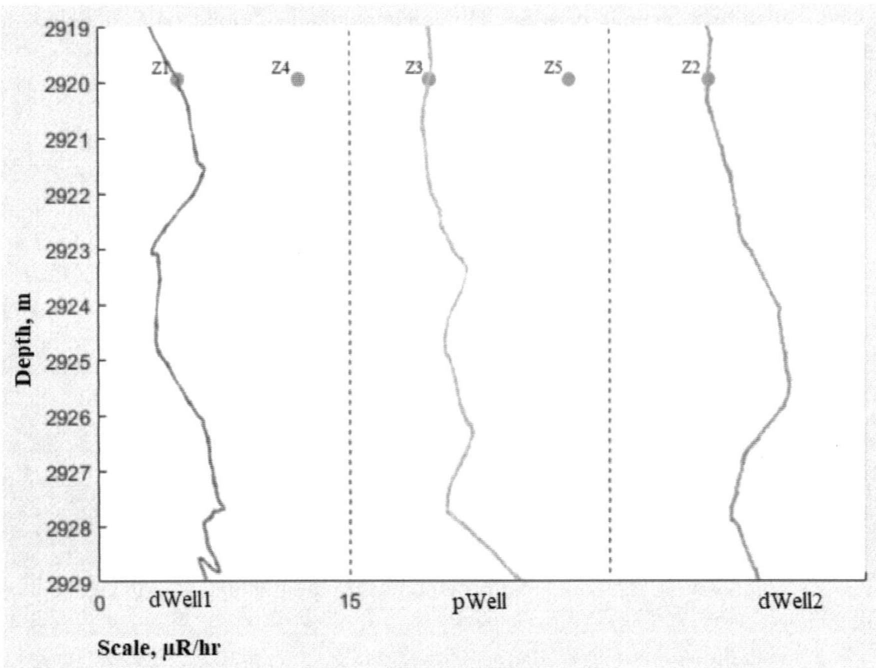

Figure 2. Fractal interpolation of wells (depth 2919-2929 m).

The method of interpolating successive random additions is based on the fact that the variance of increments (variograms) of random fractals is given as:

$$2\gamma(l) = \sum \left\{ [Z(x+l) - Z(x)]^2 \right\} = V_H l^{2H}. \quad (1)$$

where V_H is equivalent to σ .

Below are the steps used in this method of stochastic interpolation, with the correlation structure of fractal Brownian motion:

- a variance is calculated that shows the changes in the data from the received wells;
- values are calculated in the middle of the interval between wells using linear interpolation or kriging (linear regression);
- a random number is added to the approximate interpolated values taken from a normal (Gaussian) distribution with a mean value and variance:

$$\sigma_1 = \frac{\sigma_0}{2^H} \cdot \text{ and } \quad \sigma_1^2 = \frac{\sigma_0^2}{2^{2H}}. \quad (2)$$

This equation follows the power law:

$$\gamma(rl) = r^{2H}\gamma(l), \quad (3)$$

where $r = \frac{1}{2}$.

Then:

$$\gamma(\frac{1}{2}l) = \frac{\gamma(l)}{2^{2H}}. \quad (4)$$

Iterations must be repeated with all interpolated values until the optimal value is obtained. For each n-th iteration, a random value is added to each interpolated value with variance σ_n^2:

$$\sigma_n^2 = \frac{\sigma_{n-1}^2}{2^{2nH}} = \frac{\sigma_0^2}{2^n 2^{2H}}. \tag{5}$$

Next, iterations of equations with a value $R_{j,i}$ are given that denote a random number taken from a normal distribution with a mean value and variance.

Iteration 1:

$$\begin{aligned}
&\sigma_1 = \frac{\sigma_0}{2^H};\\
&Z_{1,1} = Z_{1,0} + \sigma_1 R_{1,1};\\
&Z_{2,1} = Z_{2,0} + \sigma_1 R_{2,1};\\
&Z_{3,1} = \left[\left(\frac{Z_{1,0} + Z_{2,0}}{2}\right)\right] + \sigma_1 R_{3,1}.
\end{aligned} \tag{6}$$

Iteration 2:

$$\begin{aligned}
&\sigma_2 = \frac{\sigma_1}{2^H};\\
&Z_{1,2} = Z_{1,1} + \sigma_2 R_{1,2};\\
&Z_{2,2} = Z_{2,1} + \sigma_2 R_{2,2};\\
&Z_{3,2} = Z_{3,1} + \sigma_2 R_{3,2};\\
&Z_{4,2} = \left[\left(\frac{Z_{1,1} + Z_{3,1}}{2}\right)\right] + \sigma_2 R_{4,2};\\
&Z_{5,2} = \left[\left(\frac{Z_{2,1} + Z_{3,1}}{2}\right)\right] + \sigma_2 R_{5,2}.
\end{aligned} \tag{7}$$

Iteration 3:

$$\begin{aligned}
&\sigma_3 = \frac{\sigma_2}{2^H};\\
&Z_{1,3} = Z_{1,2} + \sigma_3 R_{1,3};\\
&Z_{2,3} = Z_{3,2} + \sigma_3 R_{3,3};\\
&Z_{3,2} = Z_{3,1} + \sigma_2 R_{3,3};\\
&Z_{4,3} = Z_{4,2} + \sigma_3 R_{4,3};\\
&Z_{5,3} = Z_{5,2} + \sigma_3 R_{5,3};\\
&Z_{6,3} = \left[\left(\frac{Z_{1,2} + Z_{4,2}}{2}\right)\right] + \sigma_3 R_{6,3};
\end{aligned} \tag{8}$$

$$Z_{7,3} = \left[\left(\frac{Z_{4,2} + Z_{3,2}}{2}\right)\right] + \sigma_3 R_{7,3};$$

$$Z_{8,3} = \left[\left(\frac{Z_{3,2} + Z_{5,2}}{2}\right)\right] + \sigma_3 R_{8,3};$$

$$Z_{9,3} = \left[\left(\frac{Z_{5,2} + Z_{2,2}}{2}\right)\right] + \sigma_3 R_{9,3}.$$

Iteration 4:

$$\sigma_4 = \frac{\sigma_3}{2^H};$$
$$Z_{1,4} = Z_{1,3} + \sigma_4 R_{1,4};$$
$$Z_{2,4} = Z_{2,3} + \sigma_4 R_{2,4};$$
$$Z_{3,4} = Z_{3,4} + \sigma_4 R_{4,4};$$
$$Z_{4,4} = Z_{4,4} + \sigma_4 R_{4,4};$$
$$Z_{5,4} = Z_{5,4} + \sigma_4 R_{5,4};$$
$$Z_{6,4} = Z_{6,4} + \sigma_4 R_{6,4};$$
$$Z_{7,4} = Z_{7,4} + \sigma_4 R_{7,4};$$
$$Z_{8,4} = Z_{8,3} + \sigma_4 R_{8,4}; \tag{9}$$
$$Z_{9,4} = Z_{9,3} + \sigma_4 R_{9,4};$$
$$Z_{10,4} = \left[\left(\frac{Z_{1,3} + Z_{6,3}}{2}\right)\right] + \sigma_4 R_{10,4};$$
$$Z_{12,4} = \left[\left(\frac{Z_{4,3} + Z_{7,3}}{2}\right)\right] + \sigma_4 R_{12,4};$$
$$Z_{13,4} = \left[\left(\frac{Z_{7,3} + Z_{3,3}}{2}\right)\right] + \sigma_4 R_{13,4};$$
$$Z_{14,4} = \left[\left(\frac{Z_{5,3} + Z_{9,3}}{2}\right)\right] + \sigma_4 R_{14,4}.$$

Where: σ_i - root-mean-square deviation for the given iteration, $R_{j,i}$ - a random number taken from the normal distribution for the j-well, i – iteration number, $Z_{j,i}$ - data sources of the j-well, i iteration.

This method introduces small changes to the initial data (input values), the final result of distribution corresponds to the input values and during the construction the correlation structure is preserved

In the study, the Archie [3], Simandoux [4], Waxman [9] models were used to evaluate reservoir petrophysical parameters from logs, both at input wells and in generated pseudo-wells. The application of these models depends on the type of rock lithology. The Archie model is used mainly for pure reservoir rocks with an effective shale content of less than ten percent. The Waxman-Smits and Simandoux models are applied to rocky rocks (sandy sands) with an effective shale content of more than ten percent.

Four modules of programs were developed using geostatistical, fractal geometry and methods of interpreting wells. In these programs, geostatistical methods for constructing variograms, kriging, and conditional simulation were implemented, following the recommendations given in the references.

The method of R/S-analysis of fractal geometry and the method of successive random additions was implemented in these programs, following the recommendations in [5-8].

The Archie, Simandoux and Waxman-Smits models designed to study wells have been implemented. Program modules were written in the MATLAB language and are named VARGRAM, GEOFRACT, GEOSTAT1 and GEOSTAT2, respectively. The developed programs in the MATLAB language can be transferred to the source code C++ using the built-in generator.

4. Conclusion

The formation description technique developed in this study offers many applications for solving problems that arise in the study of hydrocarbon reservoirs. Graphs constructed at a given depth are important for well experts because their study can give a broad idea of the distribution of gamma radiation, resistivity and porosity in the area of missing wells, which allows a petrophysical study of hydrocarbon reservoirs.

Application of mathematical modelling with the use of computer technologies allowed developing a computer program for the analysis of well data and subsequent interpretation in the form of graphs, sections of the reservoir, areal maps. The study revealed an increase in the accuracy of generated data with an increase in the wells in question.

References

[1] Savinykh Yu A, Savinykh R I, Ganyaev V P, Bogdanov V L and Muzipov Kh N *A Method for Preventing the Deposition of Paraffin in an Oil Well* Patent for invention RUS 2263765 19.04.2004.
[2] Muzipov Kh N 2005 Intensification of Oil Inflows by Means of sound Waves PhD Thesis (Tyumen)
[3] Sep. – Oct. 2013 IOSR Journal of Applied Geology and Geophysics 1 4 pp 10-4
[4] Simandoux P 1963 Dielectric Measurements in Porous Media and Application to Shaly Formations Review de I'institutFrancais du Petrole, Supplementary issue
[5] Berteig V 1989 Predictions of Hydrocarbon PV with Uncertainties (SPERE) p 2-5
[6] Delhome J P Kriging in the Hydrosciences, Reservoir Characterization 1
[7] Galloway W E and Hobday D K 1983 Terrianeous Depositional Systems (New York: Springer-Verlag)
[8] Haldorsen H H 1986 Simulation Parameter Assignment and the Problem of Scale in Reservoir Engineering ed Lake and Carroll (Orlando: Acad. Press)
[9] Hewett T A and Behrens R A 1990 Conditional Simulation of Reservoir Heterogeneity with Fractals (SPEFE,) p 217-25
[10] Hurst H E 1951 Long Term Storage Capacity of Reservoirs (ASCE)
[11] Volokitin Ya E and Khabarov A V 2009 Complex methodology for estimating the oil saturation coefficient of heterogeneous reservoirs. Karotazhnik (Well Logger) J. (Tver: International Association for Scientific, Engineering and Business Cooperation in Well Logging and Operations in Wells) 12 pp 143-66

IOP Conf. Series: Earth and Environmental Science **181** (2018) 012031 doi:10.1088/1755-1315/181/1/012031

Salvaging of waste drilling muds by means of injection and mechanical ways

A N Shuvaev and M V Panova

Industrial University of Tyumen, Tyumen

E-mail: anshuvaev46@mail.ru

Abstract: Over the past decades, in the process of oil and gas production in Western Siberia, one of the most common methods of salvaging of waste drilling muds is the introduction of cement into the salvaging, followed by agitation by an excavator. This method does not ensure the uniform distribution of any detoxifier preparation and contradicts the logic in the field of excavator application. More than two hundred years, this machine belongs to a group of diggers, not mixers. This article presents modern methods of obtaining homogeneous mixtures, borrowed from the foundations, road and hydraulic engineering.

1. Introduction

The greatest danger for objects of the natural environment is the production and technological waste of drilling in the well construction process, which are accumulated and stored directly on the drilling site. The waste drilling mud is a drilled rock, which is flushed to the surface from the bottom of the well by the drilling mud. The rock separated from the drilling fluid is drill cuttings.

Drilling of wells is carried out mostly in sedimentary deposits, in which clay rocks are the most common. Their share is 65-80%. In their composition they contain a wide range of pollutants of mineral and organic nature, represented by materials and chemical reagents used for the preparation and processing of drilling muds, except from oil, oil products and pollutants of mineral nature.

The mineralogical composition of the waste drilling mud is determined by the lithological composition of the drilled rocks and can change significantly as the well deepens.

The chemical composition of drill cuttings depends on a number of factors: the composition of the drilling fluid, the used drilling mud, the physical and chemical properties of the rock, etc. The main components of the waste drilling mud, which can have a negative effect on the microflora of soils and other environmental objects, are oil, mineral salts (especially chlorides), surfactants and heavy metals.

The share of waste drilling fluids in drilled areas is up to 98%.

Waste drilling muds is an aggregate-stable colloidal system stabilized by reagents - surfactants (sodium carboxy methyl cellulose, rubber composition-2, sulfanole, higher fatty alcohols, etc.); they can vary over a wide range, which greatly complicates their clarification and purifying. The composition of waste drilling mud can contain suspended substances from 2500 to 25000 mg/l, oil and petroleum products from 50 to 150 mg/l and oxidizable organic matter (chemical oxygen demand) from 1000 to 3500 mg/l [2]. Drilling fluids also have a high content of hard metals.

2. Research

Content from this work may be used under the terms of the Creative Commons Attribution 3.0 licence. Any further distribution of this work must maintain attribution to the author(s) and the title of the work, journal citation and DOI.
Published under licence by IOP Publishing Ltd

The proposed general list of works on the processing of salvaging with waste drilling muds consists of the following stages:

1. Field studies of the salvaging by remote and subsurface vertical exploration and chemical studies;
2. Construction of a technical site and development of measures to ensure local stability of technological machines;
3. Transportation and storage of materials and chemicals;
4. Technological operations for preparation of mixtures of chemical preparations and their introduction, distribution and mixing in the waste drilling mud salvaging;
5. Quality control of obtaining a homogeneous mixture by geo-radar survey and control drilling with sampling for chemical studies;
6. Control monitoring by vertical exploration through the estimated time of toxicity elimination.

3. Results and discussion

3.1. Field studies of waste drilling mud salvaging

The existing methods for studying soil masses are based on the study of cuttings taken during the drilling of wells. Therefore, the reliability of the data depends on the frequency of sampling and chemical analysis. This method does not allow to determine the qualitative and quantitative composition as well as homogeneity of waste drilling mud with a maximum degree of probability and accuracy, which in turn makes it impossible to select effective preparations for toxicity elimination and their dosage.

During the last 15 years geophysical non-destructive methods for studying ground masses have been used in construction practice. Remote and subsurface vertical exploration is one of the most promising geophysical methods (Figure 1). The principle of their action consists in the emission of electromagnetic waves and the analysis of their reflection from various boundaries. And remote vertical exploration to a depth of up to 15 meters can be carried out both with the help of space vehicles and drones. This method allows to organize the search for buried and hidden salvaging locations of waste drilling mud and to determine their approximate qualitative composition. And a subsurface vertical exploration with control sampling gives complete information on the composition of waste drilling mud. The analysis of the dielectric constant of the salvaging massif determines the homogeneity of the waste drilling mud, the location and presence of layers and interlayers as well as the basic physical parameters (Figure 2). This allows to select the optimal and effective methods of production technology, to manage the process of introduction and mixing components, as well as assess the quality of the performed work. For greater reliability, these data are refined based on the results of the control drilling.

Figure 1. Geo-radar survey.

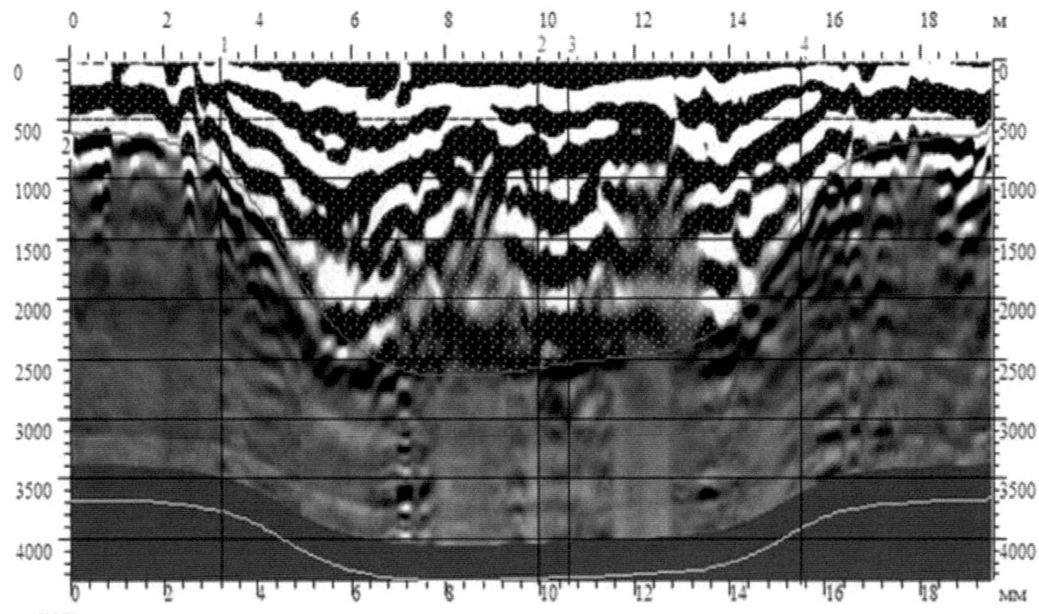

Figure 2. Radargram of the longitudinal section of the waste drilling mud salvaging.

Geo-radar survey is performed in longitudinal and transverse directions. Dimensions of grids depend on the parameters of the salvaging and are from 1 to 5 meters. In the absence of passability on the surface of the salvaging the device moves on a sledge or in a boat at a speed of 1 to 5 km/h. The following parameters and properties of the salvaging product are determined by decoding:

- bundles and lenses in terms of density and mineral component;
- volumetric distribution of moisture,
- leakage of liquid outside the salvaging;
- Density-Moisture-Viscosity-Shear Stability diagrams.

3.2. Construction of a technical site and development of measures to ensure the local stability of technological machines

At this stage, works on the equipping the work site are performed: warehouses, cabins, workshops, personal hygiene equipment, access roads, etc. If there is no provision for accessibility of vehicles, inventory portable shields are installed to move the equipment along the surface of the salvaging.

3.3. Introduction and distribution (mixing) of compounds in the salvaging

The technological part is based on classical solutions from the field of foundation engineering, strengthening of soils in the road sector and hydraulic engineering construction [1]. In Fig. 3 classification processing technologies of waste drilling muds is presented.

3.4. Injection technology.

It is divided in turn into jet and pressure (cup) ones [2]. Currently, in the world and domestic practice, there are many injection methods for introducing chemical reagents and chemicals into soils to change their physical and mechanical properties. Injection technology was patented in 1794 by the French engineer S. Berigny. It is a kind of deep anchoring of soils by injection of treatment or fixing solutions through pre-drilled wells or injectors.

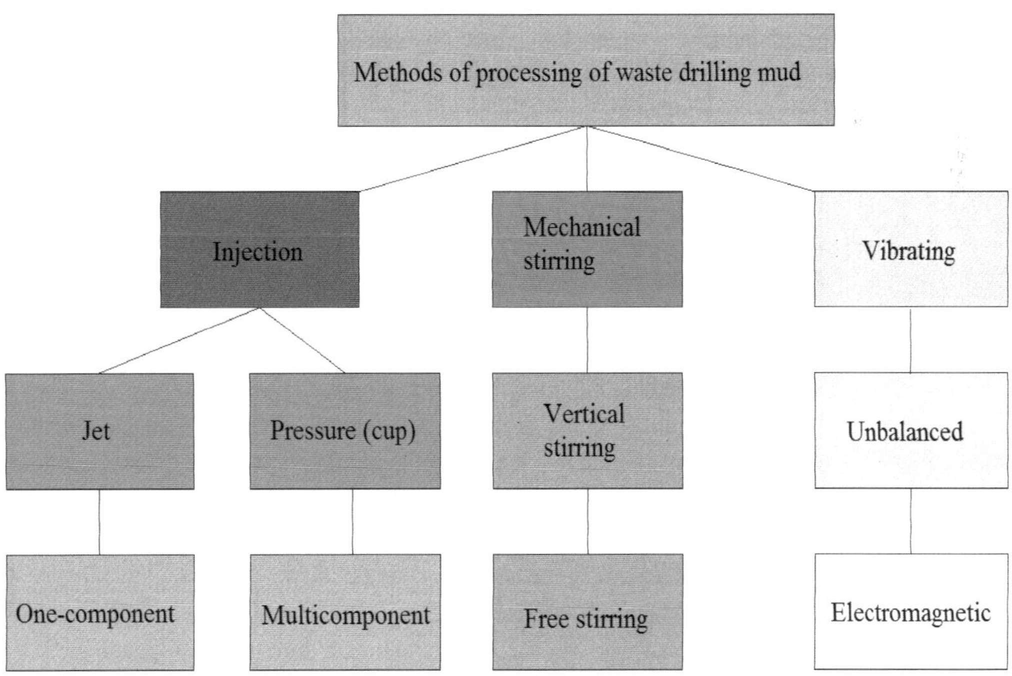

Figure 3. Classification of processing technology of waste drillin mud.

IOP Conf. Series: Earth and Environmental Science **181** (2018) 012031 doi:10.1088/1755-1315/181/1/012031

3.5. Jet-grouting technology.

It is based on the use of the energy of a high-pressure jet of a solution at a pressure of 40-100 MPa for the destruction of soil and its simultaneous mixing. The pressure is created by means of a monitor equipped with one or two nozzles through which a solution is fed into the ground in a direction perpendicular to the well axis. Simultaneously with the feeding of the solution, the monitor rotates and it moves along the well. There are three types of jet fastening, determined by the simultaneous use of various components:

- one-component - the jet of chemical solution is applied at a pressure of 40-60 MPa, the radius of the fixing zone in clay soils does not exceed 0.6-0.8 m;
- two-component - the jet of chemical solution is supplied in the air cage, the radius of the treatment zone is up to 3.0 m;
- multicomponent - the jet of water in the air cage is used for soil erosion, and the processing chemical solution is supplied with significantly lower pressure into the eroded cavity, the radius of the zone exceeds 5 meters. Injection pressure determines the energy of the jet and the radius of its action. The recommended value of injection pressure is 450 atm. (bar).

The equipment used to carry out the injection works is made by different, both foreign and domestic manufacturers and has a significant difference in technical and operational parameters. At the same time, the main technological units are drilling equipment, a mixing station, a solution pump and high pressure hoses.

Drilling equipment is used for wells of a certain diameter and depth for the installation of injectors.

Figure 4. Drilling equipment:
a) large-sized installation, b) transportable installation.

3.6. Mechanical stirring.

When mechanical stirring, the blade body is inserted into the salvaging of the waste drilling mud with a simultaneous dosage of the compound and additives. During the rotation of the blades, the distribution of the compound and the formation of a homogeneous mixture occur. The feeding of the compound under dry and wet methods of treatment is carried out under pressure through a flexible pipeline directly into the working zone of the mixer milling cutter. Admissible technological parameters for a dry process of waste drilling mud treatment should be maintained within the following limits: binder consumption 100-400 kg/m^3; the pressure in the hose inlet 0.2-0.4 kPa; the pressure in the hose outlet 0.2-0.5 kPa; rotation rate of the mixer milling cutter 100-200 rpm. Acceptable values of technological parameters when treatment of the waste drilling mud in the salvaging with the wet method: binder consumption 100-250 l/m^3; pressure in the hose inlet 2.0 MPa;

IOP Conf. Series: Earth and Environmental Science **181** (2018) 012031 doi:10.1088/1755-1315/181/1/012031

pressure in the hose outlet 0-1.0 MPa; rotation rate of the mixer milling cutter 10-20 rpm; the lifting rate of the mixer milling cutter 0.1-0.5 m/s.

4. Conclusion

Technologies for treatment of the waste drilling mud with the use of methods of jet injection, mechanical and vibro-mixing with productivity from 20 to 1000 m^3 per hour are developed. The use of remote and subsurface vertical exploration makes it possible to study in detail the waste drilling mud salvaging, carry out their quality control and monitoring.

References

[1] Rzhanitsin B A 1986 *Chemical anchoring of soils in construction* (Moscow: Stroyizdat) pp 42-45

[2] Voronkevich S V 2005 *Fundamentals of technical land reclamation* (Moscow: Scientific World) pp 272-6

Application of thawed and peat soils for construction of road-bed for automobile roads

A N Shuvaev, A B Shabarov and M V Panova

Industrial University of Tyumen, 38 Volodarskogo st., Tyumen, 625000, Russia

E-mail: anshuvaev46@mail.ru

Abstract: The presence in the permafrost zones of loose rocks unsuitable for the erection of embankment set a number of problems for science before converting them into a group of suitable rocks. The practice of transport construction in the specific conditions of Siberia for the last two decades has accumulated experience and expanded the list of local materials used in the erection of embankments, namely frozen clay and peat soils. The strength, reliability and durability of such soil massifs depend on their water-heat regime and stress-strain state. In the developed physics-mechanical model, the principal moment that distinguishes it from the known ones is a sufficiently detailed account of the mutual influence of the temperature field, the distribution of moisture, ice content, as well as stresses and deformations. Theoretical premises were tested in experimental construction.

1. Introduction

A vast region in Russia is occupied by permafrost soils (over 11 million km^2, which is almost 63% of the territory) with unique climatic and ground-geological conditions which are unique in complexity and variety and create significant difficulties in the design and construction of both roads and railways. The complexity of conditions is determined by the almost fully absence of stone materials and the widespread distribution of silty, fine dispersed and wetland clay and peat soils. The road construction requires large material costs, from 3 to 5 times higher than in the middle part of Russia. The current normative documents (Industry Building Code 84-89 "Research, design and construction of highways in permafrost regions" and the Interstate Standard GOST 33149-2014 "Roads of common use for motorways, the rules for designing highways in difficult conditions") are in most cases not ideal and impeccable, especially since the first ones were created 30-40 years ago, while the latter do not go beyond the framework of SNiP 2.05.02-85, developed more than 50 years ago and updated in the Design and Construction Specifications 34.13330.2012. [1]. The reasons for the roads failure in these conditions are not so much the impact of vehicles, but the natural factors effect on the road. These problems are the most complex and have not been yet adequately studied. Over the past 50 years, the practice of road construction in Siberia has shown that when roadbed construction an expansion of the use of local materials such as wetlands of frozen and thawed clay and peat soils should be considered as the most effective ways to reduce costs and increase in the pace of roads construction. An acute problem is the forecasting of the formation of the embankment and the adoption of measures to ensure the stability and durability of the roadbed on the basis of constructive-technological and organizational solutions.

Content from this work may be used under the terms of the Creative Commons Attribution 3.0 licence. Any further distribution of this work must maintain attribution to the author(s) and the title of the work, journal citation and DOI.
Published under licence by IOP Publishing Ltd

2. Research

The operating procedures have the following drawbacks: one-layer structures are considered, one-dimensional problems are solved, only final precipitations are considered, not taking into account that they go in time. As a result, not enough reliable calculations are obtained.

3. Results and discussion

The aim of this work is the development of a complex model for forecasting the water-heat regime and the stress-strain state of the wetland thawed and frozen clay and peat soils of the disturbed structure in the embankment body during its formation at the stage of construction and operation, taking into account the model universality and the analysis simplicity of individual determinants phenomena.

Physical principles of modeling are: the law of conservation of energy (heat balance) in the form of an equation of thermal conductivity, taking into account the heat of phase transitions, the equation of the moisture balance in different phases, determining the saturation in different parts of the investigated region, and also the equations determining the stresses and deformation.

For an arbitrary small but finite volume V, the soil components averaged over the volume are: the ground densities $p_1=\frac{m_1}{V}$, water $p_2=\frac{m_2}{V}$, ice $p_3=\frac{m_3}{V}$ and gas $p_4=\frac{m_4}{V}$ are found from the integral law of conservation of mass [2]:

$$\int_v \frac{dp_i}{dt}dv = -\int_s p_i \, v_i^n ds + \int_v \sum_j J_{ji} dv \qquad (1)$$

where: $J_{ij} = -J_{ji}$ have the dimension $kg/m^3 \cdot s$ and characterize the intensity of the phase transition of the mass from j to the ith component per unit volume of the medium per unit time, that is J_{32} characterizes the intensity of ice melting, J_{23} - hardening of water; J_{42} -condensation of steam; J_{24} - evaporation of water; $p_j \, v_i^n$ are the projections onto the external normal to the surface S of the phase fluxes, that is $p_i \, v_i^n = (J_i)^n$; $p_i = \frac{m_i}{v}$.;.

The change in the averaged mass density of the soil skeleton *p<[,* water p_2 and ice p_3 can occur due to water migration, phase transitions and changes in soil porosity.

Velocities $\bar{v}$ and flows $\bar{J_i}$ of matter in the solid phase: for the soil skeleton $\bar{J_1}=p_1\,\bar{v_1}$ and ice $\bar{J_3}=p_3\,\bar{v_3}$ are exclusively related to the deformation motion and are determined by the equations of the stress-strain state of the soil [3,4]:

$$\bar{v_1}=\bar{v_3}=\bar{v}_{def}(x,y,z,\sigma_i,\tau_{ij},t) \qquad (2)$$

where: $\sigma_i(i = x, y, z)$, τ_{ij} (ij = x, y, z) are the normal and tangential stresses in the soil, determined by external loads on the roadbed, the pressure of the overlying layers and the soil properties.

The flow of liquid moisture in a colloidal capillary-porous body in the general case is due to diffusion, convective and effusion flows [3] and is determined by Onsager's linear relations:

$$\bar{J_2}=p_2\,\bar{v_2} = -\lambda_{2m}\nabla\theta - \lambda_{2m}^T\nabla T - k_2\nabla P \qquad (3)$$

where: $\nabla\theta = \frac{1}{C_m}(\nabla w)_T$ – gradient of moisture transfer;

$C_m = \left(\frac{dw}{d\theta}\right)_T$ – isothermal specific moisture capacity;

$\lambda_{2m} = a_{2m} p_1 C_m$ - coefficient of moisture conductivity;

$\lambda_{2m}^T = \frac{\lambda_m}{C_m} \delta$ - coefficient of thermal conductivity

For each reference volume V at each time t, the humidity and ice content are calculated from the mass conservation equations (1):

$$W = \frac{m_2}{m_1} = \frac{p_2}{p_1} \quad \textit{and ice content} \quad L = \frac{m_3}{m_1} = \frac{p_3}{p_1}, \tag{4}$$

The model of the heat and moisture regime of the soil in the integral form is written using (4) as:

$$\int_v \frac{dp_1}{dt} dV = -\int_s p_1 (\bar{v}_1)^n ds \tag{5}$$

$$\int_v \frac{dp_2}{dt} dV = -\int_s [\nabla\lambda_{2m}(\nabla\theta)^n + \lambda_{2m}^T(\nabla T)^n + k_2(\nabla P)^n] dS + \int_v J_{32} dV + \int_v J_{42} dV \tag{6}$$

$$\int_v \frac{dp_3}{dt} dV = -\int_s J p_3 (\bar{v}_3)^n ds - \int_v J_{32} dV \tag{7}$$

$$\int_v \frac{dp_4}{dt} dV = -\int_s [\lambda_{4m}(\nabla\theta)^n + \lambda_{4m}^T(\nabla T)^n + k_4(\nabla P)^n] dS - \int_v J_{42dv} \tag{8}$$

The heat balance equation can be represented as:

$$\int_v \frac{d(pH')}{dt} dV = -\int_s \lambda(\nabla T)^n dS, \tag{9}$$

where $H = \int_0^T C_{ef} dT$ is the equivalent enthalpy of the soil mixture, determined taking into account the presence of skeleton soil material, ice, unfrozen water and steam-air mixture.

The initial data for the analysis of the heat and humidity and stress-strain state of the soil are given at the initial time t = 0 in the form: $T = T_o$ (x, y, z, o); $p_i = p_j$ (x, y, z, o); where v j =1, 2, 3, 4 (9) in the whole considered calculation area. The boundary conditions for heat and moisture transfer are given on the outer surface by [4] and at a distance to the depth from the surface. Calculation of the humidity temperature in non-stationary conditions on the external surface is determined by:

$$\lambda_{3F} \frac{p_F}{b_F} \left(\frac{dW}{dn}\right)_F + \lambda_{3F} b_F \left(\frac{dT}{dn}\right)_n + \lambda_3 \frac{p_F}{b_F} (W_F - W_B) = 0 \tag{10}$$

Equations (5-9) are a mathematical integral form of recording the basic laws of frozen ground physics - the equations of conservation of mass and energy. The equations of conservation of mass and energy in differential form are written according to [3]:

$$p_0 \frac{du_i}{d\tau} = -d_i V \left[\bar{J}_{i\,dif} + p_i k \frac{k_i}{\eta_i} (\nabla P - p \nabla h) \right] + I_i \tag{11}$$

The water pressure in the soil can be determined from the equation of soil consolidation in the form of V.A. Florin for the control volume V:

$$\int_v \frac{\partial H}{\partial t} dV = \int_v \left(\nabla \frac{1}{\omega} \frac{\partial H^*}{\partial t} + \frac{1}{2\gamma\omega} \frac{\partial \sigma_z^*}{\partial t} \right) dV + \int_S \frac{1+e}{2\gamma\omega \left(\frac{de}{d\theta}\right)} k \nabla H^n dS \tag{12}$$

$$\omega = 1 - \frac{\beta(1+e)}{2\frac{(de)}{d\theta}} ;H^* u\, \sigma_z^*$$

where the water pressure and the sum of the stresses in the soil skeleton at the end of consolidation, i.e. in a stabilized state.

The quantities H^* and θ^* do not change in the future and the equation of consolidation according to V.A. Florin [4] has the form:

$$\frac{dH}{dt} = C_v \nabla^2 \mathrm{H} \tag{13}$$

where: $C_v = (1+e)(1+\xi)\, R/2\gamma\omega a$ - coefficient of consolidation.

Under the influence of stresses o_s the volume v, which coincides with the control volume at the time instant t, decreases. In this case, the soil skeleton v<\ is assumed to be constant. The pore volume v_n = v - v^ decreases according to the law of damped hereditary creep on ΔV_{def} [3]:

$$\Delta V_{def} = \varepsilon_{v\ def} V = \frac{e(t_1)-e(t_2)}{1+e(t_1)} V, \tag{14}$$

The total change in volume v during the time $\Delta t = t_2 - \Delta t_1$ due to the action of the stresses σ_Σ and the phase transition is determined by the following formula:

$$\Delta V_p = \Delta V_{def} + \Delta V_{phase\ tr.} \tag{15}$$

The coefficient of volume strain or $\varepsilon_V = \frac{\Delta V}{V}$ is equal to the sum of the relative longitudinal strains [2, 3]:

$$\varepsilon_V = \varepsilon_z + \varepsilon_x + \varepsilon_y = (1+\mu_y)\varepsilon_z, \tag{16}$$

where: $\varepsilon_V = \mu_y \varepsilon_z;\ \mu_H = a\left(\frac{y}{b'/2}\right)^n$ when Z > Z_o is the coefficient taking into account the possibility of transverse soil deformations; $\varepsilon_x = 0$ under the assumption that there is no deformation in the X direction along the road axis.

Equations (16) and (15) yield expressions for the relative strains ε_x, ε_z, ε_y of the linear strains S_z and S_y in the directions Z and Y and the deformation rates v_{1z} and v_{1y}:

$$\varepsilon_z = \frac{\varepsilon_V}{1+\mu_y} = \frac{\Delta V_{def} + \Delta V_{phase\ tr}}{V(1+\mu_y)};\ \varepsilon_y = \mu_y \varepsilon_z; \tag{17}$$

$$S_z(t,z,y) = \int_0^z \varepsilon_z(t,z,y)dz; S_y = \int_0^y \varepsilon_y(t,z,y)dy; \tag{18}$$

or with the condition of constancy of ground movements in the plane of the surface of the embankment, it can be approximately taken:

$$V_{1z} = \varepsilon_z \frac{\Delta z}{\Delta t}; V_{1y} = \varepsilon_y \frac{\Delta y}{\Delta t};$$
$$\bar{v}_1 = \bar{v}_1 \left(\varepsilon_z \frac{\Delta Z}{\Delta t}, \varepsilon_y \frac{\Delta y}{\Delta t}; \right). \quad (19)$$

To study the processes of stabilizing the embankments of clay soils erected in winter during 2009-2013, the Department of Highways and Aerodromes of Tyumen State Architectural and Construction University conducted field research on cluster sites and highways of the OOO TNK-Uvat field network. The research task consisted of:
1 – study of the dynamics of settlements of soil embankments erected of clay soil at a negative temperature (in winter);
2 – study of the dynamics of the temperature mode and physics-mechanical characteristics of the soils during thawing (in summer).

Samples were collected at the study sites to determine the physics-mechanical characteristics of the soil. To control the settlement of the embankments, metal marks were set at the stage of the embankment erection (Figure 1).

Figure 1. Stamps setting.

To determine the vertical and horizontal distribution of temperature fields, the temperature sensors DS18S20, located at the base of the roadbed along the axis, the break points and on the surface of the deposited layers, were used. The deformations of the embankments of loam (Figure 2) with moisture content of 10%, 19% and 27% with freezing-thawing are significantly different when compared with a sandy embankment. When frozen, its deformations occur, which magnitude is more from 2.5 to 3 times. In the initial stage of thawing, there is a sharp swelling and an increase in deformations of up to 20%, followed by attenuation. Compared to sand, a residual deformation equal to half of the total one

IOP Conf. Series: Earth and Environmental Science **181** (2018) 012032 doi:10.1088/1755-1315/181/1/012032

is observed. The curve Ky is inversely proportional to the strain curve. The initial value of Ky is 0.97 then it drops to 0.925 and stabilizes. At the stage of swelling, a decrease in Ky continues, after which it assumes a constant value, but less than tolerated 0.92. Thus, after one cycle of freezing-thawing of cohesive soil, the embankment is decompressed to values less than those permitted by regulatory documents.

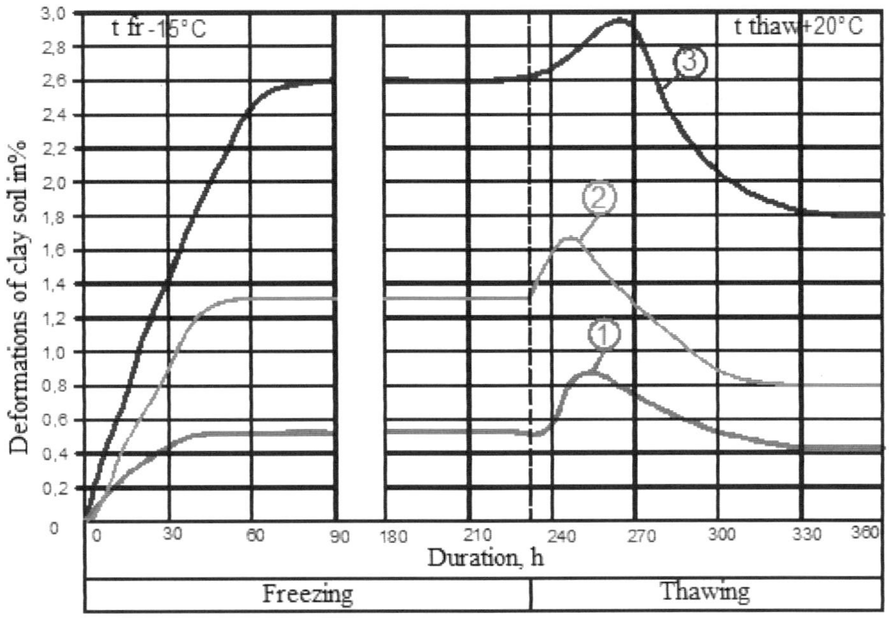

Figure 2. Deformations of embankment of loam during freezing and thawing
1- humidity -10%; 2 - humidity of 19%; 3 - humidity 27%.

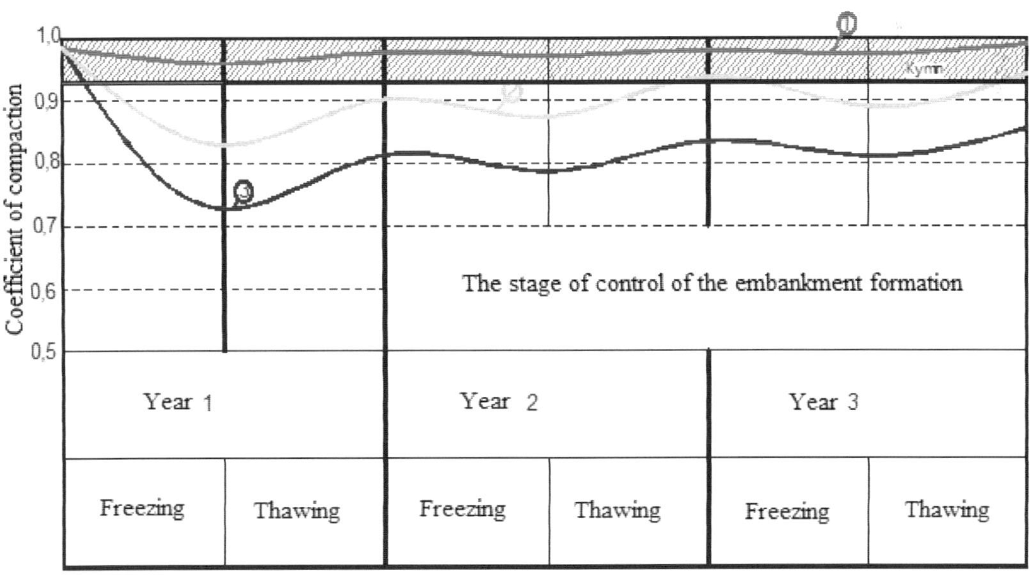

Figure 3. Dynamics of soil compaction factor for the embankment
1 - fine sand W=7%, 2 - loam W=19%; 3 - loam W=27%.

Figure 3 shows the dynamics of the coefficient of compaction of various soils. Analysis shows that sandy soils embankments are formed after the first freeze-thaw cycle. Fluctuations in density, and at the same time Ky from 1 to 0.88 for loam with a moisture content of 19%, occur within 2 years. In the first freeze-thaw cycle, decomposition of the soil to Ky=0.8 occurs. For loam with a moisture content of 27% the coefficient of compaction is reduced to 0.72.

The formation of a soil embankment with a humidity of 19% occurs during 2 freeze-thaw cycles, and loam with a moisture content of 27% for more than 3 years.

4. Conclusion

The main factors influencing the formation of the roadbed from "substandard" soils were determined and substantiated; an algorithm and computer program were developed; experimental field and laboratory studies were conducted; the monitoring of the built-up road sections in territories occupied by permafrost soils was carried out.

The developed physical and mathematical model allows calculating the variation in time and space of the main parameters of frozen and dehydrated soils: humidity, w(x, y, z, t); the ice content, L (x, y, z, t); temperature, T(x, y, z, t), normal and tangential stresses in the soil σ_i(x, y, z, t), τ_{ij} (x, y, z, t), and also deformations ε_i(x, y, z, t). A quasi-three-dimensional approach to the mathematical modeling of the roadbed has allowed us to obtain computational equations with two independent variables y and z, and a time t, which allows us to analyze the development of processes in time. The conducted field research and monitoring showed good convergence with theoretical assumptions.

References

[1] Shuvaev A N 1997 *Roadbed from frozen soils* (Moscow: Nedra) 123-34
[2] Nigmatulin R I 1987 *Dynamics of multiphase media* (Moscow: Nauka) I 342-56
[3] Tsytovich N 1973 *Mechanics of frozen soils* (Moscow: Vysshaya shkola) 346-63, 412-23
[4] Puzakov N 1960 *Water-thermal mode of the roadbed and highways* (Moscow: Avtotransizdat) 72-8

IOP Conf. Series: Earth and Environmental Science **181** (2018) 012033 doi:10.1088/1755-1315/181/1/012033

Optimization of the gas drying process in the production of hydrogen by training the fuzzy Sugeno model

V G Logachev and A Yu Sobol

Industrial University of Tyumen, 38 Volodarskogo st., Tyumen, 625000, Russia

E-mail: sobol2978@yandex.ru

Abstract: The article deals with the processes associated with obtaining and drying hydrogen under steam-carbon dioxide conversion of methane. A fuzzy Sugeno model has been created, and with the help of the training this model, a sample of process parameters has been obtained, some of which values worsen the process, and some, on the contrary, are optimal for its flow.

1. Introduction

At the moment, there is a problem of non-renewable natural resources in the world, and there are various ways to solve it, in this particular case, it is hydrogen power. There are many different ways of producing hydrogen: steam methane conversion, steam-carbon dioxide methane conversion, plasma conversion of hydrocarbons, etc. [1]. In the production of hydrogen in some ways, for example by steam-carbon dioxide conversion of methane, there is a problem of dehydration of hydrogen before transportation, or further use in production. This paper is devoted to solving the problem of hydrogen drying by the Sugeno fuzzy modeling method.

2. Research

The task is as follows: to determine the appropriate technological scheme for gas drying and the sampling of the input parameters of the technological process, create a mathematical model on the basis of this scheme and using the software package Matlab determine the values of process parameters and how they influence the flow of this technological process as well as identify the optimal values.

Hydrogen production technologies are definitely energy-intensive and cost-effective, but the physical and chemical properties of hydrogen give undoubted advantages over other types of energy resources, such as oil and natural gas. The advantage of hydrogen over natural gas is its power capacity and density: the natural gas has a power capacity of natural gas is about 40 MJ/kg, and that of hydrogen – 140 MJ/kg, but 1 kg of methane is equal to 1400 liters and 1 kg of hydrogen – 11200 liters; the latest is not a non-renewable natural gas resource [2]. Also, the production and use of hydrogen in the power sector is more environmentally friendly and makes it possible to make the power supply system decentralized.

The idea of using hydrogen in power engineering is not new. In the 80s of the XX century, hydrogen fuel engines were developed [2]. National and international programs for the development of elements of hydrogen power, including renewable energy sources, are being adopted and implemented in the USA, the EEC countries, Japan and China, and an active propaganda campaign is being carried out. In Madrid, Rome, Amsterdam, Stockholm and other European capitals there are buses on hydrogen. An electric car with a hydrogen engine was purchased by the Prime Minister of Japan, and Iceland is almost

Content from this work may be used under the terms of the Creative Commons Attribution 3.0 licence. Any further distribution of this work must maintain attribution to the author(s) and the title of the work, journal citation and DOI.
Published under licence by IOP Publishing Ltd

completely switched to hydrogen power: hydrogen engines are installed on boats and cars, houses are heated by heat sources on hydrogen.

The desire of Europe and the USA to develop alternative energy is advisable: Europe does not possess its own oil and gas resources, the USA possess those, but in a small amount. The transition to hydrogen power with the use of renewable energy will allow them to cease to depend on the oil and gas suppliers – Russia and the OPEC (Organization of Petroleum Exporting Countries) countries, as well as solve environmental problems.

In this paper, it is proposed to simulate the process of dehydration of hydrogen during steam-carbon dioxide conversion of methane. The flow diagram of the technological process of hydrogen production (Figure 1) in the paper of the author [3] describes the process of steam-carbon dioxide conversion of methane, after which hydrogen-saturated hydrogen vapor with a volume fraction of 2.6% in a mixture with hydrogen leaves.

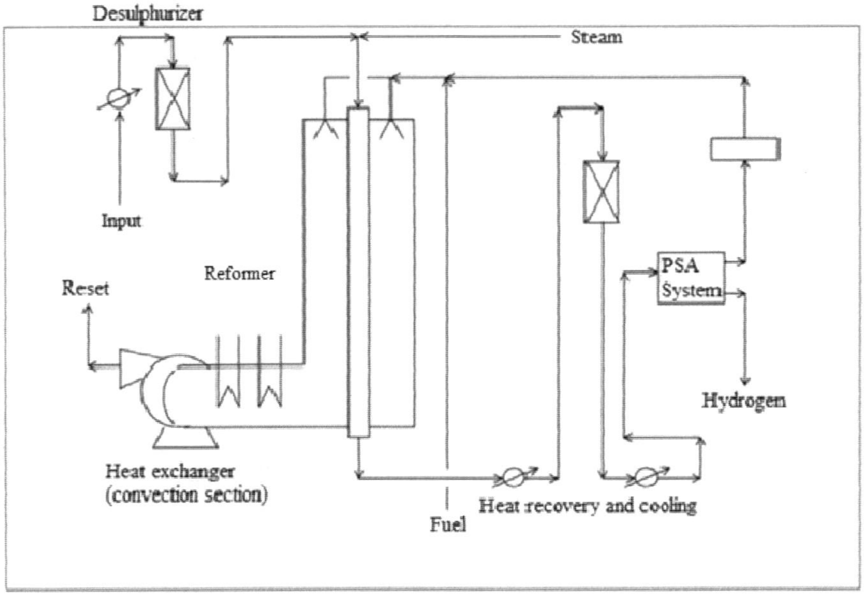

Figure 1. Technological scheme of conversion of methane with water steam to produce hydrogen (configuration with upper heating).

In the case of steam-carbon dioxide conversion of methane in the presence of hydrogen [1], the composition of the source gas is close to the methanol synthesis reaction, in the temperature range from 300°C to 340°C, oxide and carbon dioxide in the feed gas, the carbon dioxide reduction reaction proceeds simultaneously with the methanol formation one:

$$CO + 2H_2 \leftrightarrow CH_3OH \ \Delta H^O_{298} = -21.664 \text{ kcal/mole} \quad (1)$$

$$CO_2 + H_2 \leftrightarrow CO + H_2O \ \Delta H^O_{298} = 9.838 \text{ kcal/mole}$$

The last reaction under the conditions of synthesis proceeds practically up to chemical equilibrium, and as a result, carbon dioxide can influence the equilibrium concentration of methanol. The degree of influence is determined by the concentrations of the reacting components and the technological parameters of the process. In any case, the participation of carbon dioxide in the side reactions can be neglected due to the small amount of impurities formed, but the carbon dioxide reduction reaction cannot be neglected.

For this process, the structure of the drying process model of the gas in the absorber is shown in Figure 2.

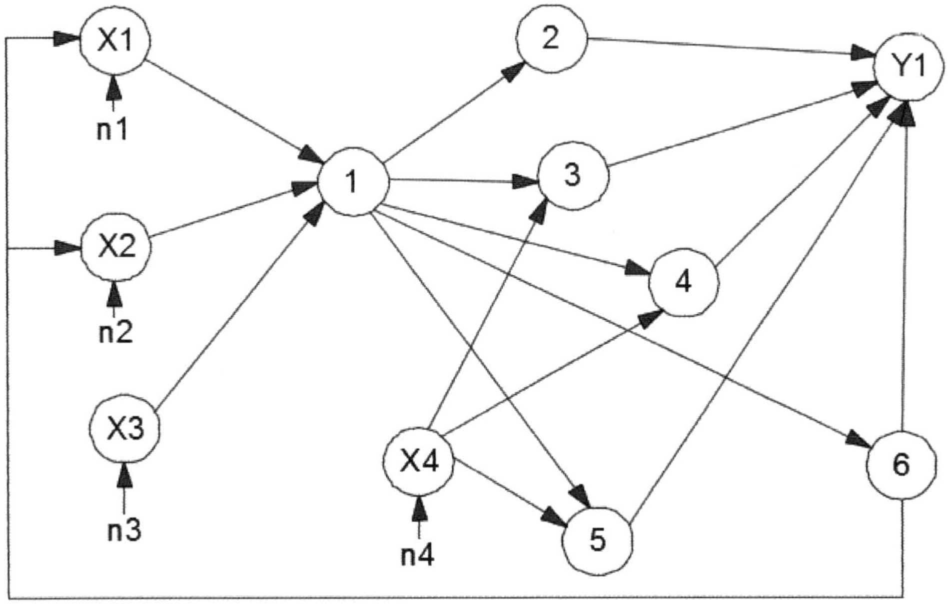

Figure 2. Model structure of the technological process.

Input variables have the following meaning: X_1 is gas flow rate; X_2 is the amount of diethylene glycol; X_3 is drying process temperature; X_4 is the drying process pressure.

Output variables: y_1 is the drying efficiency. The internal nodes of the neural network correspond to the structural elements of the absorber: 1 - input node, 2 - contact centrifugal elements, 3 - nozzle in the form of rings, 4 - filter, 5 - cartridges, 6 - diethylene glycol feeding unit.

In determining the statistical data obtained on the basis of the experience of the technological flow diagram of the author [4] for the countercurrent dehydration gas, during natural gas drying, the statistical data specified in [5] were obtained. Since the statistical data on the production of hydrogen cannot be obtained, based on the physical properties of hydrogen, statistical data were compiled for hydrogen. In view of the fact that the density of hydrogen is 8 times less than that of natural gas, it means that for 1 m^3 for drying, 8 times less diethylene glycol is needed. For the efficiency of drying a coefficient improving the process of drying natural gas (reducing the temperature of the dew point) is taken. When compiling the table, the following data were obtained (Table 1).

Table 1. Drying efficiency for hydrogen.

Gas flow rate, (H^2) thousand m^3/h.	The amount of diethylene glycol kg/m^3	Process temperature ^{0}C	Process pressure	Drying efficiency of gas
1112	4.3	14	4	59.85
960	5	13	4.6	27.01
1040	10.7	15	5.2	61.31
824	16	15	5	78.10
864	13.2	15	4.5	68.98
664	12	15	4.3	72.63
640	12.3	15	4.3	61.68
1040	9	13	4.4	90.88
1064	1.3	12	4.4	41.61

1056	1.5	12	4.4	56.2
800	3.3	13	5.1	0
816	7.1	11	5.2	58.39
1088	11	14	5.1	75.18
824	6.4	16	4.7	76.28
824	4.2	15	4.8	76.64
824	4.5	15	4.9	61.68
1016	11.4	11	5	85.04
1000	10.2	29	5.1	48.18
1000	8.5	15	4.8	73.72
1000	9.7	15	4.7	50
960	10,3	16	4.7	70.8
936	8.6	19	4.9	85.4
968	9.4	24	4.9	58.76
992	13.8	11	5	44.16
1064	6.7	22	5.1	100
1072	7.3	10	5.2	67.15
1008	12.9	15	4.8	100
936	16.8	15	4.6	81.75

3. Results and discussion

3.1. RESULTS 1.

At this stage, the technological process of steam-carbon dioxide conversion of methane was chosen for further consideration and analysis, the technological flow diagram of hydrogen production was selected (Figure 1), a mathematical model was created (Figure 2) and statistical data for further modeling (Table 1) were obtained.

Based on the initial data (Table 1), this article proposes to predict the process of hydrogen dehydration efficiency by creating and teaching a fuzzy Sugeno model containing parameters affecting the drying efficiency.

Forecasting the efficiency of gas drying is a typical task of nonlinear regression analysis. Forecasting the efficiency of gas drying is carried out according to the following process parameters: amount of diethylene glycol, gas flow rate, process temperature, process pressure, drying efficiency.

Data for calculating the efficiency of hydrogen drying were taken from the author's paper. The efficiency of hydrogen drying φ_{H_2} is calculated by the formula (1):

$$\varphi_{H_2} = \frac{100\%}{(T_{CH_4(MAX)} - T_{CH_4(MIN)})} \times T_{CH_4(\text{тек})} \quad (1)$$

where $T_{CH_4(MAX)}$, $T_{CH_4(MIN)}$, $T_{CH_4(\text{тек})}$ are temperatures (maximum dew point, minimum dew point, current temperature of natural gas).

Using the Matlab programming environment, the procedure for evaluating the information content of the process parameters on the basis of rough input-output models is described (Figure 3).

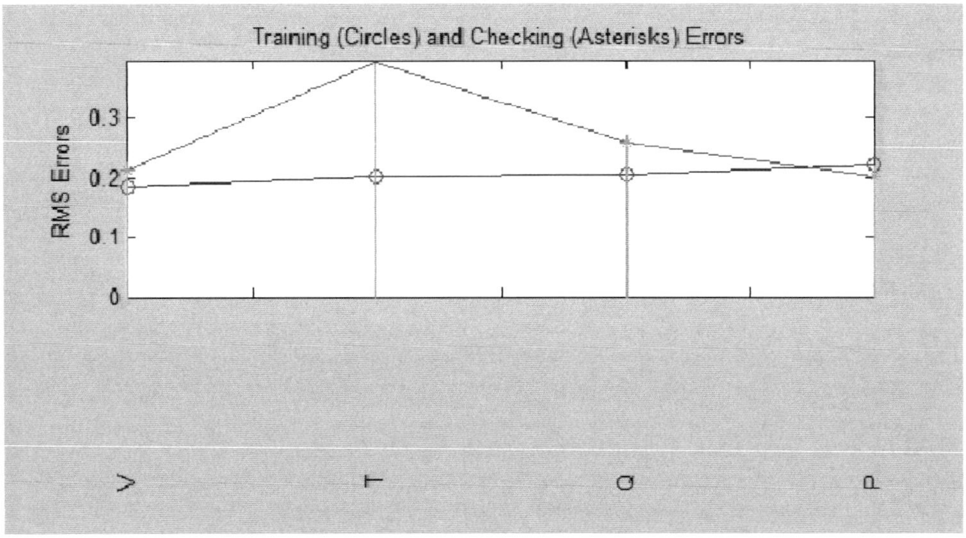

Figure 3. Rough input-output models:
amount of diethylene glycol - Q, gas flow -V, process temperature -T, process pressure - P, drying efficiency – E.

3.2. RESULTS 2.
As can be seen from the graph, the most informative sign is the gas flow rate, and the second one is the pressure of the technological process. For gas flow rate, the value of the discrepancy for the training sample is 0.1856, and for the test sample it is 0.2141.

The discrepancies in the test and training samples are of the same order, therefore, one more input variable can be added to the model. It would be logical to select the gas flow rate and pressure to compile the ANFIS model, but this directive approach does not guarantee that the chosen ANFIS model with two inputs will provide the maximum forecast accuracy.

The figure (Figure 4) shows the graph of the procedure for evaluating the information content for the pairs of variables.

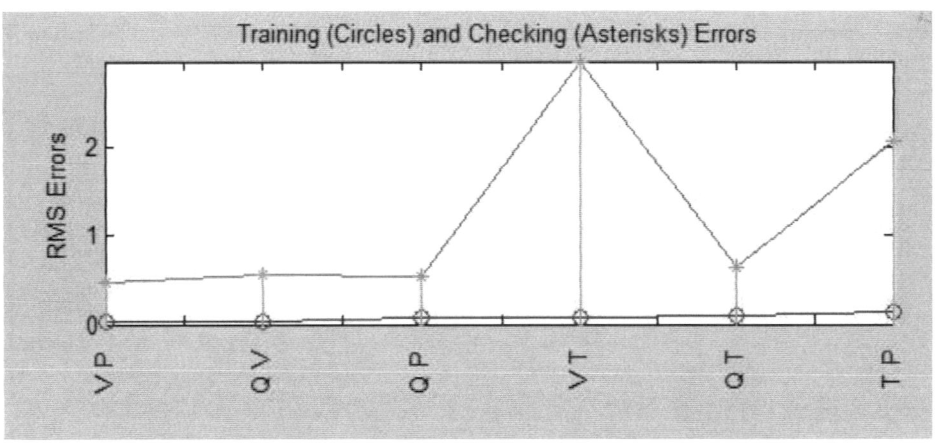

Figure 4. Testing of rough models "two inputs - one output"

3.3. RESULTS 3.
The most informative pair of features are gas flow rate and process pressure, for them the discrepancy value for the training sample is 0.0248, and for the test sample it is 0.4732. The difference in the

discrepancies on the training and test samples increases, which signals the effect of the model complication.

The figure (Figure 5) shows a graph of the procedure for evaluating the information content for triples of variables.

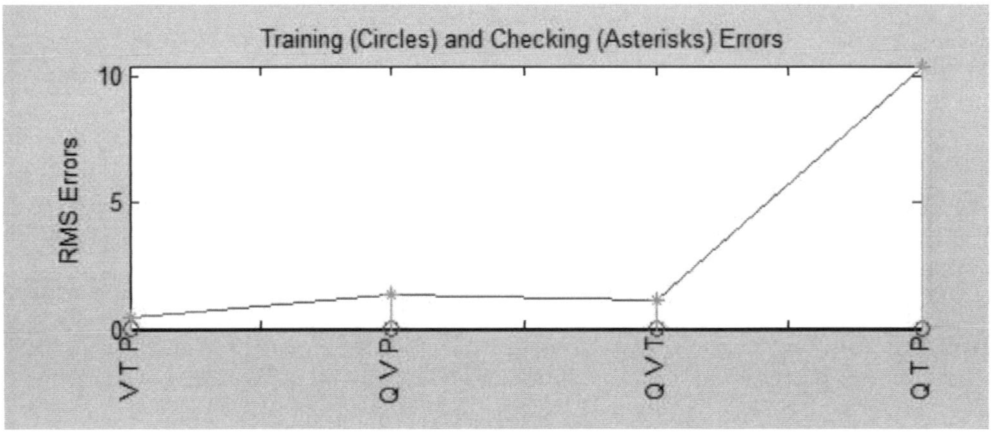

Figure 5. Testing of rough models "three inputs - one output".

3.4. RESULTS 4.

The best forecasting is provided by the model with such input variables: gas flow rate, process temperature, process pressure, for them the discrepancy value for the training sample is 0, and for the test sample it is 0.4649.

The minimal discrepancies on the training and test samples did not significantly decrease in comparison with the model "two inputs - one output". Therefore, adding one more input variable to the steam gas flow rate and the process pressure does not greatly increase the accuracy of predicting the efficiency of gas drying. To provide better generalizing properties, it is preferable to use simpler models, so for the further research the model "two inputs - one output" is selected (Figure 4).

The figure (Figure 6) shows the training results of the selected two-input model. In the previous stages, candidate models were trained using the function only during the first iteration of the ANFIS algorithm. This is done to quickly select a suitable set of input variables. When selecting input variables, it can be spend more time for training systems using several iterations.

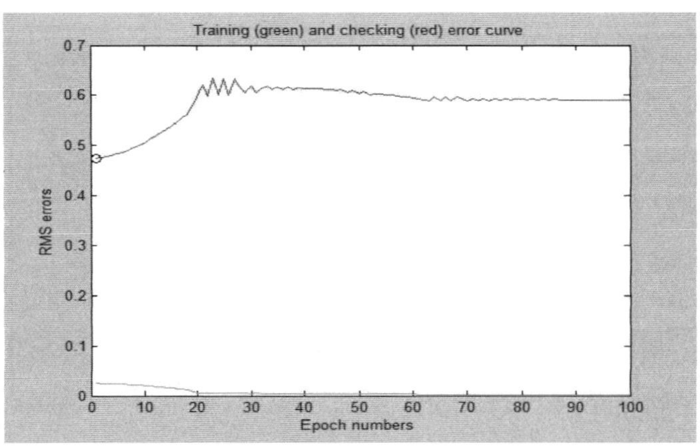

Figure 6. Testing of rough models "three inputs - one output".

3.5 RESULTS 5.

The figure (Figure 6) shows the dynamics of training a fuzzy model in the form of the dependence of learning errors (lower curve) and testing (upper curve) on the number of iterations of the algorithm. The minimum of the test error is achieved still by the 1st iteration of the algorithm. Then the test error increases, which indicates the effect of retraining, i.e. the loss of the model generalization properties. The minimum point is marked on the graph by a circle.

The figure (Figure 7) shows several kinds of input and output surfaces for the best two-input ANFIS model, which takes into account the gas flow and process pressure, which is a fuzzy model with a minimum test error.

a)

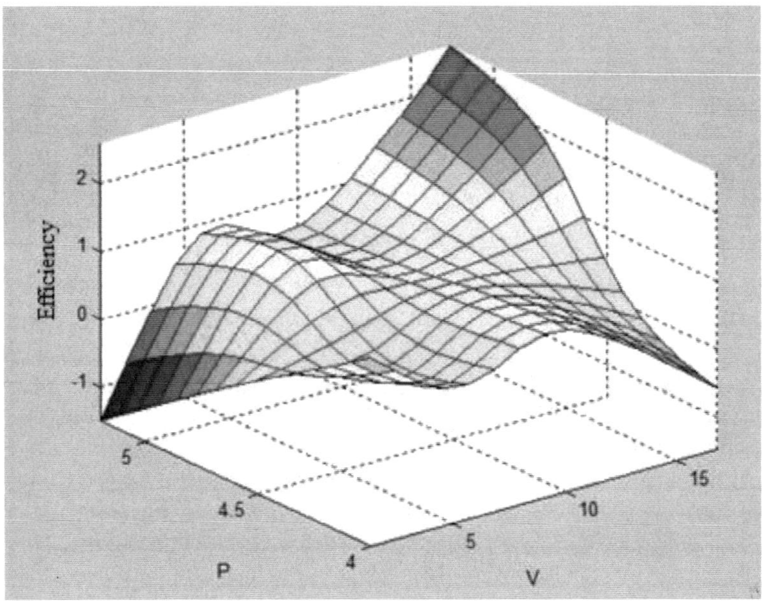

b)

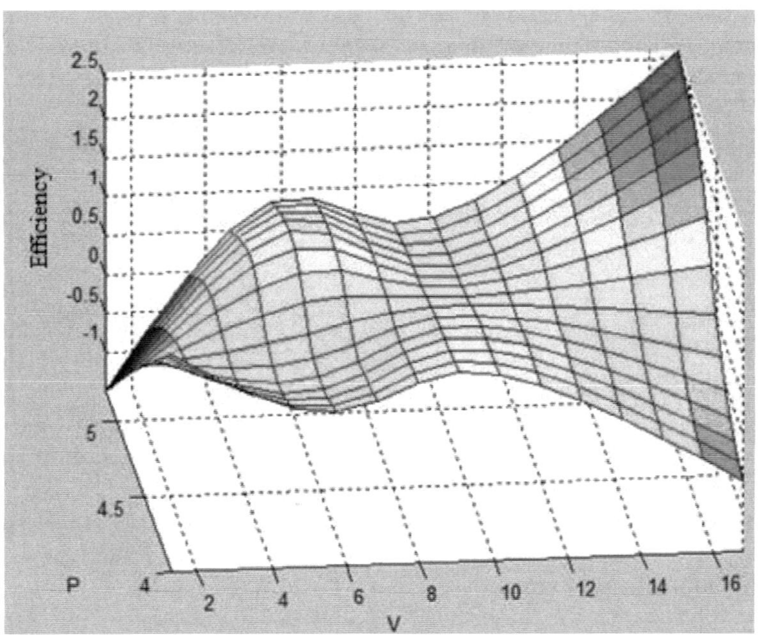

c)

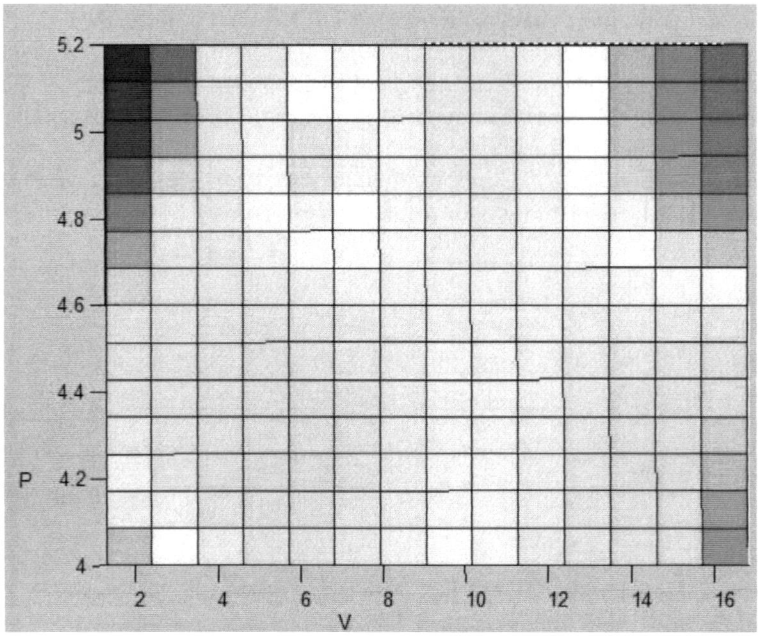

Figure 7. (a, b, c) The surface "inputs - output" of the fuzzy model after training.

3.6. RESULTS 6.

For the model depicted in figures (Figure 7 (a, b, c)), the discrepancies in the training and test samples are 0.473 and 0.225, respectively.

The sections of the yellow color of the model show a high efficiency of gas drying, close to 100%, and blue sections show ineffective drying. However, considering the far corner of red color, gives a forecast for improving the efficiency of gas drying, i.e. increasing pressure and diethylene glycol consumption during the hydrogen drying process can significantly improve its dehydration, especially since the drying efficiency was calculated for natural gas based on the dew point temperature. The dew point temperature of hydrogen, based on its physical properties, is much lower than that of natural gas.

The yellow area with the minimum amount of diethylene glycol and process pressure gives an understanding that other factors also affect this technological process, but their effect is much less than the amount of diethylene glycol and pressure during the technological process. Such unexpected behavior is explained by the absence in the training sample of the factor space data in this area.

As a result of training the fuzzy Sugeno system, it was found out that the maximum dehydration of gas in the hydrogen production is more dependent on the pressure and gas flow rate. The system showed a minimum test error for the 1st iteration.

4. Conclusion

For hydrogen dehydration, the countercurrent absorption scheme after steam-carbon dioxide conversion of methane is analyzed in this paper. A mathematical model is created in which the parameters influencing the process of hydrogen drying are gas flow rate, amount of diethylene glycol, drying process temperature, drying process pressure. For these parameters, statistical values are determined when the process passes. Based on these statistics, Sugeno fuzzy modeling is performed, the parameters are determined which values maximize the efficiency of gas drying: there are the gas flow rate and process pressure. The fuzzy model is trained, the optimal influence on the technological process is provided by the 1st iteration of the algorithm. Based on these data, the ANFIS model was obtained and analyzed in the Matlab software package.

IOP Conf. Series: Earth and Environmental Science **181** (2018) 012033 doi:10.1088/1755-1315/181/1/012033

As a result of the analysis, it was found out that the discrepancies in the training and test samples are equal to 0.473 and 0.225, respectively. The maximum efficiency of drying is achieved with the maximum increase in the gas flow and the increase in the pressure of the technological process. At maximum process pressure and minimum gas consumption, the gas drying efficiency is minimal.

References

[1] Demidov D V 2013 *Steam-carbon dioxide conversion of methane on nickel-zirconium catalysts* (Nizhny Novgorod) p 141
[2] Radchenko R V *Hydrogen in power engineering* (Ekaterinburg) p 211
[3] Sobol A Yu 2017 Hydrodynamic and mass-transfer characteristics of hydrogen after steam-carbon dioxide conversion of methane *Advances in Modern Science* (Belgorod) **5** 4 pp 66-9
[4] Sobol A Yu 2014 Absorption of carbon oxides in installations with a packed absorber *Oilfield business* (Moscow) 1 pp 42-5
[5] Laryukhin A I 2008 *Development of mathematical models of absorption drying and hydrate formation in the preparation of natural gas* (Izhevsk) p 156

IOP Conf. Series: Earth and Environmental Science **181** (2018) 012034 doi:10.1088/1755-1315/181/1/012034

Development of scientific and methodical approaches to the optimization of drilling deep wells

V M Spasibov and V V Kozlov

Industrial University of Tyumen, 38 Volodarskogo st., Tyumen, 625000, Russia

E-mail: spasvm@tsogu.ru

Abstract: When optimizing the construction of oil wells, it is economically expedient to search for optimal control over the process of deepening. An important role for operational management is the creation of a mathematical model of the process. The task was to develop a model of the deepening process (MDP) based on the analysis of the amount of energy transferred to the bottom of the well and spent on the destruction of the rock, on the basis of which it is possible to obtain particular acceptable models for optimizing the process of managing the well deepening technology.

1. Introduction

There is a lot of research on improving the structure of the well deepening management system (WDMS) with an emphasis on the "system approach" [1,2,3], with the creation of a global management system that is formed out of independent private systems - subsystems.

The main types of work in the implementation of the method for deepening the wells are shown in Figure 1.

The implementation of each type of work, presented in Figure 1, is associated with a diverse complex of tasks that are often functionally unrelated, and which differ significantly in rotary and turbine drilling.

Obviously, the creation of a global management system leads to the need for creating complex, cumbersome, difficult-to-implement process control systems (PCS). At the same time, the diversity of processes, as well as their autonomy, make the creation of a single mathematical model for the construction of a well not only an extremely complex but also an inexpedient direction. Such a model cannot be used for operational control, so the decomposition of tasks and control systems is necessary.

Content from this work may be used under the terms of the Creative Commons Attribution 3.0 licence. Any further distribution of this work must maintain attribution to the author(s) and the title of the work, journal citation and DOI.
Published under licence by IOP Publishing Ltd

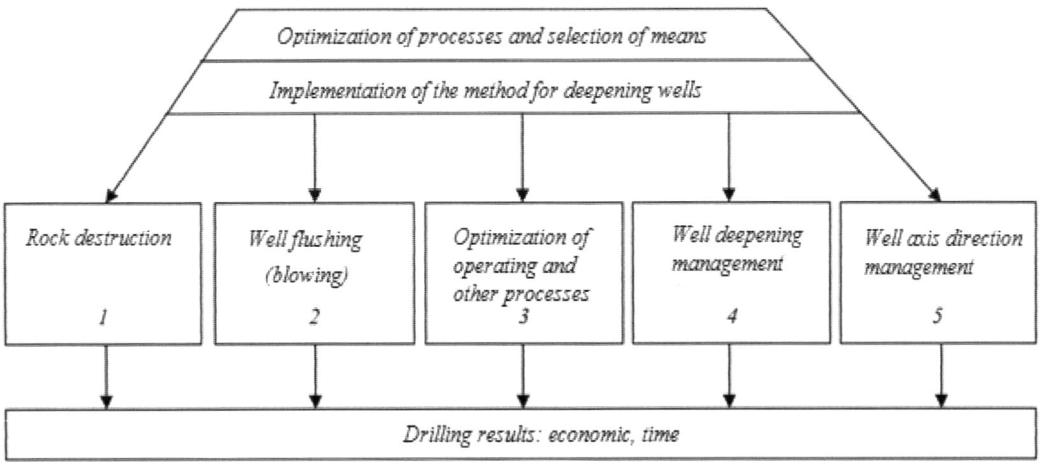

Figure 1. Scheme of work for the implementation of the method of deepening wells.

In the construction of the well, the main costs are associated with the deepening, which include the costs of washing and about a third of the cost of equipment, materials, pipes, which is more than 40% of the total cost of construction and more than 50% of the time consumption. Therefore, in the system approach to solving the optimization problem, it is economically expedient to control the deepening process. Moreover, deepening only to the productive horizon when drilling with a circle-shaped bottom with a three-cone bit, with the wear-out of its supports earlier than of its cutting structure. The greatest difficulty in optimization is precisely the deepening of the well with a great influence of external both controlled and uncontrolled factors and the most complex mathematical description, and not the descent of the column with its cementing and other well construction processes.

2. Research

The first task that requires a solution is the development of a mathematical model for managing the deepening process - MMDP for turbine drilling. The MMDP should reflect the main patterns of rock destruction, preferably of a generalized nature, taking into account the interaction of the elements of the drilling tool, especially in dynamics, hence its structure should be created on the basis of the mathematical model of the well deepening process - MDP, to meet the drilling and drilling rig conditions, to have as short shape as possible, accessible and convenient for the operational management of the well deepening.

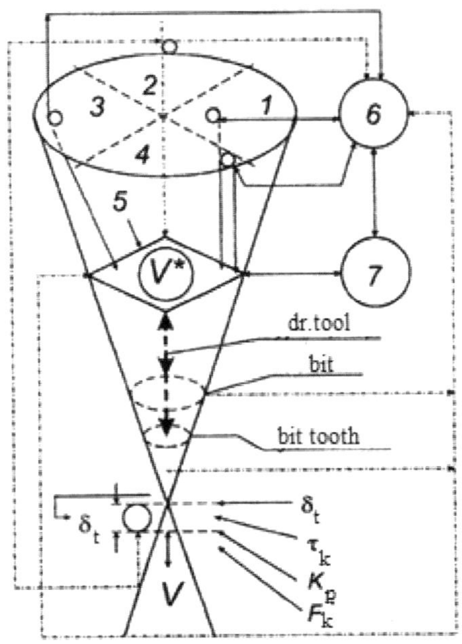

Figure 2. Scheme of connections in deepening wells and managing this process.

3. Results and discussion

3.1. Creating a model for the well construction

In order to develop a model of the deepening process, we outline the main links of the drilling process (Figure 2). At the mouth 5: 1 - drilling mode parameters controlled from the mouth; 2 - physical and mechanical properties of rocks; 3 - technical and technological parameters and mechanisms designed before drilling a well; 4 - destabilizing factors, the manifestation of which is possible during the deepening of the well; 6 - control device - controller; 7 - the operator.

At the cone base, the following things are reflected: δ_z - the magnitude of deformation of the rock with allowance for the number of bit teeth that are in simultaneous force contact with the bottomK_z; τ_K - contact time; F_K^* - the area of contact between the tooth and the rock by the end of τ_K; K_P - number of hole coverages by drilling bit cutting structure per unit of time. These are the factors determining the speed of the deepening of the well V (V^* - the rate of penetration, fixed at the mouth), the work of the entire drilling tool and the effectiveness of the structure of the control system.

In the enlarged form, as follows from Figure 2, the system consists of two main parts: a ground - control device (6,7) and a drilling tool (drill string, BHM, bit) - a controlled object. The control device performs the functions of measuring, displaying information - the parameters of the drilling mode, including from the bottom of the well, developing, in accordance with certain algorithms, control commands and providing, in accordance with them, the drill string supply by means of an operator. Moreover, according to the corresponding program, the necessary parameters of the washing liquid are set: flow rate, pressure, viscosity, density $(Q_{ZH}, P_{ZH}, \eta_{ZH}, \rho_{ZH})$.

The control object provides the required axial load on the bottom and carries out well deepening. The functions of the controlled organ are performed by a bit, and ultimately a bit tooth, which forms δ_z, τ_K, K_n, and hence V^* by means of the axial load. The control system should work in accordance with the adequately reflecting conditions and the specifics of the method of drilling a well by a mathematical model of control over the process of borehole wiring and, first of all, by the model of the deepening process (MDP).

The output parameters considered in Figure 2 depend on the power that the turbo-drill can achieve at a given flow rate of the washing liquid Q and the value of the axial load on the bottomhole - P_3 (or on the bit P). Parameters Q and P are independent control factors. However, the smooth regulation of the flow rate of the washing liquid presents significant difficulties (the need for frequency converters), therefore for turbine drilling P remains the main control action on the controlled body. The change in P is realized by feeding the upper part of the drill string (UP) to the bottom, while the speed of lowering the UP between the feeds is measured by V^*. Naturally, the characters of changes of V^* and V are different but with the elimination of errors in the measurement of V^*, associated with the vibrations of UP, etc. $V^* = V$.

The speed of drilling progress V, with the same rock for the same type of bit, is determined by δ_1 – the deepening of the rock-breaking bit's tooth top into the rock, which depends on the area F_K^*, the number of hole coverages with the cutting structure of the bit - K_n and the contact time of the bit tooth with the rock τ_k. At the same time, the value δ_1 also depends on several basic parameters:

$$\delta_1 = f(P; a_P; K_Z), \tag{1}$$

where a_P - rigidity of the rock; K_Z - the number of teeth of the bit being in simultaneous force contact with the bottom.

The value δ_1 can be determined in accordance with the kinematics of the bit and the dynamics of the drill string [1,4,5]

$$\delta_1 = r \cdot (\cos \omega_{sh} t - \cos \pi/z) - U, \tag{2}$$

where r, ω_{sh} – the radius of the bit cutter and the frequency of its rotation; z - number of teeth on peripheral crown of a cutter; U - displacement in a column during vibration process in it.

However, according to the results of many studies [5,6,7,8,9] it was not possible to obtain the calculated value of U, acceptable for its introduction into the MMDP. Moreover, it is difficult to determine the quantities entering the expressions in the process of deepening the well, so it is obvious that δ_1 is not desirable to be left in the design expression of the MMDP.

3.2. Developing a model of the well deepening process

In the work, the task was solved of developing a model of MDP, on the basis of which it is possible to obtain particular acceptable models for optimizing the process of control of the technology of the well deepening.

The rate of well deepening depends on the amount of energy transferred to the bottom and spent on the destruction of the rock. In connection with this, MDP should be built on the basis of this universal physical parameter with the use of its derivative - power - N_i. It is desirable that the basis of the MDP is the power consumed for the destruction of rocks (N_p), and during drilling it is possible to adjust the rate of well deepening using the components N_i, which is important in the operational management of the well deepening process.

On the basis of the analysis of [1, 2, 5, 8], the type of MDP was chosen, which in general form is represented as

$$N_p = \frac{P_3 \cdot \delta_1 \cdot K_Z}{\tau_{MD}}. \tag{3}$$

When drilling with the BHM, it is difficult to control only the N_p, since the power of the turbodrill - N_T is also expended on other processes occurring at the bottom of the wells, i.e. all key components of N should be present in the MDP.

The power spent on the bit operation can be determined according to the recommendation of V.S. Fedorov [5]:

$$N_i = \frac{1}{\eta_t} \cdot \left(N_{p1} + N_h + 1{,}8N_T\right) + N_{BT} + N_r \tag{4}$$

where η_t - coefficient that takes into account the power costs in the bit support and for the grinding of the rock particles at the bottom of the well; N_{p1}, N_h, N_T - the powers spent on the rock destruction, the raising of a part of the drilling tool with the weight G "due to the tooth surface of the cutters" and the friction of the teeth of cutters on the bottom of the well; N_{BT}, N_r- powers for friction of the bit on the borehole wall and in the drilling fluid.

However, not all components of N_i are expended simultaneously, for example, the N_p and N_h powers are expended in different half-periods of the axial tooth vibrations of the bit.

Figure 3 shows the change in the amplitude of such bit vibrations in time $h_3 = f(t)$ and half-periods of such vibrations: $T_1 = \tau_{md} + \tau_m$ – when the tooth of the bit is pressed into the rock during the time τ_{md}, and during the time τ_m – comes out of the indentation hole; $T_2 \cong T_H$, as it is assumed for an undeformable bottom.

Naturally, this form of vibration (or the pattern of interaction between the tooth and the bit) during the destruction of the bottom varies somewhat depending on the properties of the rock and other factors, but in the first approximation such a scheme can be adopted.

As a result of the research it was found out that the hydraulic power of the turbodrill (except for the costs in the engine itself) is consumed in various ways in each half-cycle of the axial tooth vibrations of the bit (Figure 4).

One of the differences is due to the fact that during time T_1, the main part of the power – N_P under the influence of the axial load on the bit P, consisting of the static part P_C and the dynamic part P_D , is expended on the direct destruction of the rock, whereas during T_2, N_{DP} power supplied to the bottom face, is realized mainly on the friction of the working surface of the bit on the rock under the force P_C. The other differences can be seen in Figure 4.

In accordance with the above-mentioned characteristics of N_{TG}, the components of this power are calculated in the course of turbine drilling of wells.

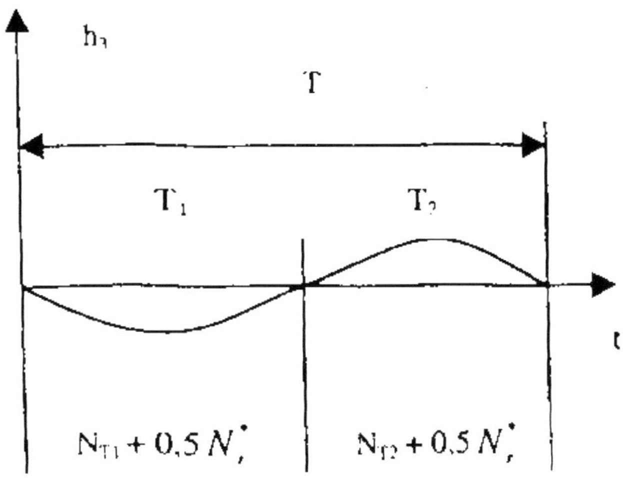

Figure 3. Scheme of amplitude variation of axial tooth vibrations of the bit and power costs N_{rH} during T_1 and T_2.

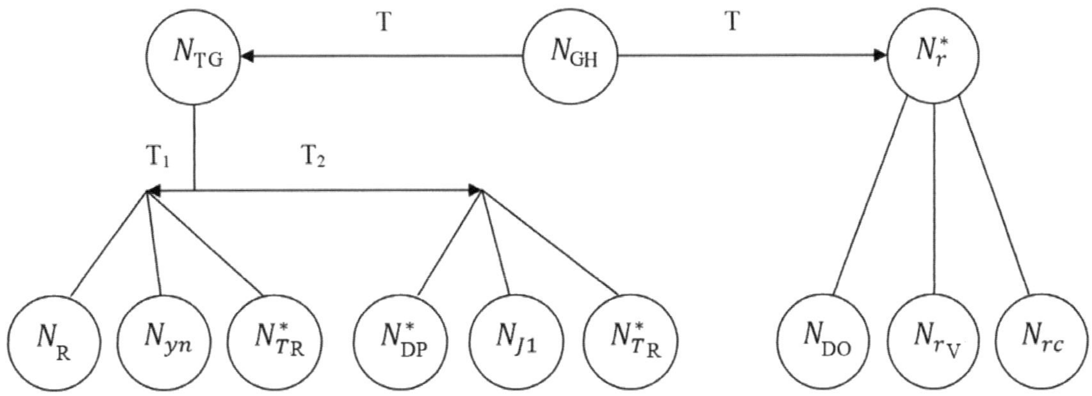

Figure 4. The scheme of the power consumption of the drilling pumps during the period *T* of axial tooth vibrations of the bit:N_{GH} – drilling pump power; N_{mg} – hydraulic power of the downhole motor expended in the half-periods T_1 and T_2 in time $t = T$; N_P, N_{UP}, N_{TP}^*– the power consumed for direct rock destruction, elastic interaction of the bit and the drill string with the bottom, and the friction of the bit on the borehole wall, in the axial and radial engine supports, during the operation of calibrators connected to the motor shaft, etc. N_{DP}^* – power supplied from the engine to the bottom of the well (not taking into account aN_{J1}); N_{J1} – power used to restore the moment of inertia of the rotor of the BHM; N_G – part of N_{GH}, spent on the destruction of the rock at the bottom of the well and its purification N_{DO}, on energy dissipation in the flow of the fluid under vibrations N_{GM}, on the hydro-resistances N_{rc}.

The accepted scheme of expenses N_i allows obtaining MDP of various completeness with an output to the resulting parameters: a deepening (elastic deformation) for one tooth-bottom interaction – δ_1, penetration per one bit rotation, mechanical penetration rate *V*, or penetration per one spudding drilling.

Below we propose a procedure for obtaining MDP, on the basis of which it is not difficult to create a MMDP without an abundance of hard-to-determine coefficients.

In accordance with the adopted cost scheme N_i, we give the main expressions and calculate the necessary components N_{TG}:

N_{UP} – the total power that returns to the drilling tool when the axial load is removed from the bottom, and is spent on the dispersion in the rock (N_{PC}) and the drilling tool (in the drill string – N_K)

$$N_{UP} \cong N_{PC} + N_K; \quad (5)$$

Power N_{UP} is determined by the elastic properties of the bottom and tools; N_{TP}^*– half the power spent on friction of the bit on the borehole wall and the washing liquid, in the elements and mechanisms of the turbodrill (in the axial and radial supports, etc.),

$$N_{TP} = 2 \cdot N_{TP}^* \quad (6)$$

N_{DP} – The power put to the well bottom at the time T_1 and expended on the interaction between a bit's cutting structure and a bottom without its intensive destruction is defined as

$$N_{DP} = 2\pi \cdot M_U \cdot P_C = N_{DP} + N_{J1}, \quad (7)$$

$$N_{DO} = N_{OCH} + N_{DP}, \quad (8)$$

where M_U – the specific moment at work of a bit at the bottom, connected with the coefficient of friction (resistance) of its cutting structure with rock – μ_{GP};N_{OCH}, N_{DP} – the cost of power for cleaning the bottom of the well and for its destruction by a jet of fluid leaving the bit nozzles; N_{GC}, N_{GM} – (hydraulic) power consumed by the hydraulic resistors in the entire hydraulic tract of the drill string and borehole, as well as the formation of hydro-impulse pressure in the flow of the washing fluid under the influence of the vibrating shaft of the BHM on the flow [4,10].

Since N_J– total power to maintain the main moment of inertia of the rotating shaft of the turbodrill, equal to $2N_{J1}$, is mainly spent on overcoming the dynamic component $N_P(N_D)$, then $N_J = 2 \cdot N_D$.

Since n decreases over timeT_1, and increases (n_2>n_1) during T_2, then due to the difference $n_1 - n_2 = \Delta n$, the turbodrill power is restored to its average value, which is constantly provided by the turbodrill.

In accordance with Figure 4, we represent the costs of the hydraulic power of the drilling pumps consumed during the time T_1 and T_2

$$N_{G1} \cong N_{G2}, \quad (9)$$

$$N_{G1} = N_P + N_{UP} + N_Z, \quad (10)$$

$$N_{G2} = N_{DP} + N_D + N_Z, \quad (11)$$

$$\mathrm{N_Z = N_{TP}^* + N_{OCH} + N_{DP} + N_{GC} + N_{GM}}. \quad (12)$$

Here N_P and N_{PC} should be calculated with the total value P_Z.

Applying equations (8) - (12), the model of the well deepening process can be represented in different variants and at different levels of detailing of the MP.

From expressions (9) - (12), we find

$$\mathrm{N_P = (N_{DP} + N_D) - N_{PC}}, \quad (13)$$

$$\mathrm{N_{P.P.} = N_P + N_{UP} = N_{DP} + N_D}, \quad (14)$$

Or if $N_{DP} \cong 2\pi M_U \cdot P_C \cdot n$ and $N_D \cong 2\pi \cdot M_U \cdot P_D \cdot n$

$$\mathrm{N_{PP} = 2\pi n \cdot P_Z \cdot M_U}. \quad (15)$$

According to the recommendation [4,5] and our improvements, to create MDP we take expression (3) and then

$$\mathrm{N_{UP} \cong \frac{P_Z \cdot \delta_1 \cdot K_Z}{1{,}8 \cdot \tau_{MD}}}, \quad (16)$$

$$\mathrm{N_{PP} = \frac{1{,}5 P_Z \cdot \delta_1 \cdot K_Z}{\tau_{eb}}}. \quad (17)$$

In accordance with the foregoing, we obtain the expanded form of the MDP

$$\mathrm{\frac{P_C \cdot K_D \cdot \delta_1 \cdot K_Z}{\tau_{MD}} = 2\pi \cdot P_C \cdot n \cdot M_U - \frac{P_C \cdot \delta_1 \cdot K_Z}{1{,}8\tau_M}}, \quad (18)$$

where δ_1 – the value of a bit tooth indentation into the rock (usually δ_1 is equal to the value of the elastic deformation of the rock); K_Z – the number of bit teeth participating in simultaneous force contact with the bottom, as recommended in [4] $K_Z = 3$; K_D – coefficient of dynamic effect of P on the bottom; τ_{MD}, τ_M – the time of the bit tooth indentation into the rock and exiting it from the indentation hole, and besides

$$\mathrm{\tau_{MD} + \tau_M = T_1} \quad (19)$$

In general, K_D = 1.0 – 1.5, but on average $K_D \leq 1.25$.

In the formula (18), the load P_C is not reduced for the convenience of the corresponding transformations and the derivation of the mathematical MMDP.

MDP by formula (18) can be converted in the required form and volume, including with acceptable accuracy for the creation of MMDP.

Let's bring the expression (18) to a simpler form, taking $\tau_{MD} \cong 1{,}1\tau_M$, $K_Z = 3$, the instantaneous radius of the bit rotation at the contact of the cutter with the bottom $R_M = (0{,}55 - 0{,}9)\ R$ and using the known calculation method M_U,

$$M_U = R_M \cdot \mu_{GP}. \quad (20)$$

Then, on average

$$\delta_{1C} = \frac{0{,}97\ R \cdot n \cdot \tau_{MD} \cdot \mu_{GP}}{K_D}, \quad (21)$$

where $\mu_{GP} = 0{,}4 - 0{,}05$ - coefficient of friction (resistance) of a bit cutting structure with a bottom at the upper limit of μ_{GP} for very soft rocks, and 0.05 for rocks of type T,TK; R - bit radius.

Time τ_{MD} can be found in accordance with the known expression for calculating the period of axial tooth vibration of the bit with an undeformable bottom

$$\tau_{MD} \cong \frac{t_Z}{2 \cdot 2\pi R \cdot n}, \quad (22)$$

where t_Z - bit cutter teeth spacing on the peripheral crown.

From the formulas (21) and (22) we obtain

$$\delta_1 \cong (7{,}6 \cdot 10^{-2} \cdot t_Z \cdot \mu_{GP})/K_D. \quad (23)$$

For the implementation of the presented MDP, it is necessary to have almost one order less coefficients difficult to determine. Note that for $Q = const$, there is no special need to operatively control the power values $N_{TP}, N_{OCH}, N_{DP}, N_{GC}, N_{GM}$, it is enough to check them periodically.

The MDP given include all physically understandable parameters and are simpler than those proposed, for example, in [1].

3.3. Mathematical models for managing the process of deepening wells

Expression (21) can be used as a MMDP, but it is not convenient for operational control purposes. Firstly, because of the small value of δ_1. Secondly, because of the need to introduce δ_1 by calculation. Thirdly, the record

$$\delta_1 = \frac{0{,}97 \cdot R \cdot n \cdot \tau_{ML} \mu_{GP} \cdot P_L}{P_C + P_L} \quad (24)$$

is unusual, although the meaning is clear from the diagrams and figures 3 and 4.

Fourthly, the time τ_{MD} is not directly controlled and is not regulated from the wellhead and this model does not reflect the number of hole coverages by the bit cutting structure K_P.

At the mouth of the well, information about *V*, and not about δ_1 is received, then it is natural that the MDP should be presented in such a way that *V* in the MMDP is explicitly related to the controlled parameters. Managed parameters should be definable, and the models themselves should reflect the physical essence of the well deepening. Moreover, it is necessary to reflect the influence of the static part of the axial load on the bit in the MDP and MMDP.

In this regard, we considered another option for obtaining MMDP. For that, we will write the equality of the specific volumetric works of rock destruction under the tooth of the bit $A_{V1} = A_{V2}$ in the form

$$\frac{N_P}{V_C F_Z} = \frac{T_U}{V_L} \quad (25)$$

Or as

$$\frac{P_Z \cdot \delta_1}{V_C \cdot F_Z \cdot \tau_{MD}} = \frac{P_Z \cdot \delta_1}{K_L \cdot \delta_1 \cdot F_K^* \cdot K_P}, \quad (26)$$

where T_U, V_L – the energy of the tooth impact on the rock and the volume of the indent hole; V_C – average value of V; K_L – coefficient, taking into account the change in the volume of the hole V_L of the destroyed rock in comparison with $\delta_1 \cdot F_K^*$.

In studies based on processing of the commercial material of the stations of the AMDP, the validity of the value of K_L was confirmed in the range of 0.35 - 0.50 - the lower limit for hard rocks, the upper one for the type MC, M. Next, we obtain V_c with the dimension m/h, (i.e., multiplying by 3600 sec), taking into account $K_P = K_{PO} \cdot n$,

$$V_c = \frac{3{,}6 \cdot 10^3 \cdot n \cdot \delta_1 \cdot K_{PO} \cdot F_K^* \cdot K_L}{F_3 \cdot \tau_{MD}}. \quad (27)$$

where K_{PO} - number of hole coverages by the bit teeth per one revolution.

The formula (27) must be transformed so that it clearly contains the control parameter P_3 and there are no parameters that need to be continuously calculated while managing the drilling.

The value of δ_1 can be determined by several methods, but in any case $\delta_1 \leq h_3$, which is due to the construction of the cone bits and follows from [4,5] – here h_3 is the amplitude of the axial tooth vibrations of the bit. We find δ_1 as

$$\delta_1 = \frac{N_P \cdot \tau_{MD}}{P \cdot K_Z}, \quad (28)$$

$$N_P = (N_T - N_1), \quad (29)$$

$$\delta_1 = \frac{P_3}{a_P}, \quad (30)$$

where N_T - the power of the turbodrill, at which it is required to find δ_1 (δ_1- is maximum at N_{TMAX}); a_P - rock stiffness,

$$N_1 = N_P + N_0 + N_K + N_{IC} + N_M + N_{UP}, \quad (31)$$

where N_P, N_0, N_K - the frictional power in the axial support of the downhole motor, the friction of the bit on the borehole wall and the washing liquid, and also on the operation of the calibrators; N_{UP} - the power consumed by the elastic interaction of the bit and the drill string with a bottom; N_{IC}, N_M - additional power costs when working of the bottom of the drilling tool on a curved well section and on the rotation of the flywheel.

By formula (31), the value δ_1 can be calculated more easily, but it is more difficult to determine a_P experimentally.

The error level in the calculation of V, as can be seen in expression (27), may depend on the values of the parameters τ_{MD} and K_{PO}. Therefore, consider the possibility of reducing such an error by refining the formula for calculating K_{PO}.

In accordance with the well-known recommendation, K_{PO} is usually defined as

$$K_{PO} = e_{SH} \cdot Z_0 \cdot R/r_{SH}, \quad (32)$$

where e_{SH}, Z_0 - the number of cutter bits and the total number of teeth on the cutter; R, r_{SH} - radius of bit and crown cutter P.

The data given in [1,4,5], as well as calculations, have shown that the energy spent on the destruction of the central part of the bottom with $R_i < R_{MG}$ is almost by an order less than those spent on the destruction of the peripheral part F_3. Usually the average value of the instantaneous well radius R_{MG} is calculated [4] within the range (0.72 - 0.55) R.

As a result of studies of the energy expended on the destruction of the peripheral and central parts of the well bottom during turbine drilling, we obtained an expression for calculating K_{PO}

$$K_{PO} \approx \frac{11{,}4R}{t_Z} \quad (33)$$

From formula (18) we find forR_M = *0,63R*

$$\delta_1 = 2{,}04 \cdot \frac{n \cdot \tau_{MD} \cdot R \cdot \mu_{GP}}{K_Z}, \quad (34)$$

make a replacement

$$F_K^* = \frac{P_Z}{K_Z \cdot P_{SH}}, \quad F_Z = \pi \cdot R^2, K_Z = 3, K_L \cong 0{,}425. \quad (35)$$

Substituting the results in (27), we obtain

$$V_C = \frac{1{,}25 \cdot 10^3 \cdot P_Z \cdot n^2 \cdot \mu_{GP}}{P_{SH} \cdot t_Z}. \quad (36)$$

The stiffness of mining works on the P_{SH} stamp can be determined in the conditions of drilling organizations according to the data of the study [5,10].

This expression can be adopted as a mathematical model for controlling the process of deepening wells.

4. Conclusion

The paper presents a method for developing a model for deepening a well MDP, in the form of a scheme for the total power of a turbodrill, in the form of an asymptotically compressed mathematical model with the possibility of calculating the turbodrill power components consumed for a period T. The MDP reflects the parameters of the bit, drillable rocks and modal parameter $P_Z = P_C \cdot K_D$.

The developed MDP allowed obtaining particular acceptable models for optimization of the process of control over the technology of well deepening.

References

[1] Balaba V I 2010 Ensuring the effectiveness and efficiency of drilling oil and gas wells on the basis of the system approach Thesis for a degree of Doctor of Engineering (Moscow) 318 p

[2] Aldred W et al 2012 Drilling automation *Oilfield Review* 2 **24** pp 18-27

[3] Zamora M and Geehan T 2013 *Developing Drilling Automation* 14 p

[4] Lyons W, Kaoyo T and Lapeyrouse N J 2013 *Formulas and Calculations for Drilling, Production and Workover of Wells* (Moscow: Premium Engineering) 344 p

[5] Fedorov V S 1991 *Scientific Bases of Drilling Modes* (Moscow: Gostoptekhizdat) 248 p

[6] Spasibov V and Kulyabin G A 2001 Construction of models for wells deepening management in turbine drilling *Higher Educational Institutions News. Oil and Gas* 6 pp 6-11

[7] Lind Yu B, Mulyakov R A and Zayrullina E I 2012 Software complex for automation of oil and gas well design *Oil Industry* (Moscow: Oil Industry Publishing House) 4 pp 35-8

[8] Tsuprikov A A 2015 Analysis of mathematical models of the mechanical speed of penetration to optimize the process of oil and gas wells drilling *Scientific Journal of KubSAU* 107 (03) pp 903-15

[9] Cheryi S G 2015 The problems of automation technological process of drilling oil and gas wells *Software @ Systems* 2 (110) pp 113-18

[10] Lukyanov E B and Strelchenko V V 1997 *Geological and Technological Research in the Process of Drilling* (Moscow: Oil and gas) 679 p

IOP Conf. Series: Earth and Environmental Science **181** (2018) 012035 doi:10.1088/1755-1315/181/1/012035

Features of wastewater treatment systems of the oil and gas field infrastructure

Elena Vialkova and Svetlana Maksimova

Industrial University of Tyumen

Abstract. When creating the oil and gas field infrastructure, the question arises of collection, storage, transportation, treatment and disposal of waste generated as a result of people's activities. One of the important tasks is the design, construction and operation of a reliable water disposal system for rotators' camps, stable under the northern conditions and including wastewater treatment plants. Technological and technical-economic analysis of existing facilities has shown that the efficiency of the existing small wastewater treatment plants is rather low, while the cost of operation is several times higher than that of stations with high capacity. This trend is also characteristic of existing technologies for treatment and disposal of sludge. In order to prevent anthropogenic impact on the nature of the North of the Tyumen Oblast (region), to increase the efficiency of small wastewater treatment plants and to reduce the costs of sludge disposal, it is necessary to apply new technological solutions and methods for treating the resulting domestic waste.

1. Introduction

The strategy for the development of the Arctic zone of Russia until 2020 suggests that the main areas for the formation of sustainable development of the Arctic include ensuring a favorable environment for residents by the rational and safe use of natural resources, preventing harm to the natural environment and the vital interests of the population. It is also about providing the population with drinking water of the required quality and reducing the negative impact on water sources due to the modern approach to the design and construction of water disposal networks. [1].

A fairly large part of the Russian territory is in the Arctic conditions (or close to the Arctic), including the North of the Tyumen region. Severe climatic conditions, frozen mobile soils, and swampy terrain create serious problems in the design, construction and operation of various facilities. When creating the infrastructure of oil and gas fields, the question arises of the collection, storage, transportation, treatment and disposal of waste generated as a result of people's activities. One of the important tasks is the design, construction and operation of a reliable domestic wastewater treatment system, stable under the northern conditions, including external networks, pumping stations and wastewater treatment plants. [1-4]

2. Overview of the problem and solutions

Construction of wastewater treatment plants in rotators' camps of the northern climatic zone has a number of features associated with low temperatures in winter, permafrost soils, a lack of access roads to construction sites, and a drastic change in the number of personnel. Shift work causes the unevenness of water consumption and water disposal, which is exacerbated by the peak nature of the flow of wastewater for treatment. Biological treatment of cold wastewater with a peak intake is a difficult task [4].

Content from this work may be used under the terms of the Creative Commons Attribution 3.0 licence. Any further distribution of this work must maintain attribution to the author(s) and the title of the work, journal citation and DOI.
Published under licence by IOP Publishing Ltd

The problem of qualitative wastewater treatment with subsequent disposal of sludge is relevant for small towns and villages in Russia. Small amounts of wastewater, as a rule, have higher concentrations of organic substances. For example, the wastewater of rotators' camps in the oil fields of the Tyumen North contain elevated concentrations of BOD, COD, nitrogen-containing substances [4-6]. Such wastewater contains technogenic pollutants: anionic surfactants, petroleum products and others [7]. There is a problem of treatment and disposal of sludge formed at small treatment plants. The necessary number of units of traditional equipment, the volume of consumables, and the cost of electricity are unreasonably high, especially under the conditions of the northern climate. To solve the problem of wastewater and sludge treatment in wastewater treatment plants of low capacity, it is necessary to introduce new energy saving methods and technologies that allow obtaining high process efficiency.

The problem of protecting natural clusters of the Arctic is directly related to the state of water sources. There is an opinion that imperfect methods of calculation, inaccurately taking into account the self-cleaning ability of water bodies, have influenced the current ecological state of water bodies, including the northern regions. Most of them are classified as "dirty" and "very dirty" and cannot cope with residual contamination of poorly treated or untreated wastewater from rotators' camps in oil and gas fields.

According to researchers [8-10], the process of self-cleaning (self-regeneration) of water bodies which compensates for the anthropogenic impact is possible only up to a certain threshold level, while preserving the buffer properties of the system, its high capacity for inactivation and transformation of pollutants. If this threshold is exceeded, the biological environment loses its ability to stabilize the water quality in a water body, and local and global changes in the aquatic ecosystem begin.

The rate of self-cleaning of the water body and decomposition of carbon-containing compounds, including surfactants and petroleum products, depends on temperature, access to oxygen, the nutrient regime of the aquatic environment, i.e. on those factors that determine its microbiological activity. In oxygen-depleted water, decomposition of carbon-containing compounds tends to slow down. Self-cleaning of water bodies from petroleum is especially slow. For 2-7 days, the content of emulsified petroleum products in water decreases by 40% at 20 degrees Celsius, and at 5 degrees Celsius by only 15%. In the presence of aquatic vegetation in model experiments, the oil film disappeared in 4-6 days at its thickness of 0.06 cm, and at 0.6 cm - in 20-22 days. [8].

From this point of view, small amounts of wastewater generated in oil and gas fields are subject to compulsory treatment and their discharge should be controlled by environmental organizations. The choice of the technological scheme for treating industrial and municipal waste waters of enterprises and infrastructure facilities is carried out on the basis of several parameters: the quality of the source water and the required degree of wastewater treatment (stress-strain state concentrations); given daily capacity of WWTPs; local geographic and climatic conditions, available resources, environmental conditions, construction experience and other conditions. In Russia, there are unified requirements for treatment plants, regardless of their capacity and operating conditions. In European countries, less stringent standards are set for small wastewater treatment facilities, and state programs for reconstruction of such structures are being developed [11,12].

The problem of disposal of liquid municipal waste of northern rotators' camps remains unresolved to the full. In the oil fields of the Tyumen North, it is often not possible to discharge municipal wastewater from temporary settlements to water bodies for several reasons: there is no suitable water body for this purpose; the water body is located too far, and laying the discharge line is not cost-effective; the water body is shallow and freezes in the winter. In a number of fields, domestic wastewater from small production sites is periodically transported by sewerage machines to central wastewater treatment plants to the nearest major population center. With the scattered placement of small treatment plants, their operation is quite complicated. [11, 13]

In this case, discharging wastewater into the reservoir pressure maintenance system in conjunction with treated production and bottom water is considered. The reservoir pressure maintenance system is a complex of process equipment necessary for preparation, transportation, and injection of a working agent into the reservoir of an oil field in order to maintain reservoir pressure and achieve maximum oil extraction rates from the reservoir. According to regulatory requirements, joint treatment and injection of formation and waste water is allowed only after complete biological treatment and disinfection of domestic wastewater at a flow rate of not more than 100 m^3/day.

However, the municipal treated wastewater may not meet the requirements of the industry standard, especially for the dissolved oxygen content downstream of the aeration biological treatment facilities. Untreated effluents contain a large amount of organic substances, creating a nutrient medium for bacteria, including sulfate-reducing bacteria. Therefore, in order to discharge wastewater of rotators' camps into the reservoir pressure maintenance system, the following measures should be envisaged: complete biological treatment of wastewater with post-treatment and mandatory decontamination at the last stage of the WWTP; additional treatment of effluents to remove residual dissolved oxygen, hydrogen sulfide and ferric iron.

Disposal of industrial and municipal liquid waste in the deep subsoil horizons is also an option. The main criterion for selecting burial sites is environmental safety of construction and operation [11, 14, 15].

Creation of areas for disposal of industrial wastewater (liquid waste) of oil and gas fields includes several stages: selection of an area and a site for construction of an underground storage facility, exploration, design and construction of a disposal site. At each stage, as well as during the operation of the site, great attention is paid to protection of groundwater. Underground disposal of petroleum-containing wastewater is a complex process which includes preparation of waste for burial, delivery and injection into absorbing wells. Geological and hydrogeological conditions of the subsoil are very complex, therefore, the project on justifying the burial of industrial wastewater cannot foresee in advance all the peculiarities of filling the underground storage with liquid waste [16,17].

In order to reduce the number of places of storage of liquid municipal and industrial waste, a method of burning treated domestic wastewater mixed with industrial and rain effluents is used for gas fields, for example, at the gas fractionation units (GFU) of TyumenNIIGigprogaz [11,16,17]. However, this method is associated with the emission of combustion products into the atmosphere and with the irretrievable use of water resources. In order to avoid the above drawbacks, according to current regulations, in oil and gas fields modern wastewater treatment plants in block-modular design must be used to ensure the required degree of treatment, taking into account the subsequent disposal of treated wastewater and sludge.

Process schemes of small towns have recently become more complex: new treatment stages are introduced; separation of the aerotank into nitrification-denitrification zones is applied; various types of reagents are added at several stages at one plant. All these measures are justified from the point of view of achieving the required treatment effects. For example, introduction of such a facility as an equalizing tank into the process scheme of a small WWTP of a shift camp makes it possible to reduce the irregularity of wastewater supply for treatment, and prevent "peak" discharges of contaminants. As a result, the volume of subsequent settling, desilting and post-treatment facilities decreases; the stability of bioreactors increases [4,11,18]. In order to achieve high quality and reduce space occupied by equipment, and, consequently, construction and operation costs, small WWTPs include membrane plants: filters, MBR reactors and others. But, as a rule, all these innovations significantly raise the cost of treatment. [19-22]

Problems with disposal of sludge formed at municipal wastewater treatment plants remain relevant. One of the promising methods, especially in low-temperature areas, is exposing the waste of the treatment plant to electromagnetic super-high frequency radiation. As a result of this short-term process, not only the amount of sludge decreases, but also its properties are significantly improved: ash content, resistivity, water-yielding capacity and others. Special attention should be paid to super-high frequency treatment due to complete disinfection of liquid domestic waste. [23-25]

3. Research

Researchers at the Department of Water Supply and Wastewater Disposal (Industrial University of Tyumen) apply a comprehensive approach to solving the identified problems of water disposal systems for rotators' camps of oil and gas fields. Scientific interests include issues relating to collection, transportation, treatment and disposal of liquid municipal waste generated at infrastructure facilities. Below are the process and technical and economic indicators of the surveyed wastewater treatment plants.

Table 1 presents comparative data for a number of rotators' camps, small towns and cities of the Tyumen region.

Table 1. Quality of wastewater entering treatment facilities on an example of settlements in the Tyumen region.

Treatment plants	Capacity, m^3/day	Suspended substances, mg/dm^3	BOD_5, mg/dm^3	COD, mg/dm^3	N-NH_4, mg/dm^3
Septic tanks of rotators' camps	Up to 5	300-6000	300-1500	600-11000	85-200
Rotators' camps of oil fields	10-250	200-1000	250-600	500-1000	70-85
Vagay (village)	400	140	390	500	52
Labytnangi (city)	3 000	200	180	250	50
Yugorsk (city)	7 000	150	120	160	18
Nefteyugansk (city)	12 000	182	131	215	50,4
Tobolsk (city)	40 000	190-200	250	400	25-27
Tyumen (city)	210 000	180-250	140-250	290-400	30-37

At the same time, the specific reduced mass of contaminants entering the treatment facility reduces significantly as the plant capacity increases (Figure 1).

In the period from 2007 to 2015, 15 operating small wastewater treatment plants for small towns in Western Siberia were studied. The capacity of wastewater treatment plants was in the range of 10 to 250 m^3/day. As a result of the study, a serious problem was identified, related to the unsatisfactory quality of biological treatment and the poor exploitation of these facilities. It turned out that the lower the plant capacity, the lower the efficiency of water treatment by main indicators (Table 2).

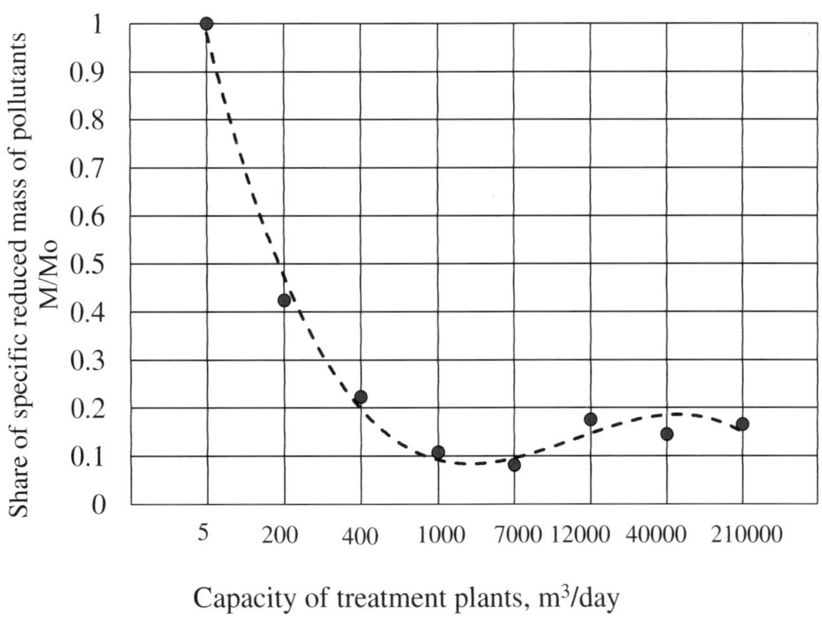

Figure 1. The change in the specific reduced mass of pollutants (cond.t/m^3) entering with wastewater, depending on the plant capacity.

Table 2. Average annual efficiency values of small WWTPs.

Treatment efficiency of WWTPs by main indicators:	WWTP capacity, m^3/day			
	10 m^3/day	40 m^3/day	120 m^3/day	250 m^3/day
Suspended substances	84%	86%	92%	97%
BOD	60%	70%	83%	85%
Ammonia nitrogen	46%	50%	56%	58%
Anionic surfactants	52%	57%	86%	87%
Petroleum products	38%	64%	85%	90%

At all the small plants studied, there are increased input concentrations of petroleum products and anionic surfactants, not typical for domestic wastewater of stations with high capacity (Table 3). At the same time, many standard designs of small WWTPs do not take into account the presence of anionic surfactants and oil products in this water, which enter the water disposal system in considerable quantities from the process of washing contaminated work clothes and with surface drains from the territories. In the wastewater of rotators' camps, the content of petroleum products is high compared to the requirements of engineering plants, and this has a significant effect on biological treatment.

Table 3. Input parameters of wastewater at WWTPs of the Tyumen region.

Indicator	WWTP 10 m^3/day	WWTP 40 m^3/day	WWTP 120 m^3/day	City WWTP up to 10 thousand m^3/day	City WWTP up to 50 thousand m^3/day	City WWTP up to 200 thousand m^3/day
Petroleum products	9. 33	8. 85	7. 48	8	6. 8	4. 9
Anionic surfactants	6. 3	8. 7	12. 5	1. 7	2. 4	3. 7

A more detailed analysis of the situation identifies many identical facts, combining which, one can draw a general conclusion that there are:

- a sharp deviation from the average concentrations for a number of contaminants (suspended substances, BOD, ammonium nitrogen, phosphates) in the incoming wastewater;
- an increase in concentrations of petroleum products associated with the appearance of surface runoffs in the domestic wastewater network and, as it turned out, with the washing of the work clothing of oil workers;
- an increase in concentrations of anionic surfactants associated with the increased use of household chemicals;
- a low temperature of incoming wastewater (during the cold period 6-100C) as a result of their long stay in the system;
- a high coefficient of hourly wastewater supply unevenness for treatment.

The consequence of this is an incorrect choice of process schemes and composition of WWTP at the design stage, and unsatisfactory operation after putting the facility into operation.

The technical and economic analysis of the operation of small WWTPs showed that the cost of cleaning 1 m^3 of wastewater is very high for plants with a capacity of 200 m^3/day or less (Figure 2). For example, for plants with a capacity of 10-20 m^3/day, the prime cost is about 8 times higher than that for a plant with a capacity of 700 m^3/day. This is due to the high costs of electricity, reagents, transportation and other operating costs. Moreover, the analysis of a number of operating wastewater treatment plants has shown that the efficiency of wastewater treatment by small plants is often much lower than that by plants with higher capacity (Figure 3).

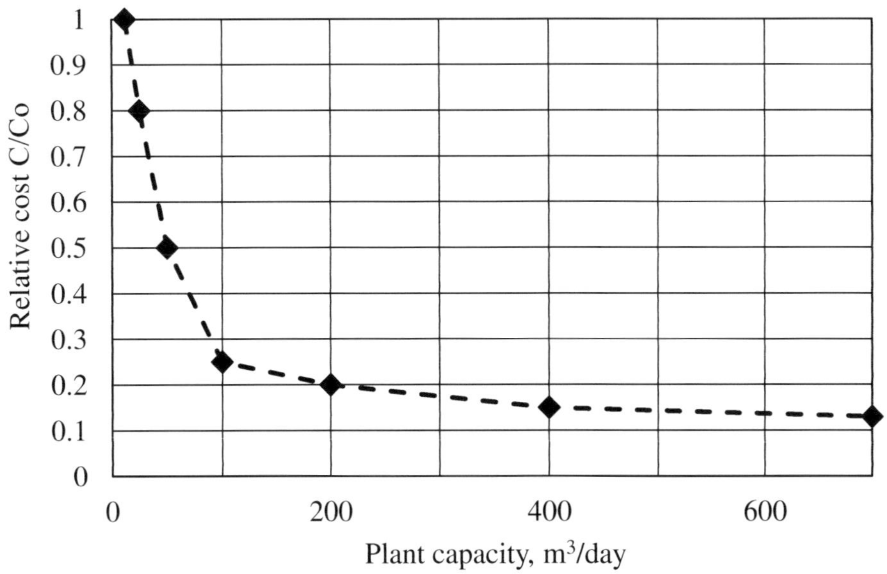

Figure 2. Dynamics of the water treatment cost reduction with an increasing plant capacity.

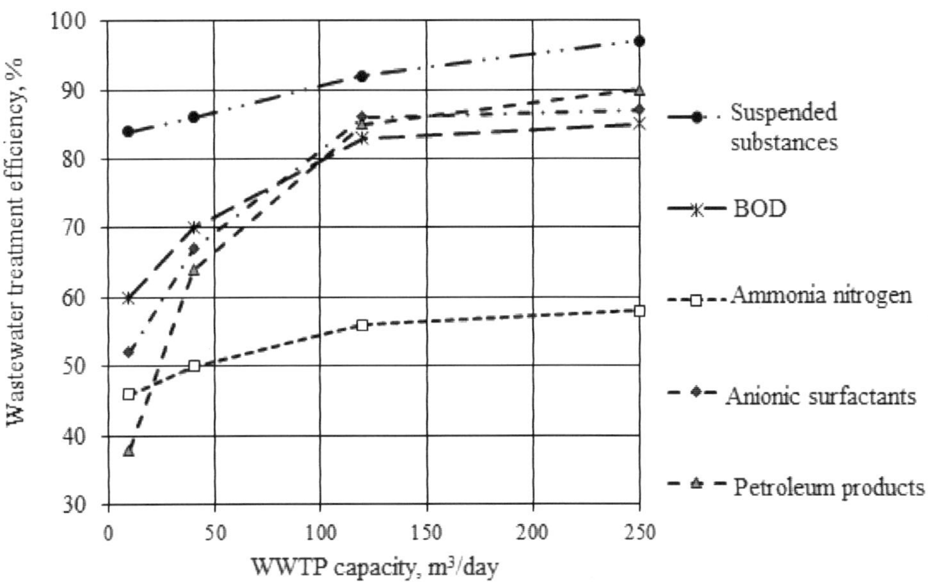

Figure 3. Efficiency of domestic wastewater treatment depending on the capacity of the treatment plant.

Process schemes of small wastewater treatment plants of the first generation were a repetition of household wastewater treatment schemes in conventional settlements: the initial wastewater goes successively to: rough water treatment facilities (coarse screens and grit chambers), biological treatment facilities (aerotanks), facilities for separating excess active sludge. Most schemes lack primary settling tanks since the high removal efficiency of suspended solids and, consequently, part of BOD can lead to inadequate organic feeding of bacteria-denitrifiers which remove nitrogen. In

IOP Conf. Series: Earth and Environmental Science **181** (2018) 012035 doi:10.1088/1755-1315/181/1/012035

construction regulations, installation of primary settling tanks is recommended at a capacity of 1000 m^3/day and above.

The introduction of chemical plants and the feeding of a coagulant to intensify the process of separation of the sludge mixture into some schemes usually excluded the return of active sludge to the aerotank. Treated wastewater after disinfection, as a rule by ultraviolet irradiation, is discharged into a water body. Sludge pads were used to dehydrate the excess active sludge. When such schemes were developed, the following factors were not taken into account: increased concentrations of contaminants in small amounts of effluents, the presence of man-made pollution in domestic wastewater, and a decrease in the temperature of wastewater in winter. A consequence of this was the low efficiency of treatment facilities. [11]

The requirements for the degree of wastewater treatment of rotators' camps led to the inclusion of post-treatment facilities in the process schemes - a second stage of biological treatment or clarifying filters with quartz sand. As the experience of operating plants designed and built using a similar scheme showed, the efficiency of the facilities also depends significantly on the quality of the incoming wastewater and temperature in the biological treatment sections.

The development of next-generation process schemes goes along the path of improving biological treatment facilities, in particular, nitrification and denitrification zones. The constant tightening of the requirements for treated wastewater discharged to water brings about more complex post-treatment facilities. At present, post-treatment of wastewater has become a multi-stage process, which includes a sorption filter in addition to the usual granular filter. In a number of schemes, a membrane bioreactor (MBR) is proposed instead of a conventional secondary settler (Figure 4). [19, 22, 26]

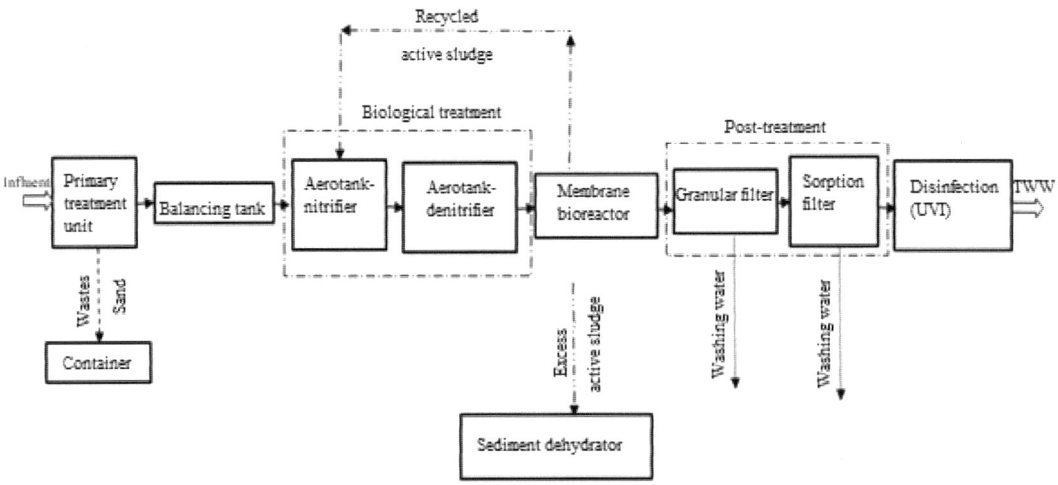

Figure 4. Process scheme of small wastewater treatment facilities with a membrane bioreactor.

Sequencing batch bioreactors (SBR-reactors) are of interest, but little used in Russia. Due to the sequencing process of biological treatment, this plant is virtually not affected by significant fluctuations in the volumes of wastewater entering the treatment facility, the composition and concentrations of pollutants. Over the last 15 years wastewater treatment technologies with SBR-reactors have been widely used in Eastern and Central European countries to modernize existing treatment facilities [20, 21, 27-29]. Along with the construction of new wastewater treatment technologies with SBR-reactors, it is possible to use existing tank facilities as SBR-reactors, such as wastewater pumping stations, settling tanks, aerotanks and other facilities. Evidently, this is the main advantage of such wastewater treatment facilities, especially in the conditions of seasonal fluctuations in the operation of rotators' camps. Taking into account the quality of the incoming wastewater from septic tank of rotators' camps and the possibility of automated process control, SBR-reactors deserve more attention of designers and developers of the infrastructure of oil and gas fields (Figures 5 and 6). [11, 27-29]

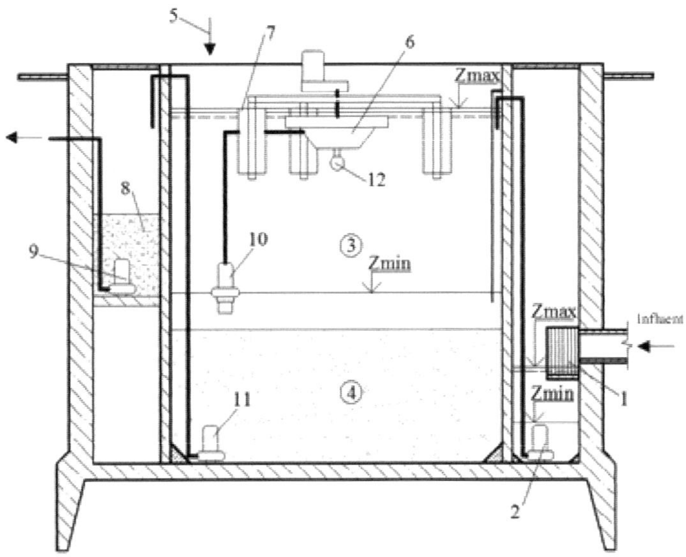

Figure 5. Transforming the existing wastewater pumping station into the SBR-technology system: 1 - screen-container; 2 - submersible wastewater pump; 3 - treated wastewater; 4 - active sludge; 5 - input of flocculant; 6 - floating turbo-aerator; 7 - pontoon blocks; 8 - settled sludge; 9 - wastewater sludge pump; 10 - decanter; 11 - excess active sludge pump; 12 - treated wastewater discharge.

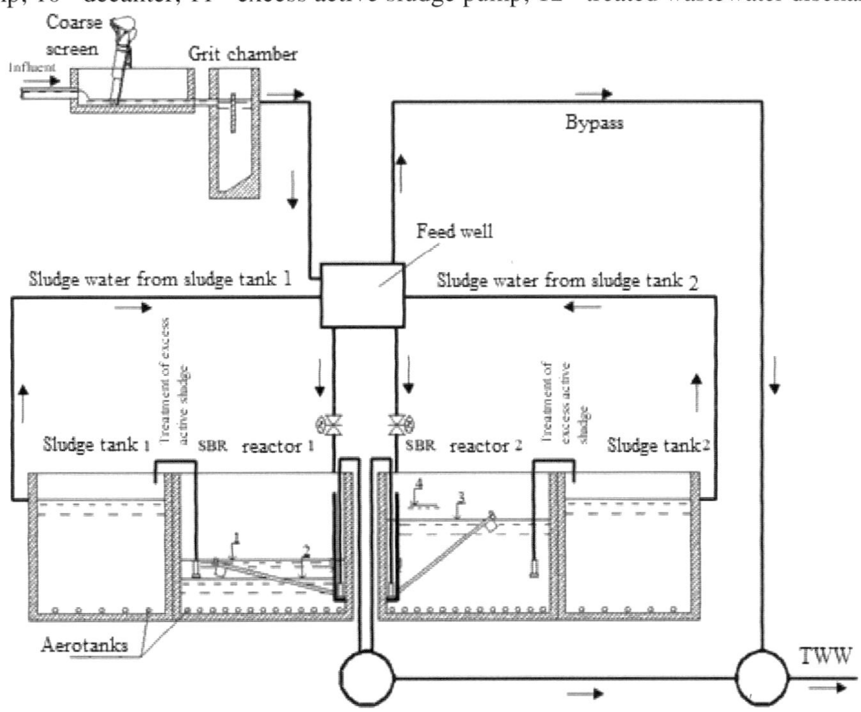

Figure 6. Process scheme of WWTPs with the Flexidiblok SBR-reactor [21]: 1- active sludge level, 2 - minimum water level, 3 - maximum water level, 4 - denitrification level

Small settlements, such as villages, urban-type settlements and rotators' camps of industrial enterprises, produce small amounts of wastewater sludge. However, this type of waste is produced daily, almost not processed properly, accumulates, and, eventually, the problem of disposal of municipal waste arises. It is especially difficult to solve this issue in winter at plants located in the Tyumen North.

IOP Conf. Series: Earth and Environmental Science **181** (2018) 012035 doi:10.1088/1755-1315/181/1/012035

Currently, there is no organization in the country that is responsible for waste disposal in the regions; there is no statistical recording of the amount of generated municipal wastewater sludge. Replacement of obsolete equipment and abandonment of low-efficiency methods of sludge processing and disposal is slow.

The amount of sludge formed depends mainly on the treatment plant's capacity (Table 4). Evidently, the greater the amount of wastewater treated, the greater the amount of sludge. However, the cost of processing 1 m3 of water and, accordingly, the sludge for small and medium-sized plants is much higher than for plants with a capacity of more than 50,000 m^3/day (Figure 7).

On the example of four operating small wastewater treatment plants of different capacities, it is possible to compare all types of waste for each facility. The results are shown in Table 5.

Table 4. Amount of wastewater sludge formed in settlements.

Settlement type	Capacity of WWTP, thousand m^3/day	Amount of wastewater sludge, m^3/day	Amount of wastewater sludge, thousand tons/year
Rotators' camps with a population of 100-500 people.	0.01-0.25	0.1-1	0.04-0.4
Settlements with a population of 1000-5000 people.	0.4-1	2-10	0.7-3.7

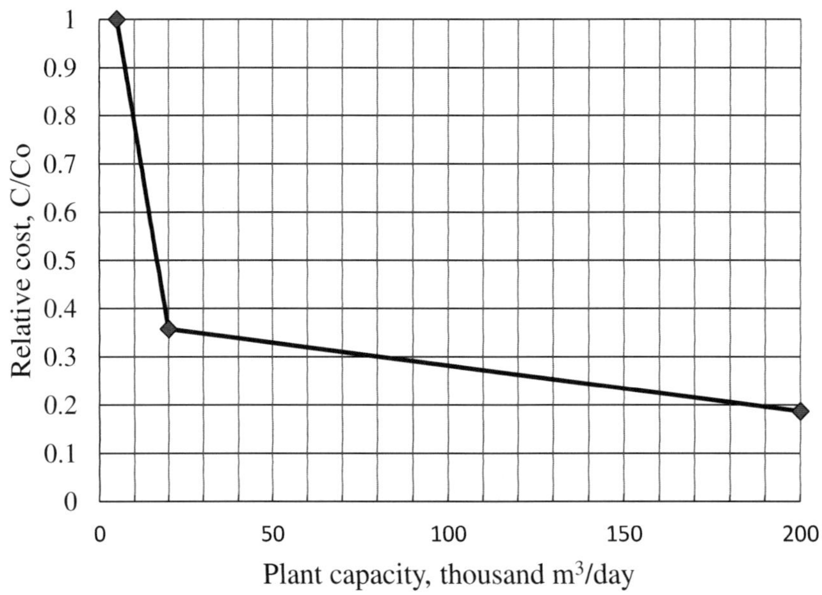

Figure 7. Dynamics of the wastewater sludge treatment cost reduction with an increasing plant capacity.

Table 5. Amount of waste collected and residue accumulated from rotator's camps in oil fields.

Waste type	Plant capacity for wastewater of rotators' camps, m^3/day			
	20	40	110	150

Wastes, $\frac{m^3/day}{m^3/year}$	$\frac{0{,}0028}{0{,}52}$	$\frac{0{,}0059}{1{,}08}$	$\frac{0{,}008}{2{,}96}$	$\frac{0{,}011}{4{,}0}$
Sediment (including sand) $\frac{m^3/day}{m^3/year}$	$\frac{0{,}0013}{0{,}47}$	$\frac{0{,}0027}{1{,}0}$	$\frac{0{,}007}{2{,}7}$	$\frac{0{,}01}{3{,}65}$
Sludge, $\frac{m^3/day}{m^3/year}$	$\frac{1{,}23}{448{,}95}$	$\frac{3{,}12}{1138{,}8}$	$\frac{6{,}83}{2493}$	$\frac{9{,}32}{3402}$

For wastewater treatment plants of low capacity, the following process scheme is used:

1) Waste that are retained on screens, as they accumulate, are transported to a landfill site or disposed of (usually by incineration).

2) Sediment from grit chambers accumulated in sink-water traps is taken to a landfill site at least twice a year.

3) Excess sludge formed in the process of complete biological treatment, after preliminary dewatering to 80% humidity with the use of flocculent (filter bags) is transported to a landfill site or disposed of (usually by incineration).

The advantage of this process scheme is that the sediment dewatered in bags is almost immediately ready to be taken away and disposed of. Disadvantages are the following: no disinfection of sediments (excluding the incineration option), large expenditures on consumables and reagents (filter bags, flocculants). In this case, no useful disposal of the municipal waste is envisaged.

There are plants the process scheme of which provides for the treatment of sediments and includes aerobic stabilizers, dehydration, accumulation and disposal.

In this case, there is a system for supplying compressed air to the aerobic stabilizer. The sediment treatment time by air is from 8 to 24 hours, the temperature is from 8°C to 35°C. The main advantage of this scheme is that expensive reagents are not used. In this case, there is also no disinfection of sediments. And at the same time there are additional costs for the supply of compressed air. As a result, such sediments are also disposed of at the landfill site.

On sites of wastewater treatment plants, it is necessary to allocate a place for temporary storage of dewatered sludge (by the volume of accumulation for 6 months). In the event of failure of the equipment for mechanical dewatering, excess active sludge is supplied to the emergency tank. The volume of the emergency tank should not be less than the volume of wet excess active sludge accumulated per month.

The sediment formed after settling the washing water of the post-treatment filters is sent to dewatering together with excess sludge or accumulated in a special tank and, if necessary, transported to a landfill site. Dewatered sediments are classified as Hazard Class IV and are subject to recycling or disposal.

4. Conclusions.

A comparative technological and technical-economic analysis of wastewater treatment plants of rotators' camps of oil and gas fields and other settlements of the Tyumen region was carried out. Based on the results obtained, it can be concluded that the technologies used for treatment and disposal of domestic wastewater of rotators' camps of oil and gas fields are inefficient and uneconomic. In order to increase the efficiency of small WWTPs and reduce costs, it is necessary to apply new technological solutions and methods for treating wastewater and processing the generated waste.

References

[1] Strategy for the Development of the Arctic Zone of the Russian Federation and the National Security Strategy up to 2020 [Electronic resource] Access mode: http://legalacts.ru/doc/strategija-razvitija-arkticheskoi-zony-rossiiskoi-federatsii-i/

[2] Tyumen Oblast [Electronic resource] Wikipedia: Free Encyclopedia Access mode: https://en.wikipedia.org/wiki/Tyumen_Oblast

[3] Features of architecture and design in northern regions [Electronic resource] Access mode: http://www.uralstroyportal.ru /articles /article1103.html

[4] Kunakhovich A A 2012 The installation of deep treatment of domestic wastewater for use in towns of the northern climatic zone *Water Supply and Sanitary Engineering* 5 pp 61-5

[5] Litueva O G 2014 Design features of water supply and sewerage systems for the development of oil deposits *Youth and Science: Coll. of Materials of the Xth Anniversary of the All-Rus. Sc.c and Tech. Conf. for Students, Post-Grad. and Young Scientists with Intern. Participation, in Honour of the 80th Anniversary of the Formation of the Krasnoyarsk Territory* (Krasnoyarsk: Siberian Federal University) [Electronic resource] Access mode: http://conf.sfu-kras.ru/sites/mn2014/directions.html

[6] Vialkova E I, Pesheva A V, Maksimova S V and Malenko N V 2012 Problems of small wastewater treatment plants in the conditions of the Tyumen North and the Ural region *VestnikStroy* 2(60) pp 58-60

[7] Pesheva A V and Vialkova E I 2012 Investigation of factors affecting the removal of oil products and anionic surfactants from domestic wastewater *Water Supply, Water Disposal and Envir. Protection Systems: III Intern. STC for Students, Post-Grad. and Young Scientists: Articles and Abstracts* (Ufa: CITO) pp 7-12

[8] Mosin O V *Self-Cleaning Ability of Water Bodies* [Electronic resource] Access mode: http://www.o8ode.ru/article/answer/method /vodoem.htm

[9] Komleva T A 2015 *Geochemical Factors of Stability of Water Systems to Anthropogenic Loads* Doctor of Chemistry Thesis (Moscow) 260 p

[10] Nikanorov A M 1990 Ecological regulation of anthropogenic impact on freshwater and estuarine ecosystems *Methodology of Ecol. Rationing: Proc. of the All-Union Conf.* 1 (Kharkov) pp 40-1

[11] Vialkova E I, Maksimova S V, Zemlyanova M V, Vorotnikova A V and Maksimov L I 2017 *Water Discharge of Infrastructure Facilities in Oil and Gas Fields in Western Siberia* (Tyumen: Industrial University of Tyumen) 199 p

[12] Tsagarakis K P, Mara D D, Nolan N J and Angelakis A N 2000 Small municipal wastewater treatment plants in Greece *Water Sc. and Tech.* 1 **41** pp 41-8

[13] Yin H B et al 2012 Multiobjective Models for central plant site selection in joint WWTPs operation of small towns *Adv. Mater. Research* **518-523** pp 2585-92

[14] Anischenko L V 2015 Possibilities of pumping wastewater into deep subsurface horizons and improving its methods *Science and Modernity* 36 pp 94-8

[15] Underground burial of industrial wastewater / B.M. Goldberg [et al.]. - Moscow: Nedra, 1994. - 282 p.

[16] Beshentsev V A and Semenova T V 2014 Protection of groundwater from pollution at injection sites during disposal of industrial wastewater (by the example of the Yamal-Nenets oil and gas producing region) *Oil and Gas Eng.* 5 pp 357-74

[17] Beshentseva O G. 2001 *Hydrological Conditions of Underground Disposal of Industrial Wastewater of Oil and Gas Complexes of Yamal-Nenets Autonomous Okrug* (Tyumen State Oil and Gas University)

[18] Bao D J et al 2014 Research on operation mechanism and application in engineering of UNITANK process *Adv. Mater. Research* **937** pp 370-4

[19] Stepanov S V, Sopkina O S, Morozova K M, Stepanov A S and Zhukova M A 2017 The effect of chemical membrane washing on biological purification processes *Water Supply and*

Sanitary Eng. 4 pp19-24

[20] Dzenis L 2005 *Technologiczne Podstawy Modernizacjimałych Oczyszczalni Sceków* (Bialystok: PAN) **29** 125 p

[21] Wiese J 2004 *Entwicklung von Strategien für Einen Integrierten Betrieb von SBR-Kläranlagen und Mischkanalisationen* (Kaiserslautern) 190 p

[22] Xing S L et al 2014 Effect of aeration conditions on the flow field in the submerged membrane bioreactor *Appl. Mech. and Mater.* **535** pp 539-46

[23] Minnigalimov R Z 2010 *Development of Technology for Processing Oil Sludge Using Energy From High-Frequency and Microwave Electromagnetic Fields* Doctor of Engineering Thesis (Ufa GUP "IPTER") 45 p

[24] Zemlyanova M V 2015 Application of ultrahigh-frequency electromagnetic radiation for treatment and neutralization of municipal wastewater sludge *Ecology and Industry of Russia* 4 pp 47-9

[25] Vialkova E, Zemlianova M and Pesheva A 2016 Using of microwave electromagnetic radiation for treatment of wastewater and sludge *Ecology & Safety* **10** pp 121-9

[26] Tomasz Janus and Bogumil Ulanicki 2015 Interface model between the bioreactor and the membrane in a membrane bioreactor for wastewater treatment *Proc. Eng.* **119** pp 1338-47

[27] Flexidiblok [Electronic resource] TopolWater Access mode: www.topolwater.com

[28] Munawar Zaman Shahruddin et al 2016 The effectiveness study of different membranes in treating industrial wastewater *MATEC Web Conf.* 69 05001

[29] Makisha N and Panteleeva I 2017 Research for waste water treatment technology with low production of excessive active sludge *MATEC Web Conf.* 106 07015

Basic concepts of formalizing the business process of design and estimate calculations in the construction and repair of oil and gas facilities

R I Khamidullin

Industrial University of Tyumen, 38, Volodarskogo St., Tyumen, 625000, Russia

E-mail: hamidullin_ri@mail.ru

Abstract. The article considers the basic concepts of formalizing the business process of design and estimate calculations in the construction and repair of oil and gas facilities. The advantages and regulation of the main points for modeling business processes for the construction of oil and gas facilities are defined, including oil and gas transportation facilities for improving the competitiveness strategy in the market and optimizing the work of the entire organization as a whole. Particular attention is paid to the choice of the method of economic and mathematical modeling of the business process on the basis of the theory of mass service for the development of our own mathematical model, which will allow qualitative evaluating and optimizing the work of design and estimate activities.

1. Introduction

The current economic crisis is pushing the leaders and analysts of oil and gas design institutes to thinking about improving the productivity of the organization to improve those business processes that most affect the economic performances. Such processes as the preparation of design and estimate documentation for a construction project, or the maintenance of marketing campaigns to attract customers and others. Practice shows that the existing approaches to presenting and improving production and technological business processes, in particular, the business process of building projects management in the oil and gas industry are little formalized.

Today, analysts conduct business process reengineering identifying weaknesses in the transformation and modernization of existing business processes. Based on the research conducted [1] it was proved that reengineering of economically important business processes is not optimal and is not particularly successful in long-term process management. The graph in Figure 1 shows the success probability distribution, depending on the duration of the reengineering. On X, the timing of the reengineering of the business process is indicated, on Y - the relative popularity of the probability, in other words - the probability of falling into a particular period of an accidental successful or unsuccessful process.

Content from this work may be used under the terms of the Creative Commons Attribution 3.0 licence. Any further distribution of this work must maintain attribution to the author(s) and the title of the work, journal citation and DOI.
Published under licence by IOP Publishing Ltd

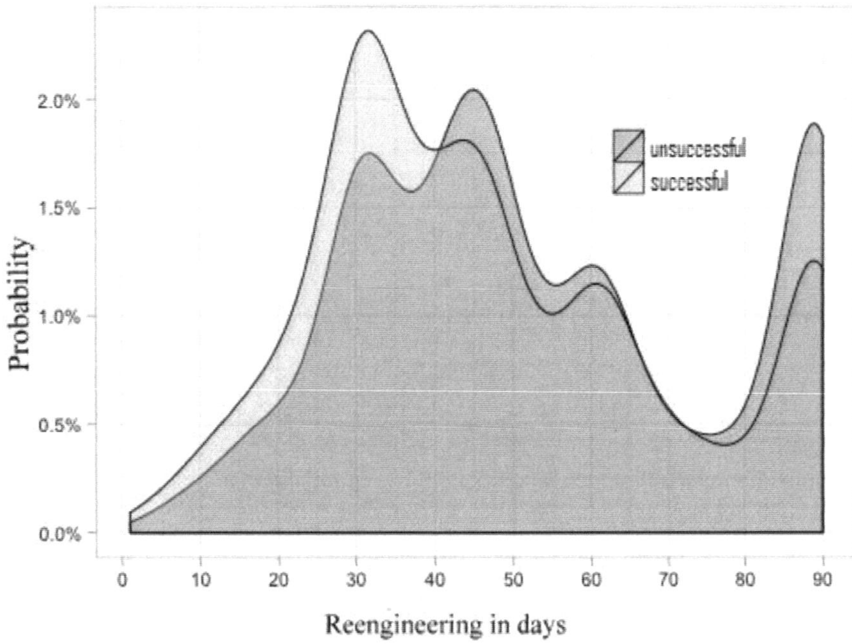

Figure 1. The success of the reengineering of the business process (for example, the development of design and estimate documentation).

As you can see, reengineering only in short periods of time provides efficient management of processes: the longer the use of this method is, the lower the quality of the construction project as a whole is. In the effective management of the project activities of the oil and gas industry instead of reengineering, there is a need for an ordered economic and mathematical model of optimization. Optimization will give a powerful impetus to the effective operation of those weak points that affect the decrease in the performance of the organization as a whole.

Determination of factors of optimization and management of actual competitiveness for the performance of the above-described work is fundamentally dependent on the specific situation in which it is located [2]. However, there are general approaches to the formulation of factors and some general framework in which management strategies and optimization exist. In particular, through the mechanism of competition, the selection of the most effective and promising construction projects takes place, which have competitive advantages and the possibility, on this basis, of long-term development.

Therefore, it is important to identify and evaluate the importance of conducting the project activity of the oil and gas industry as a business process, mathematical modeling of this business process with the designation of optimality criteria and the subsequent development of a universal program tool for automating the development of design and estimate documentation are relevant. At the same time, an important requirement is to improve the quality of the design and estimate department, taking into account the competitiveness in the construction and contracting services market to improve business performance indicators.

The work objective is to describe the basic concepts of formalizing the business process of compiling design and estimate documentation to assess the significance of optimization and automation of this business process in the oil and gas industry. The concepts will give the description and key points in the construction of a full-fledged economic-mathematical model of the business process.

IOP Conf. Series: Earth and Environmental Science **181** (2018) 012036 doi:10.1088/1755-1315/181/1/012036

2. Materials and methods

The successful operation of modern construction, repair of oil and gas facilities, including oil and gas transportation facilities, is complex, which requires new forms and methods of management. Undoubtedly, an additional factor contributing to the improvement of the production component of this sphere is the development of market relations and the enlarged emergence of a competitive environment, so attention is also paid to the economic efficiency and optimization of the work of all departments and the organization as a whole.

Ignoring these factors can lead to a situation when a system of management in the organization is in a precarious position, and the system can become inefficient.

Successful optimization of problems in the current conditions of economic development and bypassing economic problems should acquire new and more optimal conditions under which the production component will give the best result. A number of tasks of continuous increase in the efficiency of the work of design organizations become particularly acute incentives for organizational improvement, and should be guided by the need to adapt to all changes in the external environment.

Studies were carried out on the main economic difficulties and the problems that arise in the preparation of design and estimate documentation and the calculation of the cost of repairs in the oil and gas industry. The design and estimate documentation has always been a particularly important element of any construction process or various repair work, where planning of construction activities has always been strictly regulated, because for each project being developed, one writes its own estimate documentation, in which all costs are rigidly ruled, as well as instructions, and there are other standards and basic rules for budgeting.

Also, an analysis of solutions to problems arising in the business structure was conducted. The notion of the wording of the definition of a business process and consideration of the main approaches for identifying business processes where the concept of a mathematical model and a mathematical model of the business process is introduced was also important. The study made it possible to determine the basic concepts for creating a mathematical model for optimizing the business process under study using economic and mathematical methodology taking into account the author's additions.

To solve the problems, the authors relied on the concepts of Harrington D., Esseling K.S. and Nimvegen H. In these concepts there is a presentation of the theory of optimal solutions, where a significant place is given to mathematical modeling of economic processes [3,4]. To build a mathematical model, one need to have a clear idea of the purpose of the functioning and formalization of the system under study, as well as to have full information about the existing limitations.

It should also be noted that many authors of scientific papers tried to solve the problem of optimizing business processes based on various economic and mathematical models, such as Petri nets, graph theory, network models, models under uncertainty, etc. In the course of the detailed analysis, downsides in the use of the above described models were identified and the need for studying such a method in modeling business processes - the theory of mass servicing systems.

A queuing system (QS) is a system that is designed to serve incoming requests, that is, a system in which, on the one hand, there are massive requests or requirements for the performance of some services, and on the other hand, satisfaction of these requests is achieved [5,6]. Requests that come to service in the system (orders, tasks, customers, etc.) form a stream of requests. The elements of the QS that handle the requests are called service channels. In most cases, the time intervals between the time of receipt of applications and/or the times for servicing requests in the QS are random variables. In other words, in most cases it is not known exactly in advance when the next request will be received and how long it will take it to be serviced. Therefore, queuing theory is based on the mathematical apparatus of probability theory and mathematical statistics. The use of QS provides such advantages as:

- taking into account the relationship between the input and output parameters of the system;
- optimization of business processes on the set parameters in real-time.

Based on the chosen method, the business process will be formalized, which is created by a team of people with different qualifications, experience, habits, education, preferences and personal qualities.

IOP Conf. Series: Earth and Environmental Science **181** (2018) 012036 doi:10.1088/1755-1315/181/1/012036

The business process model of design and estimate activity is built in order for these people to effectively share knowledge and jointly make decisions in the course of creating a system that will give a new effect in the performance of their job responsibilities. After all, the model is the language of communication between the parties involved in creating a computer system of the model: customers, experts, architects, etc. It will be organized in such a way that each side perceiving the simulated system from their own point of view could effectively contribute to a common understanding of the subject area [7,8].

3. Results and discussion

The scientific literature describes two methods for distinguishing the business processes under study: the principle of customer-oriented chains and product chains. Therefore, our study reflects the process of identifying an end-to-end business process for drawing up design and estimate documentation for the construction and repair of oil and gas facilities on the basis of the principle of product chains. Thus, the business process of calculating and drawing up the design estimates has been singled out, which permeates the activities of such structural departments of the organization as the sales department, the production cycle itself: the construction and finishing department, the structural subdivisions in the form of material and technical supply, and the financing and verification of received documents. Each business design process has an input and an output. At the imput, there are orders from direct customers for carrying out a particular construction service. Outputs are generated by ready-made budget documentation intended for specific clients who coordinate further actions with the management of the organization. It should be noted that the resulting business management process can also be described as inter-functional on the above mentioned characteristics.

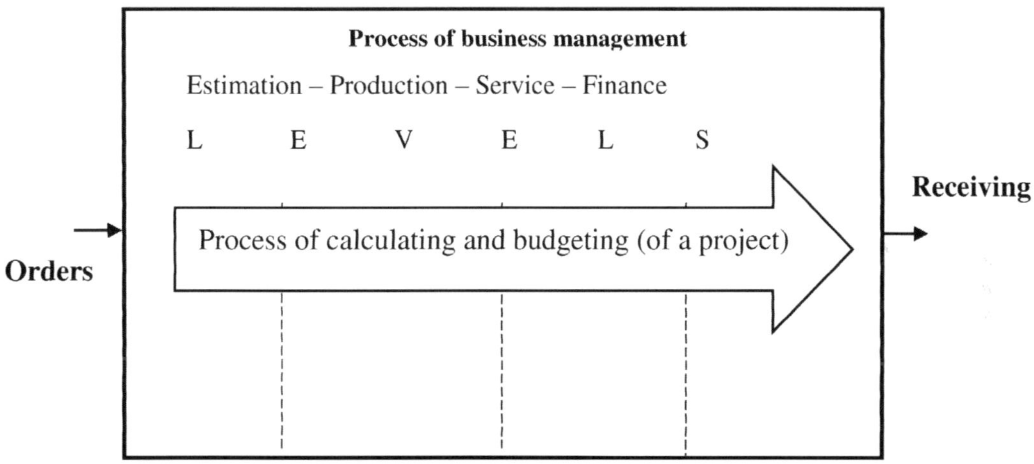

Figure 2. Scheme of the cross-functional business process for the preparation of design and estimate documentation.

As Figure 2 shows, this scheme of mapping the business process as interfunctional yields a lot of conclusions about the superior integrity of the functioning of the design and budget activity and the interrelationship of its business processes, that is, all of them are interrelated and in general their target orientation is predetermined by potential customers. Therefore, the project organization can be represented as a certain production system where requests for construction services are received at the imput, and the final version of the estimate documentation is sent to the customers of the organization to confirm the services at the output. The production level that permeates the activity of all key departments of the construction organization from the moment of the customer's request to the organization (sales department) until the approval and signing of the final estimate documentation for the performance of certain construction and repair services for the oil and gas facilities is an indispensable and central element in the developed system.

IOP Conf. Series: Earth and Environmental Science **181** (2018) 012036 doi:10.1088/1755-1315/181/1/012036

The design organization performing the design estimates is a very real production system in which it can be represented in the form of a queuing system at different levels (workplace, department), since the business process of the organization is focused on servicing orders in form of a certain production and realization of the rendered service, which is of value to the potential client. Figure 3 gives a visual representation in the form of a schematic illustration of this work as a queuing system.

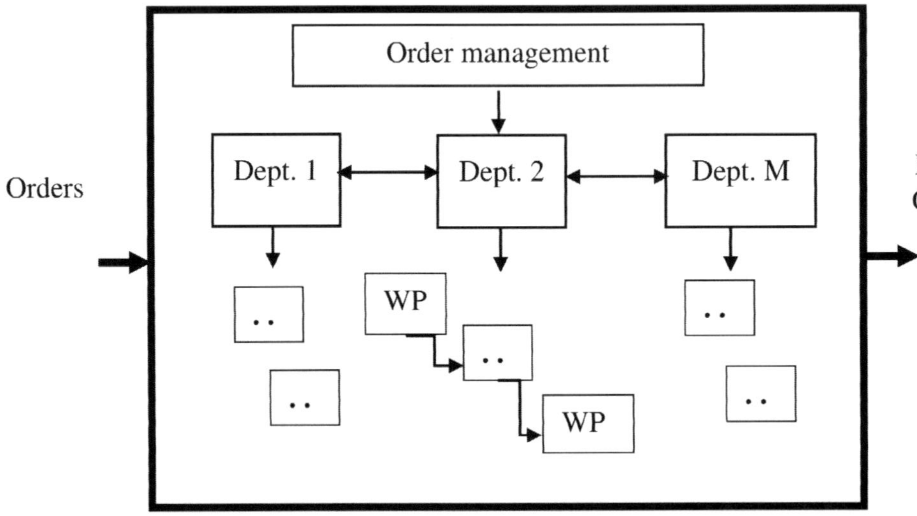

Figure 3. Project organization as a queuing system.

All orders for construction services come from potential clients and customers. Order maintenance starts from one department, ends in another, all departments are interconnected. At the output of this system, the received service is formed. At the same time, each department and workplace is also a queuing system (QS). The essential difference can only be that the existing stream of orders in this case is formed within the system, and the output streams form the results of servicing the given orders within each department (works, orders, documents, etc.).

The very formalization of the model of the business process under study will consist of a chain of sequential actions that will describe the process from all the sides: starting with the input of the business process to the output and regulating services. The input of the business process as a QS will be the number of orders (requests) for the execution of construction services, in particular, the preparation of estimates for the evaluation of ongoing services, where all the requirements of a particular customer are taken into account within the business process. The servicing department of this business process in the QS is represented by a personnel facility with their workplaces and personal computers that fulfill the orders received to compile an estimate with a given quality and a certain intensity. The queue of a business process as a QS is defined as the maximum possible number of all orders at the input in excess of those already in the process of producing the documentation. Under the output of the business process as the QS already compiled and executed design and estimate documentation will be understood, which is determined by quantitative indicators that also meet the requirements of the customer, according to the regulations of the organization.

One should not forget about the application of some restrictions, when developing this model of the business process. With some approximation, we can assume that the current order flow at the input of the simulated business process is a Poisson flow. In addition, it should be noted from the theory of probability and the theory of random processes that the use of a Poisson flow provides the hardest requirements for existing service systems. The very model of business process operations will also take into account the limit on the length of the queue. For a given business process of the specified type, there can be no more than n orders in the existing queue. It follows that if an order arrives for servicing by a business process at the moment when the same n orders already exist in the queue, then it will not be serviced and, respectively, will not be allowed in the queue. Also, the model of the business process of drawing up the estimate documentation of a construction organization can be represented by a number of indicators that can be described by mathematical expressions used to calculate the characteristics of individual operations, that is, the business functions of this business process that can be considered as open queuing systems [9]:

- business process probability indicators;
- quantitative indicators of the business process;
- economic indicators of the business process.

Therefore, it should be noted that the main task of the developed model (based on the theory of queuing) should be reduced to establishing a rational and optimal ratio between the number of requirements entering the design department and to consider the number of servicing devices in which the total costs for maintenance and losses from the servicing channels idle would be minimal.

4. Conclusion

The development of an optimal model for the compilation of any design and estimate documentation solves not only the problems in formalizing weak points in the conduct of business, but also helps in the subsequent process of creating an automated system that is iterative, so the design model allows for sequential refinements [10]. Ideally, the business process model is built in such a way that when it is detailed, the previously constructed more general elements of the model are not changed, but only new ones are added and improved, with the connection of the work of other departments and the formalization of the activity of the organization as a whole.

The peculiarity of modeling, automation and expertise of budget documents in the oil and gas sector is receiving by the customer not only the comments to estimates, but also the corrected estimated construction cost. In this case, feedback is required from the design institutes. For each observation, the design institute corrects the estimates, sends them for a second examination, and only after the removal of all comments the customer is sent a positive opinion and a ready-made package of documents. For any oil and gas company, it is important to be able to correctly assess the economic efficiency of projects at each stage of their implementation. At the same time, the accuracy of the results, the depth of the project, the amount of resources involved depend not only on the stage of the project, but also on the tasks to be performed by the assessment, the requirements and limitations of the experts conducting such an assessment.

As part of the improvement of this model, a methodology will be systematized and developed, which will include a step-by-step analysis of the investigated business process in order to determine the optimal parameters. Therefore, the computer implementation or in other words the simulation model of this QS will be an algorithm that reflects the behavior of the QS, that is, all changes in the QS state will be reflected in time for given streams of applications that arrive at the inputs of the system, because the essence of the simulation model is to describe existing processes as they would be in real life.

References

[1] Hamidullin R I and Senkevich L B 2017 About necessity of mathematical modeling of the business process of cost estimate calculations in the construction of oil and gas facilities *Higher Educational Institutions News Neft I Gaz* **6** 139-45

[2] Conforti R, Leoni de M, La Rosa M, Aalst van der W M P and Hofstede ter A H M 2015 A recommendation system for predicting risks across multiple business process instances. *Decision Support Systems* **69** 1-19

[3] Dumas M, Aalst, van der, W M P and Hofstede, ter, A H M 2005 Introduction *Process-Aware Information Systems: Bridging People and Software Through Process Technology* (Hoboken: Wiley-Interscience) pp 3-20

[4] Harrington D 2002 *Optimization of Business Processes. Documenting, Analysis, Management, Optimization* (Saint Petersburg: ABC) 205 p

[5] Hofacker I and Vetschera R 2001 Algorithmical approaches to business process design *Computers and Operations Research* 28 pp 1253-75

[6] Powell S G, Schwaninger M and Trimble C 2001 Measurement and control of business processes *Syst. Dyn. Rev.* 1 **17** 63-91

[7] Tiwari A, Vergidis K and Majeed B 2006 Evolutionary multi-objective optimisation of business processes *Proceedings of IEEE Congress on Evolutionary Computation* (Vancouver, Canada) 3091-7

[8] Tiwari A, Vergidis K and Turner C J 2010 Evolutionary multi-objective optimisation of business processes *Advances in Intelligent and Soft Computing: Soft Computing in Industrial Applications* (Springer: Heidelberg) 293-301

[9] Turner C J, Tiwari A, Olaiya R and Xu Y 2012 Business process mining: From theory to practice *Business Process Management Journal* **18**(3) 493–512

[10] Hamidullin R I and Senkevich LB 2015 Automation of the work price calculation engineer construction of estimated LLC «Tobolstroyservis» *Fundamental Research* **11**(1) 110-4